21st Conference on Power System Engineering 2022

MATEC Web of Conferences Volume 367 (2022)

Pilsen, Czech Republic
7-9 September 2022

Editors:

Martin Pelikan
Michal Volf
Michaela Vackova

ISBN: 978-1-7138-6275-8

Printed from e-media with permission by:

Curran Associates, Inc.
57 Morehouse Lane
Red Hook, NY 12571

Some format issues inherent in the e-media version may also appear in this print version.

No changes have been made to the content of these proceedings. There may be changes to pagination and minor adjustments for aesthetics.

Printed with permission by Curran Associates, Inc. (2023)

For additional information, please contact EDP Sciences – Web of Conferences at the address below.

EDP Sciences – Web of Conferences
17, Avenue du Hoggar
Parc d'Activité de Courtabœuf
BP 112
F-91944 Les Ulis Cedex A
France

Phone: +33 (0) 1 69 18 75 75
Fax: +33 (0) 1 69 28 84 91

contact-edps@webofconferences.org

Additional copies of this publication are available from:

Curran Associates, Inc.
57 Morehouse Lane
Red Hook, NY 12571 USA
Phone: 845-758-0400
Fax: 845-758-2633
Email: curran@proceedings.com
Web: www.proceedings.com

TABLE OF CONTENTS

21st Conference on Power System Engineering

Martin Pelikan[1*], *Michal* Volf[1] and *Michaela* Vackova[1]

[1]University of West Bohemia in Pilsen, Univerzitni 22, 306 14 Pilsen, Czech Republic.

Abstract. The Conference on Power System Engineering has a long-term tradition. It is an annual meeting of professors and experts from departments and universities where power system engineering is taught. The 21st year of this event was held in Pilsen (Czech Republic) between 7th and 9th September 2022 and was hosted by Department of Power System Engineering of Faculty of Mechanical Engineering, University of West Bohemia in Pilsen.

1 Preface

There were 44 delegates from academic and industrial field who came from the Czech Republic, the Slovak Republic, Germany and Hungary. The conference was opened by the dean of the Faculty of Mechanical Engineering, University of West Bohemia. The organizers received 23 contributions. 23 contributions have gone through a professional review for the publication. All presentations brought contributions to the fields of equipment for classic and nuclear power plants including their operation, fluid mechanics and flow simulations, thermodynamics, heat transfer, diagnostics of power machines and equipment etc.

The social evening with a programme gave the delegates an opportunity to meet their colleagues, share and discuss their experience and results of their research projects in less formal environment. The participants could also visit the large complex of Plasy Monastery. The programme of the conference included also a student section, in which especially Ph.D. students have an opportunity to present their contributions or posters.

The conference organizers would like to thank all the industrial and academic partners of the conference for their financial support. Financial support came from Doosan Škoda Power, ČEZ, Howden, Techsoft Engineering, TechSim Engineering, NUM solution and Helago-CZ.

* Corresponding author: mpelikan@kke.zcu.cz

Fig. 1. Logos of sponsoring organizations.

Statement of Peer review

In submitting conference proceedings to *Web of Conferences*, the editors of the proceedings certify to the Publisher that

1. They adhere to its **Policy on Publishing Integrity** in order to safeguard good scientific practice in publishing.
2. All articles have been subjected to peer review administered by the proceedings editors.
3. Reviews have been conducted by expert referees, who have been requested to provide unbiased and constructive comments aimed, whenever possible, at improving the work.
4. Proceedings editors have taken all reasonable steps to ensure the quality of the materials they publish and their decision to accept or reject a paper for publication has been based only on the merits of the work and the relevance to the journal.

Title, date and place of the conference

21st Conference on Power System Engineering
September 7 - 9, 2022 in Pilsen, Czech Republic

Proceedings editor(s): Martin Pelikan, Michal Volf, Michaela Vackova

Date and editor's signature 30.9.2022

MATEC Web of Conferences **367**, 00002 (2022) https://doi.org/10.1051/matecconf/202236700002
Power System Engineering 2022

Experimental measurement of cooler pressure loss of Formula student car

Stefan Bajić[1], *Martin* Pelikán[1*], and *Kateřina* Rostás[2]

[1]University of West Bohemia, Faculty of Mechanical Engineering, Department of Power System Engineering, Univerzitní 22, Pilsen, Czech Republic
[2]University of West Bohemia, Faculty of Electrical Engineering, Department of Materials and Technology, Univerzitní 26, Pilsen, Czech Republic

Abstract. To achieve the maximum performance of an engine while maintain safety of the race car, it is essential to cool it properly. Therefore, this article deals with the experimental measurement of the cooling modules pressure drop of the formula student car. Leading to the right choice of the fan behind the radiator. Series of measurements were taken at the UWB in the wind tunnel of the department of Power System Engineering.

1 Introduction

Every formula student car must be built according to strict rules which are controlled before the race. The cooling system of the entire car falls under these rules, of course. There are few specifications relating to the cooling system such as that there must be a firewall to separate the driver from all the parts of the fuel supply, the engine oil, any high voltage system, and liquid cooling systems. The other rule says that cooling or lubrication system must be sealed. The main rule relating to our research is that no power device may be used to move or remove air from under the car except fans designed exclusively for cooling. To properly design or select these fans, it is necessary to know the pressure drop of the cooling systems radiator.

In the formula student car from the University of West Bohemia is used the engine from the YAMAHA R6 – 2013 with the power of 91 kW. With this amount of the engine power, it is necessary to use the water-cooling system. To understand why, the next chapter describes different types of cooling systems and its use. [1],[2]

1.1 Cooling system types

Today, there are two types of cooling systems that are the most used. The first one is air cooling system and the second one is water cooling system.

*Corresponding author: mpelikan@kke.zcu.cz

1.1.1 Air cooling systems

This system is mainly used in small engines approximately up to 15 – 20 kW. It refers to scooter or motorbike engines and to airplane engines. This means that the system is used in engines that are directly exposed to the air. To help the heat to dissipate to atmospheric air by conduction the outer surface of the engines cylinder and its head, these are provided with the fins. See **Fig.** ***1***. [3],[4]

Fig. 1. Air cooling system [3]

1.1.2 Water cooling systems

Since the formula student of UWB is using more powerful engine it is necessary to use the water-cooling system.

This system is used in the engines in all vehicle types with combustion engine such as cars, trucks etc. The cooling medium is water, which is provided around the engine in the water jackets. The water absorbs heat from the engine while its kept constantly in the motion by the centrifugal water pump. The hot water is then cooling in the radiator and by the fan and by the air flow developed from the motion of the vehicle. The cooled water then circulates again trough the water cooling jackets. [4],[5]

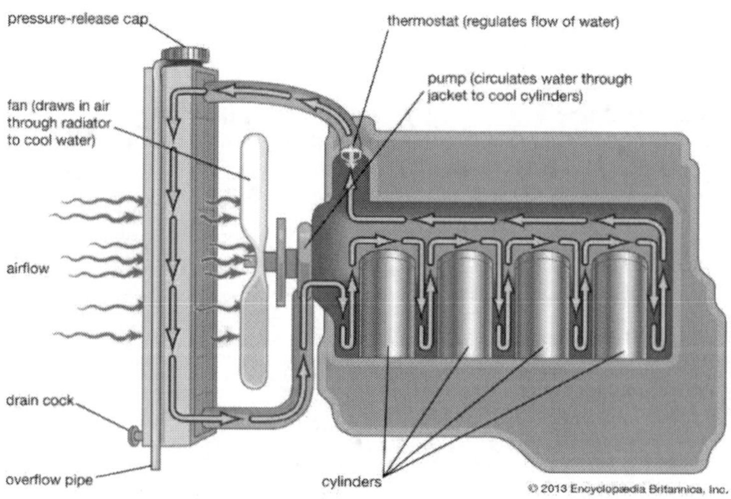

Fig. 2. Water cooling system [4]

To keep the engine safe from cracks in the cylinder blocks, radiator, and piping which it is important to prevent the cooling water from freezing. Therefore, the antifreeze mixture is added to cooling water. This measure is keeping water in the liquid state and protecting the engine in colder days when the temperatures go below the zero.

Components of the water-cooling system are thermostat valve, water pump, water jackets, antifreeze mixtures. And two components that are essentially for the research are the fan and the radiator.

The fan is provided behind the radiator to help cover the pressure drop and to blow the air over it for cooling purpose. This part is chosen after the pressure drop of the radiator is known.

The radiator consists of an upper and lower tank with honeycomb core between them. This is the part which produces the greatest pressure drop. [4],[5]

2 Experimental measurements

The measurements were taken at the department of Power System Engineering of the UWB. For the experiment the wind tunnel (**Fig.** *3*) that we have available was used.

Since the speeds of the car on the racetrack for student formulas are not very high, due to a very technical track and safety of the students of course, the measuring of the pressure drop of the radiator was the most important during the flow velocity between 40 – 80 km/h. These flow velocities are provided by the wind tunnel without any difficulty.

The measurements, however, where taken for a bigger range of flow velocities. This range was from 4 m/s up to 25 m/s which refers to 14.4 – 90 km/h. This was done to get more complete view of the pressure drop trend. In the end the two out of three measurements were used for the evaluation. The third one was not successfully completed due to technical issues in the last measurement series.

During the measurements the temperature in the laboratory was 26.4 C and the barometric pressure was 96.97 kPa. These conditions correspond to the race which is held in the Czech Republic.

Fig. 3. Wind tunnel

2.1 Set up

Before the measurement was taken the whole system needed the proper set up. This means that the measuring volume had to be modified so the part of radiator could be placed in it. After the placing the radiator into the measuring volume, it was essential to seal all the gaps, to prevent the air leakage. After the sealing and placing the pressure sampling the measuring set up (**Fig.** *4*) was ready for the experiment.

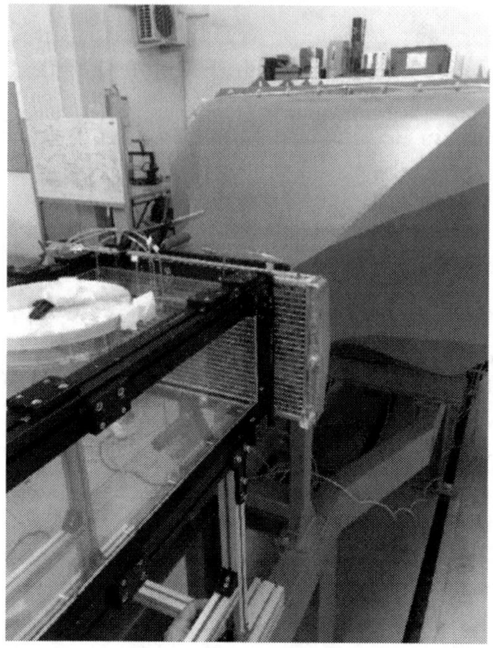

Fig. 4. Sealed radiator

The pressure samplings were placed in front of the radiator and behind it. So, the pressure drop is measured right behind the honeycomb of the radiator. These samplings were placed 70 mm from both sides of the radiator (**Fig. 5**). This was important for choosing the fan according to pressure drop of the radiator.

Fig. 5. Sampling distance

2.2 Devices

All the devices used during the experimental measurement are listed below. Also, all of them were provided by the department of Power System Engineering.

The used wind tunnel is able to produce the flow velocity from 0 – 90 m/s. Its cross section of the measuring volume is 200×300 mm.

The name of the pressure system used to measure the pressure drop is NetScanner 9116 (

Fig. 6). It is a pneumatic intelligent pressure scanner with 16 sensors and micro-processor technology. The accuracy of this system is ± 0.05 % FS and it is designed for measuring with the gas and for the wind tunnel.[6]

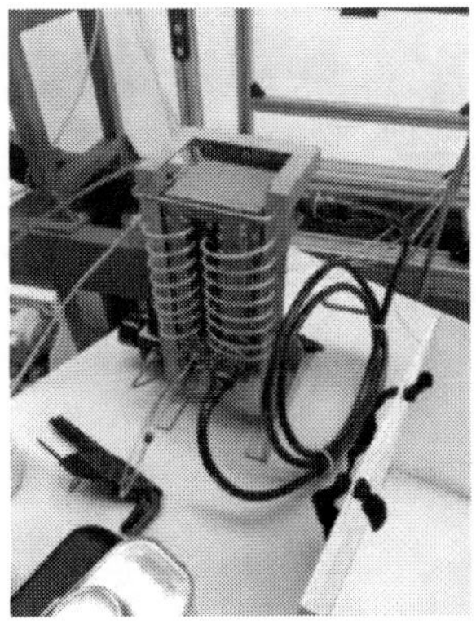

Fig. 6. Pressure system

Traverzer_2D & Pressure was the software used to control the pressure system and wind tunnel. This software is made of two parts. The first part is provided by the supplier of the wind tunnel. The second part of the software was specially made for collecting, processing and display of the data. This second part of the software was developed by one of the co-workers at the Department of Power System Engineering. [7]

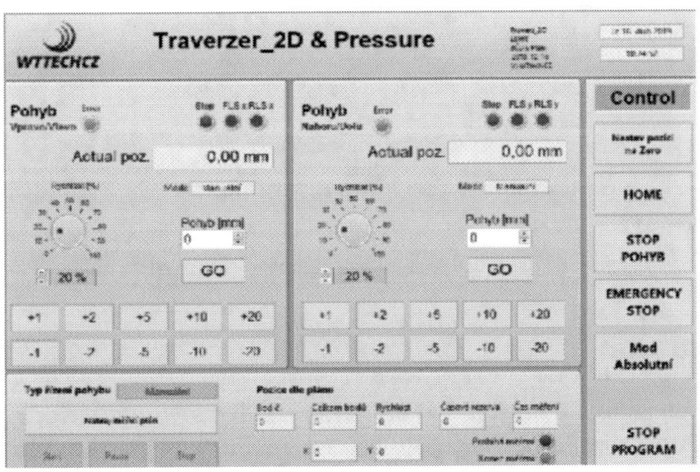

Fig. 7. Screenshot of the Traverzer_2D & Pressure [7]

3 Evaluation of the measurements

For creating the characteristics of the pressure drop at different flow velocities, the values from the measurement series were used. The pressure system directly measures the value of the pressure drop between the first and the last sampling. Also, it should be mentioned that the measurements were taken with used radiator which was damaged by the rocks during the races. This fact caused higher pressure drop then when the calculations were performed in previous researches few years ago.

Following figure (**Fig. 8**) shows the dependence of pressure drop on flow velocity. As expected with the growing speed the difference between two pressure samplings increases. This characteristic of pressure drop will help to choose the fan for the future formula student cars at the UWB.

This fan should be chosen so it covers the pressure drop and therefore the whole cooling module works properly.

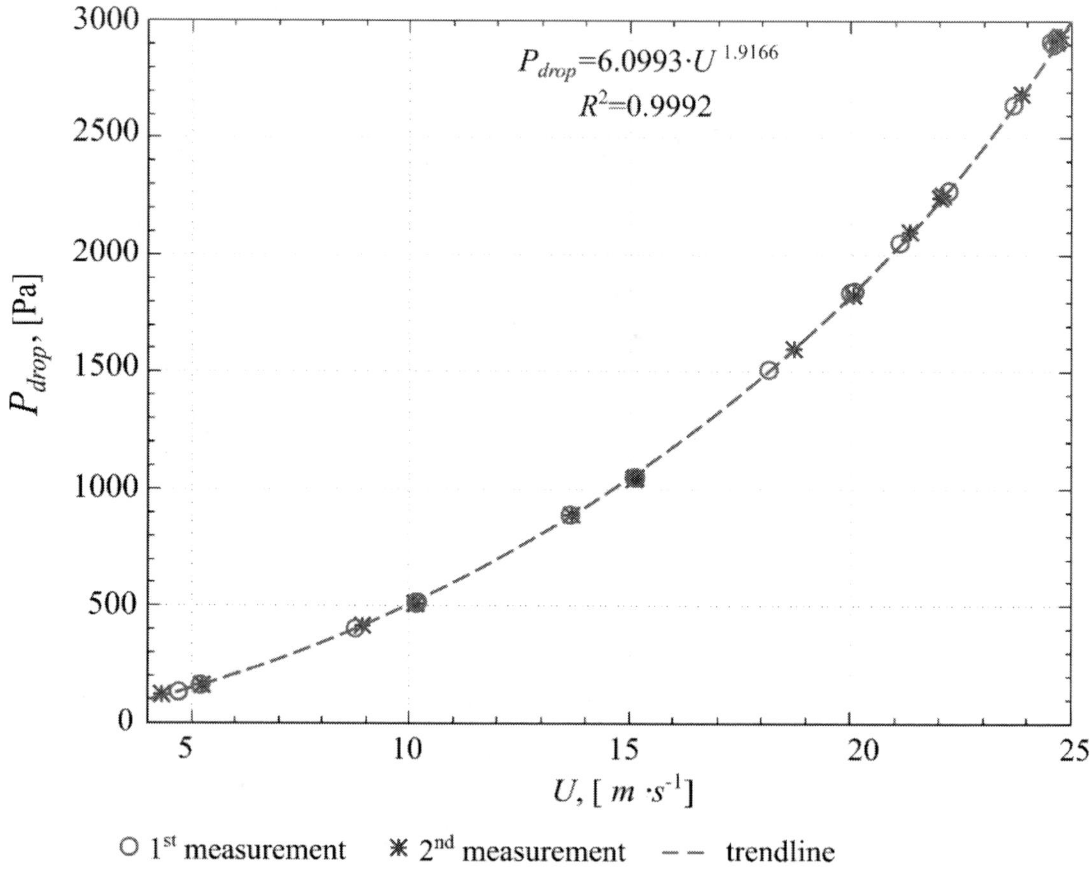

Fig. 8. Pressure loss of the radiator – experimental measurement

4 Conclusion

This research should contribute to the future students with developing new racing car in the UWB Racing Team, especially to those who will be responsible for designing the cooling module. This research will continue with numerical simulation to confirm the trend from experimental data. Since the numerical simulations are more flexible regarding to boundary conditions, they will be performed with different temperatures of the air. This should help with choosing the fan to cover the different range of the pressure drop e.g., when racing in hotter weather. If the fan is chosen wrong and it doesn't cover the pressure drop it can potentially cause overheating of the engine. The radiator will be also scanned by 3D scanner and compared with the model without any damage from the race.

Acknowledgements

This article was written with the financial support of the project Improving the quality of internal grant schemes at the University of West Bohemia in Pilsen, reg. no. CZ.02.2.69/0.0/0.0/19_073/0016931.

The presented work was also financially supported by student project SGS-2022-023 (Research and development of power machines and equipment).

References

1. Lisa Van Den Berg a Brandon Lofaro. Formula SAE Cooling System Design [online]. Mechanical Engineering Department California Polytechnic State University, 2014 [cit. 2022-05-30]. Available at: https://digitalcommons.calpoly.edu/cgi/viewcontent.cgi?article=1229&context=mesp M. Ben Rabha, M.F. Boujmil, M. Saadoun, B. Bessaïs, Eur. Phys. J. Appl. Phys. (to be published)
2. SAE International. FORMULA SAE Rules 2022: Version 2.1. 2021
3. WHAT IS AIR COOLING SYSTEM AND HOW IT WORKS IN VEHICLE. Https://www.theengineerspost.com [online]. [cit. 2022-04-25]. Available at: https://www.theengineerspost.com/air-cooling-system-in-engine/
4. Types of Cooling System In Engine. Https://www.theengineerspost.com/ [online]. [cit. 2022-04-25]. Available at: https://www.theengineerspost.com/cooling-system-for-ic-engine/
5. Cooling Systems in Automobiles & Cars. International Journal of Engineering and Advanced Technology (IJEAT) [online]. 2013 [cit. 2022-04-25]. ISSN 2249 – 8958. Available at: https://www.ijeat.org/wp-content/uploads/papers/v2i4/D1447042413.pdf
6. TE connectivity, “product-CAT-SCS0002.datasheet,” 9116, 08/15/2022.
7. HORÁČEK, Vít. Measuring the velocity field in test section of wind tunnel by pressure probes [online]. Pilsen, 2019 [cit. 2022-08-15]. Available at: https://dspace5.zcu.cz/bitstream/11025/38038/1/Horacek_Bakalarka.pdf. Bachelor thesis. University of West Bohemia.

MATEC Web of Conferences 367, 00003 (2022) https://doi.org/10.1051/matecconf/202236700003
Power System Engineering 2022

An Analysis of the Bi-modal Size Distribution of the Coarse Droplets in a Steam Turbine

Ondřej Bartoš[1*], *Adam* Huněk [1] and *Petr* Pavlíček [2]

[1] CTU in Prague, Energy Engineering Department, Technická 4, Prague 6, 166 07, Czechia
[2] University of West Bohemia, Department of Power System Engineering, Univerzitní 22, Pilsen, Czechia.

Abstract. The atomization of liquid is a widely studied problem in the field of aerosol research. Over the past year, research at CTU has focused on the formation of coarse droplets in a wind tunnel which simulates the conditions present in a steam turbine. An analysis of the droplets formed in the wind tunnel indicates, for the majority of the measurements, the bi-modal droplets size distribution. The aim of this paper is to suggest a simplified explanation of this bi-modal distribution. The CFD simulation of the flow in the vicinity of the liquid film break up was used to determine the relative velocity between the formed droplet and the bulk flow. The preliminary explanation of the bi-modal distribution is based on the first droplets mode corelating with the formation in the bulk flow and the second mode correlating with the droplets formed in the blade's wake.

1 Introduction

The liquid film atomization is an important technical problem in many practical applications, but a general theoretical description of the whole process is still absent. Water film atomization can also be found in low pressure parts of steam turbines. The coarse water droplets are formed from the water film on the blades and casings. This process is connected to undesirable effects on the reliability and efficiency of steam turbines [1, 2]. Related effects such as additional energy losses, blade erosion or the reduction of the blade's lifetime can be further noted. The study of liquid atomization, movement and its potential impact of the formed droplets on the moving blades in turbines at high speed has an important role in the never-ending effort to enhance energy production efficiency and to decrease its environmental impact.

Two populations of droplets are present in steam turbines. Fine droplets are formed by nucleation and condensation as expanding steam passes the saturation line. These fine droplets fall mostly within a diametral range of 0.1 – 1 μm. These fine droplets are transported, mostly by turbulent diffusion, on the blades and casings. Subsequently, the captured droplets can create a liquid film which tears off in the form of coarse droplets. The expected diametral range of coarse droplets in steam turbines ranges between 1 μm to 1 mm. The measurement of the size distribution function of the coarse droplets in the nozzle shows

* Corresponding author: ondrej.bartos@fs.cvut.cz

the bi-modal function of the coarse droplets. The aim of this contribution is to suggest some explanation of this phenomena.

2 Droplet formation

The following chapter is devoted to the analysis of the liquid film breakup. The current theory is based on the works of Rayleght, Weber and Ohnesorge [4]. The theory suggested several regimes of liquid film atomization. One can distinguish high-speed liquid breakup among Rayleight-type, Membrane-type and Fiber-type atomization. The photogrammetric measurements carried out in the wind tunnel show that the atomization of the trailing edge of the profile is very stochastic and may be visible in all of these atomization modes (Fig.1).

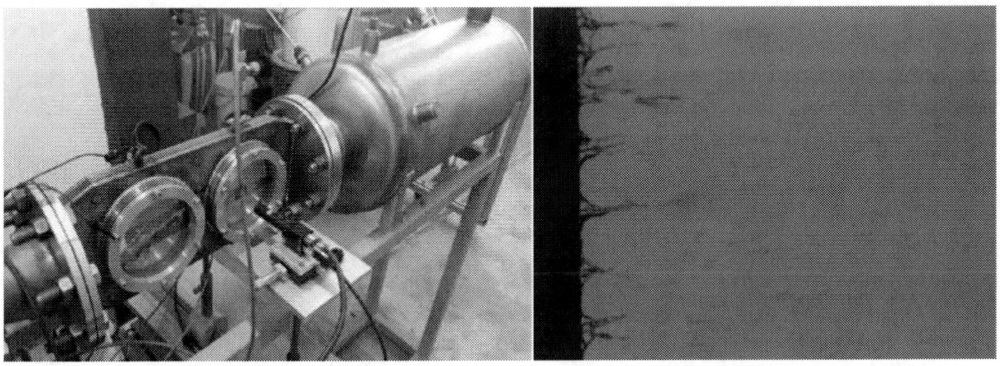

Fig. 1. The wind tunnel and the water film breakup on the trailing edge of the blade.

The mentioned theoretic modes are described by the non-dimensional Weber number as defined by:

$$We = \frac{\rho_g {w_r}^2 d}{\sigma} \tag{1}$$

where (ρ_g) is vapour density, (σ) is surface tension of the water, (w_r) is the relative velocity between the vapour and the droplets and (d) is the diameter of the droplets.
For the above-mentioned reason, the description of the "chaotic" atomization is still not sufficient. One of the possible parameters which can be determine is the maximum droplet diameter as defined by the critical Weber number We_{crit}. The critical Weber number is the non-dimensional ratio between drag forces and surface tension. The critical value is not constant for any cases, but for the low-viscosity liquid suddenly exposed in the high velocity air steam was estimated We_{crit}=13 [3, 4]. It is necessary to note that there are more values for We_{crit} in literature in a range of 10-40 [1, 3].

3 Measurement

The experiment was carried out in the wind tunnel with two blades possessing differing surface properties that are placed in the air flow. The first blade was coated with electrical insulating and hydrophobic paint and the second was polished stainless steel. The overall view of the tunnel is in Fig. 1.

3.1 The wind tunnel

The motivation for the wind tunnel assembly arose from a previous measurement [6] of the coarse droplets in the steam turbine under operational conditions. The measurement was performed in three steam turbines in Czechia. The amount of the acquired coarse droplets was very low due to their low number density level. Due to difficulties connected with measurements in an operating steam turbine, a wind tunnel was designed and manufactured for the analysis of the coarse droplet formation from the liquid films. The aim of the design was to create, as closely as possible, the conditions in a steam turbine while utilising the advantages of working in the laboratory.

The wind tunnel is designed as a classical CD nozzle, but the planned operational regime is mainly subsonic or transonic. An aerofoil NACA0008 is placed 50 mm behind the nozzle throat. The aerofoil simulates the blade in the turbine, and it is possible to remove or replace it. There is a groove on the aerofoil which supplies liquid to the surface. The liquid is pumped to the aerofoil through the dosing pump with a flow between 1 ml/min to 500 ml/min. The tunnel is equipped with four large optical windows that provide good visual access for the measurements. It is possible to operate the tunnel with steam in a continuous mode or with compressed air in a periodic mode.

3.2 The size distribution measurement

The droplet distribution function measurement was performed on a Spraytec instrument (Malvern, Inc.). Both blades were tested under the same conditions in the wind tunnel. The measuring position of the laser beam was about 20 cm behind the trailing edge of the blade. Due to the principle of measurement (scattering of light on the droplets), the conditions for evaluation were not always met, therefore the results are calculated only for the states when the conditions for evaluation were acceptable. In Fig. 2, the volumetric or mass fraction distribution is shown. It is possible to observe the bi-modal distribution.

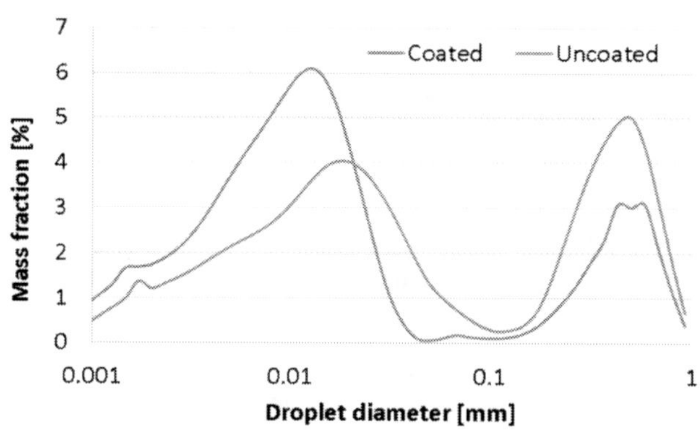

Fig. 2. Mean size distribution of the droplets for both blades.

4 Data processing

4.1 Droplet analysis

The collected data of the size distribution function of the droplets were divided into two groups and the Sauter mean diameter was computed for both of them. The edge diameter between the small and large droplets was estimated at 60 μm (approximate saddle between the modes).

$$D_{32} = \frac{\Sigma d^3}{\Sigma d^2} \tag{2}$$

Table 1. contains six values of the Sauter mean diameter for three measurements of coated and three measurements of uncoated blades. The Sauter mean diameter is used for comparing each separate measurement. The rate is the mass ration between small and big droplets. The introduced distributions are representative examples of the typical measurement.

Table 1. Sauter mean diameter and the droplet rate.

	Diametre D32		Mass rate	
	Small [μm]	Big [μm]	Small [%]	Big [%]
Coated 1	5.3	386.2	72.4	27.6
Coated 2	5.0	458.8	88.3	11.7
Coated 3	4.8	353.3	75.6	24.4
Uncoated 1	7.9	332.2	59.8	40.2
Uncoated 2	5.8	136.8	75.9	24.1
Uncoated 3	5.2	209.4	84.2	15.8

4.2 Flow field

Information regarding the relative velocity in the vicinity of the trailing edge is not possible to measure directly. The local velocity from the CFD simulation is used to determine the relative velocity in order to calculate the Weber number. The velocity of the liquid film is expected to be negligible. A mesh with 1 114 764 polyhedral cells with a prismatic boundary layer was created. This mesh had an increased density near the NACA profile. SST k-ω model of turbulence was used for the simulation. Pressure boundary conditions were used, both at the inlet and outlet. The inlet pressure was known accurately. However, the outlet pressure was less exact since the flow exiting from the nozzle enters the pipe system before exiting into the atmosphere. The outlet pressure was considered atmospheric (101 325 Pa) assuming that the pressure losses in the pipe system are negligible. The details of the simulation are recorded in [7].
Fig. 3 and Fig. 4 show the velocity of the flow both along and perpendicular to the blade respectively.

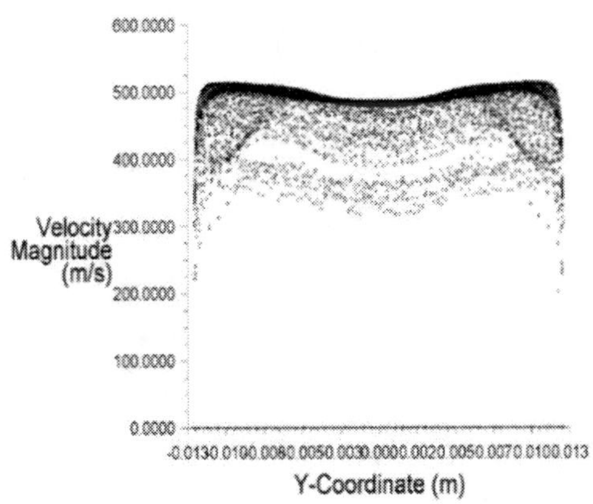

Fig.3. Velocity profile for the Y coordinate (along the channel trailing edge of the profile)

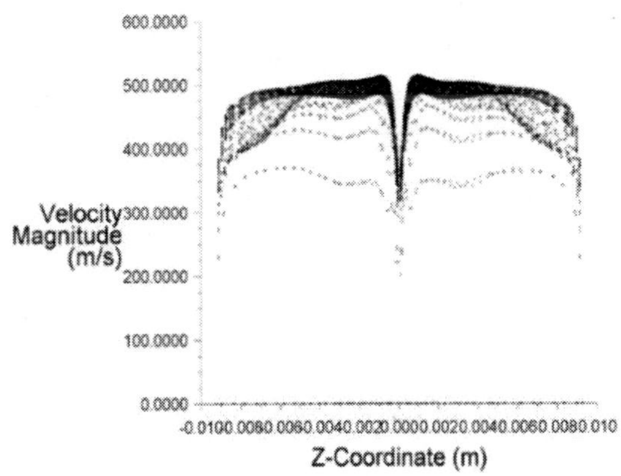

Fig. 4. Velocity profile for the Z coordinate (perpendicular to the channel)

5 Results

The bulk flow in the nozzle is generally transonic. As is possible to see in Fig. 5 and Fig. 6 close to the profile's trailing edge the flow velocity is approximately 300m/s. The critical Weber number describes the maximum droplet diameter, for the analysis the D_{95} was used. The value D_{95} is the diameter for which 95% by mass is contained in the smaller diameter. In Table 2. are written values of D_{95} and Weber numbers for the group of small droplets. The total mass fraction of these droplets is around 75% (Table 1.). The rest of the droplet mass is represented by the small number, at least 3 orders of magnitude, of droplets with big diameter D_{32} is around 200-300μm and D_{95} is around 0.5mm. The existence of these droplets has to be explained. The process of the liquid film is stochastic and also the flow of the carrier gas around the obstacle is connected with the wake and turbulence. On Fig. 5 and Fig. 6 is shown the turbulence intensity and velocity in the channel.

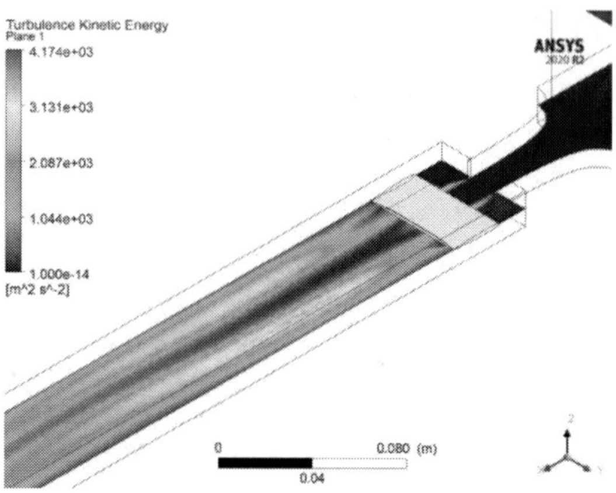

Fig. 5. Turbulent kinetic energy in the channel.

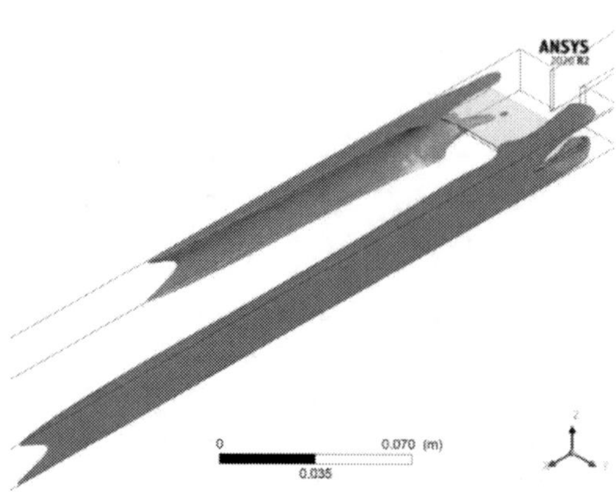

Fig. 6. Velocity areas lower than 60 m/s.

The CFD computation provides the statistical mean velocity, and the random fluctuations are possible to estimate according to kinetic energy turbulence. For the purpose of this paper, a reverse logic was utilised. The value of the local velocity fluctuation was found to meet a Weber number similar to the small droplets ($We \approx 40$) and, together, the velocity from the turbulent kinetic energy. The estimated value $w_r = 60$ m/s was suggested. The Weber numbers for the big droplets with diameter D_{95} are shown in Table 2. One can expect that this part of the water went to the vortex region with a smaller relative velocity and is accelerated slowly. This leads to satisfying the Weber theory, although the droplet is bigger than it should be.

Table 2. Table of the D_{95} for both droplet modes and the values of We at different velocities.

	Droplet diameter D_{95} [μm]		We (300 m/s)	We (60 m/s)
	small	large	small	large
Coated 1	23	500	44	38
Coated 2	12	500	22	38
Coated 3	20	580	37	44
Uncoated 1	32	580	59	44
Uncoated 2	20	200	37	15
Uncoated 3	10	430	19	32

6 Conclusions

The process of liquid atomization from the liquid film on the obstacle in transonic flow is an extremely complex problem. The purpose of this paper is not to explain all the details of the liquid atomization or to provide a detailed study of the flow field. The main aim is to suggest an explanation relating to the second mode of the droplet's existence in the size distribution function. The influence of the different blade's coating is not significant, but there is a small improvement for a coated blade because the number of large droplets is slightly lower [8]. The result of the paper can help to better understand the erosion processes in the steam turbines, as well as, in other technical applications. Future research incorporating detailed analysis is still necessary.

We gratefully acknowledge the support of the Grant Agency of the Czech Technical University in Prague, Grant No. SGS 21/154 and the National Centre for Energy TN01000007.

References

1. M.J. Moore, C.H. Sieverding: Two-phase flow in turbines and separators, Theory, Instrumentation, Engineering, Hemisphere Publishing corporation, 1976.
2. M. Hoznedl at al. Experimental research on the flow at the last stage of a 1090 MW steam turbine. Proceedings of the Institution of Mechanical Engineers, Part A: Journal of Power and Energy. 2018, 232(5), 515-524. DOI: 10.1177/0957650917749692. ISSN 0957-6509.
3. J. Valha: Proudění mokré páry a její erozivní účinky v průtočné části parní turbíny. SVUSS, Sborník referátů, 1978
4. H. Liu, Science and engineering of droplets: fundamentals and applications. Norwich, NY: Noyes Publications, c2000. ISBN 0-8155-1436-0.
5. J. O. Hinze, "Fundamental of the Hydrodynamic Mechanism of Splitting in Dispersion Processes," AIChE Journal, Vol. 1, No. 3, 1955, pp. 289-295.
6. O. Bartoš, X. Cai, M. Kolovratník, EPJ Web of Conferences. 2014, vol. 67, ISSN 2100-014X
7. O. Bartoš, P. Pavlíček, L. Měšťanová. An experimental study of coarse droplet formation. In: AIP Conference Proceedings. Melville, NY: AIP Publishing, 2019. Volume 2189. vol. 1. ISBN 978-0-7354-1936-0.
8. O. Bartoš, J. Vlasák: The influence of surface properties on droplet formation. AIP Conference Proceedings 2323, 060002 (2021); https://doi.org/10.1063/5.0041515 Published Online: 08 March 2021

Surface Structure Analysis of X12CrNiMoV12-3 and X6CrNiTi18-10 Steel Samples Processed by Laser Shot Peening (LSP)

David Bricín[1,2*], *Hynek* Gilík[2]

[1]Centrum výzkumu Rez s.r.o., Morseova 1245/6, 301 00, Pilsen, Czech Republic
[2]Department of Material Science and Technology, University of West Bohemia, Univerzitni 8, 301 00 Pilsen, Czech Republic

Abstract. The aim of the experiment was to verify the influence of the chosen strategy of laser shot peening (LSP) on the structure and surface properties of samples made of martensitic steel X12CrNiMo12 and austenitic steel X6CrNiTi18-10. The analyses performed consisted of surface topography analysis of the samples, using a confocal microscope, Olympus LEX|T OLS 5000 and metallographic analysis using a Carl Zeiss Observer Z1m light microscope and a Tescan Mira 3 scanning electron microscope. The metallographic analysis was complemented by the measurement of microhardness HV0.01, which was performed using a Struers DuraScan semi-automatic hardness tester. The LSP process introduced plastic deformation into the surface layer of both materials analysed. The plasma and shock wave generated during the process further affected the surface roughness of both tested materials, which was reflected by an increase in the value of the roughness parameter Rz. Microhardness measurements showed that the influence of the LSP process can increase the surface hardness of the material. Metallographic analysis using light microscopy methods failed to identify differences between the structure near the shot peened surface and the non-peened surface, probably due to the coarse grain structure of both materials analysed.

1 Introduction

Laser Shot Peening (LSP) technology is one of the surface treatments of materials that significantly harden their surface layer, thereby reducing their susceptibility to failure by initiating cracks from their surface [1-2]. This makes the technology applicable to the most exposed areas of dynamically and statically stressed machine components, such as turbine blade parts or welded joints [3]. The technology works on the principle of the interaction of a pulsed laser beam with the surface of the material to form a high-pressure plasma [3-4]. When plasma is generated, heat and shock waves are generated and propagate into the material being shot and its surroundings. The heat generated in this process, however, can reduce the effect of the LSP process. For this reason, a thin ablative layer in the form of, for

* Corresponding author: david.bricin@cvrez.cz, bricda@kmm.zcu.cz

example, black adhesive tape is applied to the surface of the material to be shot to protect the surface of the material to be shot against thermal effects. In addition to the ablative tape, a stream of water, the so-called cover layer, flows over the surface of the material to be shot. This uniform layer is designed to trap the emerging plasma and thus increase the time for which the emerging plasma acts on the surface of the material, see Fig. 1. In addition to the above, the type of laser beam used and its wavelength, the size of the beam spot, the pulse frequency, the pulse overlap, the firing strategy, and the properties and type of the material to be shot also affect the results achieved when shooting the material surface using the LSP method [3, 5]. Thus, the aim of this paper was to verify the ability of the selected laser pulse shooting strategy, to achieve the desired surface hardening of two types of steels, namely steel with austenitic structure X6CrNiTi18-10 and steel with martensitic structure X12CrNiMoV12-3, which are used for components in the energy industry such as turbine blades and piping systems.

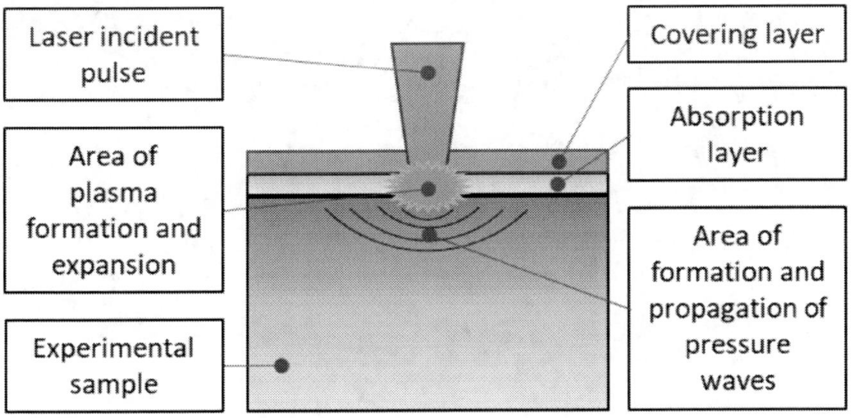

Fig. 1. The principle of the LSP process. When the laser strikes, the absorber layer is vaporized. A plasma is formed between the sample and the cover layer, flowing water, which creates a shock pressure wave propagating into the sample material volume

2 Experimental materials and methods

Two types of material were selected for the experiment, namely the austenitic refractory chromium-nickel steel stabilized with titanium X6CrNiTi18-10 and chromium-nickel-manganese-vanadium steel X12CrNiMoV12-3. Their chemical composition, recorded by EDS analysis, is given below, see Tab. 1. The initial microstructure of the samples taken is then shown in the metallographic images, see Figs. 2-3. The first of these materials, X6CrNiTi18-10 steel, is commonly used for the manufacture of piping systems. The second material, X12CrNiMoV12-3 steel, is commonly used for the manufacture of steam turbine blades.

Table 1. Chemical composition of analysed materials

Element/material	**Fe [hm.%]**	**Si [hm.%]**	**Mn [hm.%]**	**Cr [hm.%]**	**Mo [hm.%]**	**Ni [hm.%]**	**V [hm.%]**	**Ti [hm.%]**
X6CrNiTi18-10	68.4±0.13	0.6±0.0	1.9±0.0	18.7±0.1	-	10.1±0.13	-	0.3±0.0
X12CrNiMoV12-3	81.5±0.3	0.2±0.0	0.8±0.0	12.6±0.1	2±0.1	2.7±0.16	0.3±0.0	-

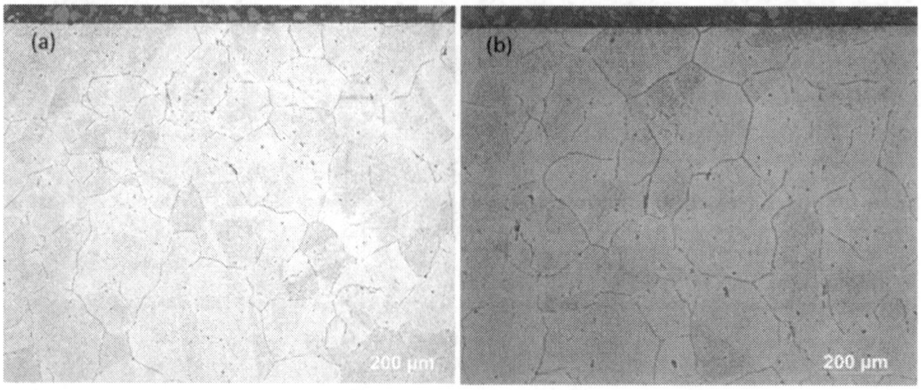

Fig. 2. Microstructure of X6CrNiTi18-10 steel samples. (a) section in the direction of firing; (b) section in the direction perpendicular to the direction of firing

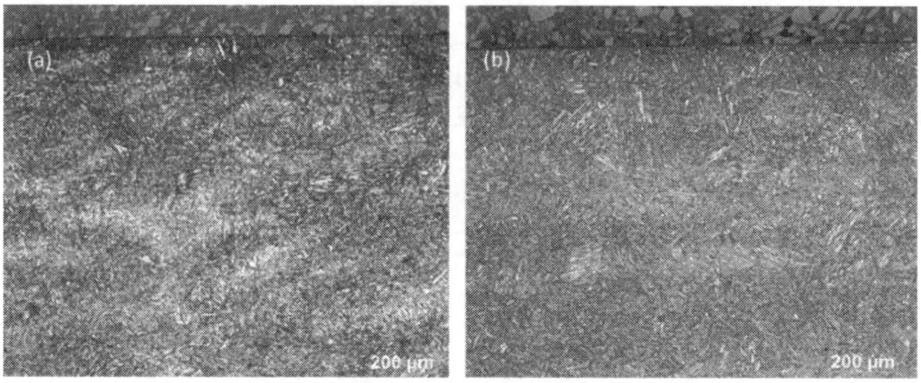

Fig. 3. Microstructure of X12CrNiMoV12-3 steel samples. (a) section in the direction of firing; (b) section in the direction perpendicular to the direction of firing

The materials for this experiment were supplied in the form of sheet metal plates with a nominal thickness of 6.5 mm for the X12CrNiMoV12-3 material and a thickness of 5.5 mm for the X6CrNiTi18-10 material. The delivered material was shot peened using a solid-state diode laser with Bivoj marking in the HiLASE centre (FZU, Dolní Břežany, CZE). This laser operates at a wavelength of 1030 nm. In this case, a pulse energy of 3 J and a duration of 30 ns for each pulse was chosen to focus on the sample surface. When shooting the surface of the delivered materials, a strategy was chosen in which the previous pulse was 52.8 % covered by the next pulse. The overlap between each pulse sequence was then 77.8%. Vinyl tape was used as a protective ablation layer. After surface shot peening of the samples, two samples were taken from each sheet of metal. The first sample was taken in the direction of the laser pulse shot sequence (longitudinal direction), the second in the direction perpendicular to it. An Olympus Lext OLS5100 confocal laser scanning microscope (Olympus Czech Group, s.r.o., Prague, CZE) was used to analyse the surface topography of the samples before and after the LSP shot process. A CarlZeiss Observer Z1m light microscope (Carl Zeiss s.r.o., Prague, CZE) and a Tescan Mira 3 scanning electron microscope (TESCAN ORSAY HOLDING, a.s., Brno, CZE) were used to analyse the differences in the microstructure of the sample surface. To induce the microstructure of the X6CrNiTi18-10 material, an etchant consisting of nitric acid (HNO_3), Hydrofluoric acid (HF) and glycerine ($C_3H_8O_3$) was used in a 2:2:1 ratio. A Vilella-Bain etchant was then used to induce the microstructure of the X12CrNiMoV12-3 material. The metallographic analysis was complemented by the measurement of the microhardness progression HV0.01, which was performed using a Struers DuraScan semi-automatic hardness tester (Struers GmbH, Roztoky u Prahy, CZE). The first indentation was performed for each sample at a distance of

50 μm from the edge of the sample. The distance between indentations was set to 50 μm. Hardness measurements were stopped in the distance where the hardness measured before the experiment was reached at the centre of each sample.

3 Experimental results and discussion

The LSP shot peening of the surface of the samples caused changes in the surface topography and roughness of the analysed samples, see Fig. 4 and Tab. 2. From the images, see Fig. 4, it can be seen that the surface of the samples was more affected by the austenitic steel, where the shot peening caused a more pronounced surface deformation, which was manifested by an increase in the variance of the values of the height profile of their surface compared to the martensitic steel samples and the surfaces of the samples that were not affected by LSP shot peening. These changes in surface topography are then reflected by the change in surface roughness of the samples after the LSP process, see Tab. 2.

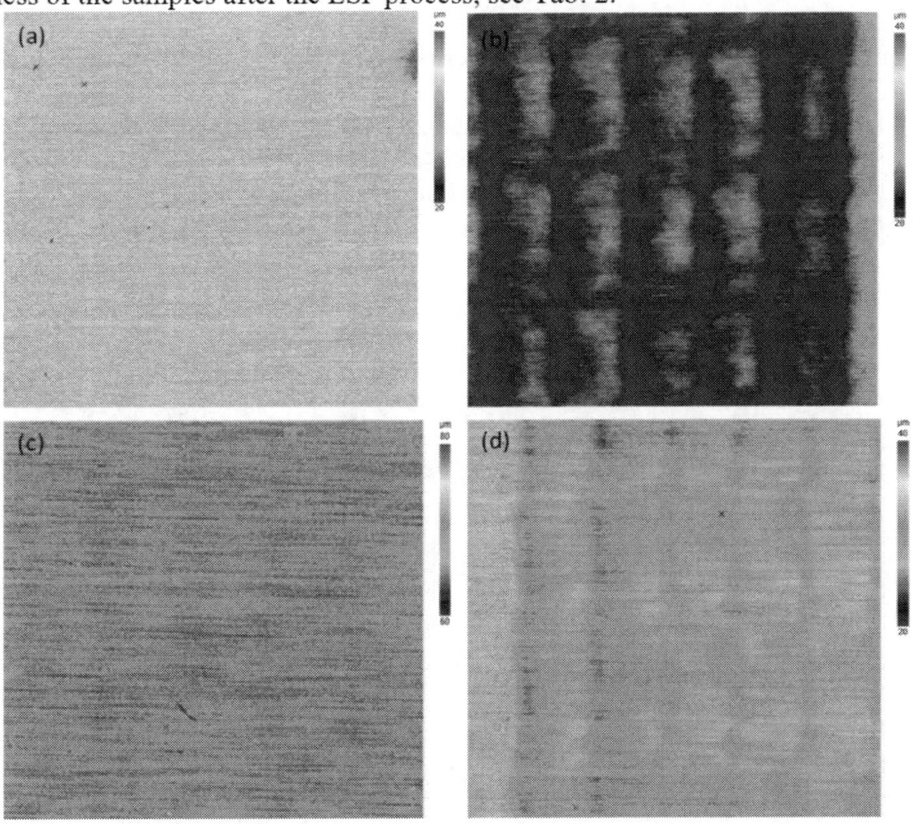

Fig. 4. (a) Height profile of the sample surface before LSP shooting, material X6CrNiTi18-10; (b) Height profile of the sample surface after LSP shooting, material X6CrNiTi18-10; (c) Height profile of the sample surface before LSP shooting, material X12CrNiMoV12-3; (d) Height profile of the sample surface after LSP shooting, material X12CrNiMoV12-3

In the case of the samples analysed, no significant change in their arithmetic mean surface roughness value Ra was observed, see Tab. 2. The fact that there was no significant change in this surface roughness parameter of the analysed samples was related to the fact that their surface was protected by ablative tape. Its use reduces the susceptibility of the sample surface material to melting and evaporation, which leads to the formation of surface craters, which can increase the surface roughness of the samples [6]. The presence of surface microcraters is more indicative of the presence of the mean depth of roughness parameter Rz. In the case of samples made of X6CrNi18-10 material, after LSP shot peening, the value of this parameter increased by about 10 % in the longitudinal direction and by about 54% in the transverse direction. For the X12CRNiMoV12-3 samples, the value of this parameter increased by about 2 % in the longitudinal direction and by about 29 % in the transverse direction. The size and number of depressions or craters could be further influenced by the pulse energy and duration [7].

Table 2. Linear profile roughness

Material	Longitudinal direction		Transverse direction	
	Ra[μm]	Rz[μm]	Ra[μm]	Rz[μm]
X6CrNiTi18-10	1.3±0.03	13±0.84	1.2±0.05	9.5±0.57
X6CrNiTi18-10 (LSP)	1.4±0.09	14.3±1.07	2.2±0.36	14.6±1.91
X12CrNiMoV12-3	1.7±0.04	16.2±0.74	1.6±0.19	13.3±0.96
X12CrNiMoV12-3 (LSP)	1.6±0.04	16.5±0.81	1.8±0.21	17.2±2.59

The pulse energy input causes changes in the microstructure of the material and changes in the applied residual stress. These changes are then reflected in a change in the hardness of the surface layer of the samples, see Figure 5. In the case of the X6CrNiTi18-10 austenitic steel samples, the hardness measurements determined the depth of substrate influence on the sample in the shooting direction to be approximately 0.6 mm, and in the direction perpendicular to this direction to be approximately 1.2 mm. In the case of samples made of martensitic steel X12CrNiMoV12-3, the depth of influence in the shooting direction was about 0.2 mm, in the perpendicular direction about 0.4 mm.

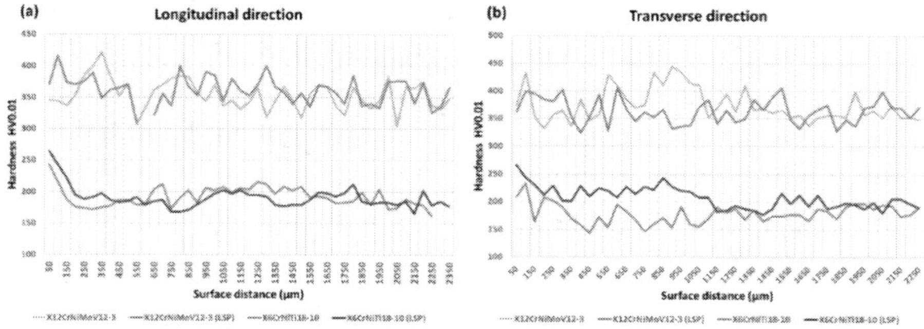

Fig. 5. Comparison of microhardness HV0.01 (a) section in the direction of firing; (b) section in the direction perpendicular to the direction of firing

It is known from the literature that the waveform of the imposed residual compressive stresses by a single shot, is uniform in the volume of the material, with a small depth gradient [8]. However, when multiple shots are taken, as shown by the simulation results in [9], a non-uniform distribution of the applied residual stresses occurs. This non-uniformity can then become apparent when measuring the microhardness of the material as in the case of this study. The highest hardness was achieved at or near the surface of the material. For austenitic steel X6CrNiTi18-10, the highest hardness value was recorded at a distance of 50 μm from the sample surface, in both directions analysed. In the shooting direction, a hardness of 264HV0.01 was measured at the surface of the sample shot by the LSP method. Compared to the unaffected sample surface, this was an increase in hardness of approximately 9 % in this region. In the direction perpendicular to the shooting direction, a hardness of 266HV0.01 was measured on the LSP-treated sample surface. Compared to the unaffected sample surface, this was an increase in hardness of approximately 28 % in this region.

For martensitic steel X12CrNiMoV12-3, the highest hardness was recorded for the specimens shot by the LSP method at a distance of 0.1 mm from the specimen surface for the cut in the direction of shooting and at a distance of 0.35 mm from the specimen surface for the perpendicular direction. In the shooting direction, 416HV0.01 hardness was measured in this area from the surface of the LSP specimen. Compared to the unaffected sample surface, this was an increase in hardness of approximately 21 % in this region. In the direction perpendicular to the direction of firing, a hardness of 353HV0.01 was measured in this area from the LSP specimen surface. Compared to the unaffected sample surface, this was an increase in hardness of approximately 5 % in this region. The introduced energy can then affect the microstructure of the surface of the shot peened material to a large extent. See Figs. 6-7 below for the surface structure of the samples recorded using an optical microscope.

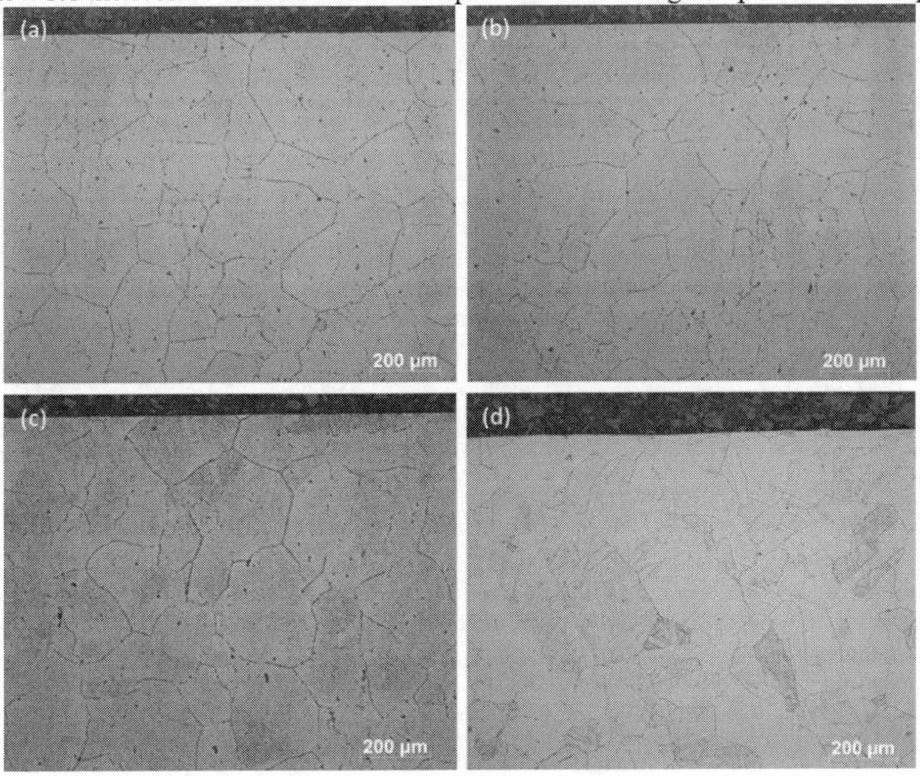

Fig. 6. Comparison of the surface structure of X6CrNiTi18-10 austenitic steel samples before and after LSP shot peening. (a) Structure of the unaffected surface, longitudinal direction- in the direction of shooting; (b) Structure of the surface after LSP shot, longitudinal direction- in the direction of shooting; (c) Structure of the unaffected surface, transverse direction- perpendicular to the direction of shooting; (d) Structure of the surface after LSP shot, transverse direction- perpendicular to the direction of shooting

From the metallographic images, taken by light microscopy (LM), it can be seen that the structure of the X6CrNiTi18-10 austenitic steel samples has not changed significantly due to the LSP process, see Fig. 6.

In a study [10], which investigated the microstructural changes in the surface of austenitic steels due to their LSP shot peening, it was found that grain refinement can occur during the LSP process, which mainly involves plastic deformation of the material surface by twinning and strain-induced martensitic transformation. However, none of the above was observed in the material structure in this case. The absence of significant microstructural changes due to the LSP process is related to the coarseness of the grains of the material used. As it was found the coarse grain structure of the material is able to absorb the generated shock wave in its volume without successfully refining the microstructure of the material [11]. This is also because in coarse-grained material there are far fewer suitably oriented planes of slip relative to the external load in which dislocations could move, and so it cannot experience such a significant increase in their density, which would result in a noticeable microstructural change, as in the case of bulk fine-grained austenitic steels.

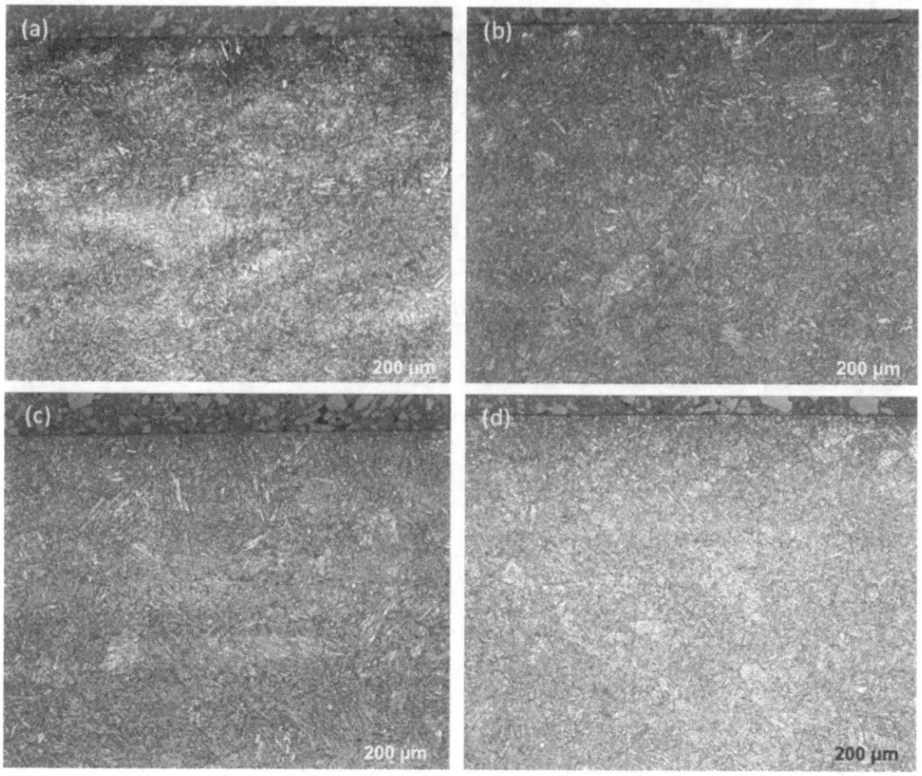

Fig. 7. Comparison of the surface structure of the martensitic steel X12CrNiMoV12-3 samples before and after LSP shot peening. (a) Structure of the unaffected surface, longitudinal direction-perpendicular to the shooting direction; (b) Structure of the surface after sniping, longitudinal direction- perpendicular to the shooting direction; (c) Structure of the unaffected surface, perpendicular direction- perpendicular to the shooting direction; (d) Structure of the surface after LSP shot peening, perpendicular direction- perpendicular to the shooting direction

Similarly, no significant structural change was captured by LM methods in the case of the shot peened surface of the martensitic steel X12CrNiMoV12-3, see Figure 7. As it has been found that shot peening of the surface of martensitic steels by LSP can lead to refinement of the martensite laths, reduction in the size of the original carbides and precipitation of nano-carbides in their surface [12-13]. However, these differences cannot be observed by LM methods.

Below are metallographic images that were obtained by analysing the structure of the shot peened samples using an electron backscatter detector (EBSD), see Figures 8-9. Figures 8(a) and 9(a) show the reconstructed grain boundaries of the two analysed materials. From the image, see Figure 8(a), it can be seen that some of the grains of the austenitic steel X6CrNiTi18-10 bear traces of plastic deformation by twinning. As noted, see ref. [10], this phenomenon can occur in the case of surface shooting of the following types of material. In the structure of the martensitic steel X12CrNiMoV12-3, the described refinement of the martensite lath structure was not observed, see Fig. 9(a), as in the case of the previously cited study, and see Ref. [12]. This is due to the fact that at such low magnification these structures are beyond the resolution of the method used. Figures 8(b) and 9(b) then show ipf maps of the surface structure of the analysed materials. The results show that the grain orientation is coincidental for both analysed materials. Therefore, the LSP process did not cause significant changes in their orientation. Using a component designed to determine the fraction of recrystallized structure, it was possible to observe the depth of plastic deformation in the surface structure of the X6CrNiTi18-10 austenitic steel samples, see Figure 8(c). The depth of this plastic deformation was not uniform over the entire surface. Its value reached values of units to tens of micrometres. In the case of the X12CrNiMoV12-3 martensitic steel samples, this area could not be visualized, probably because of the low resolution of the method, at the low magnification used. The last images document the microstructure of the samples when the component is used to show areas where significant plastic deformation of the material structure has occurred, see Figure 8(d) and Figure 9(c). From figure 9(c), it can again be seen that any changes in the structure of the X12CrNiMoV12-3 martensitic steel are below the resolution of this method at its low magnification. For the austenitic steel samples, see Fig. 8(d), it can be seen that, in addition to plastic deformation in the surface layer, traces of the metallographic preparation operations have become visible.

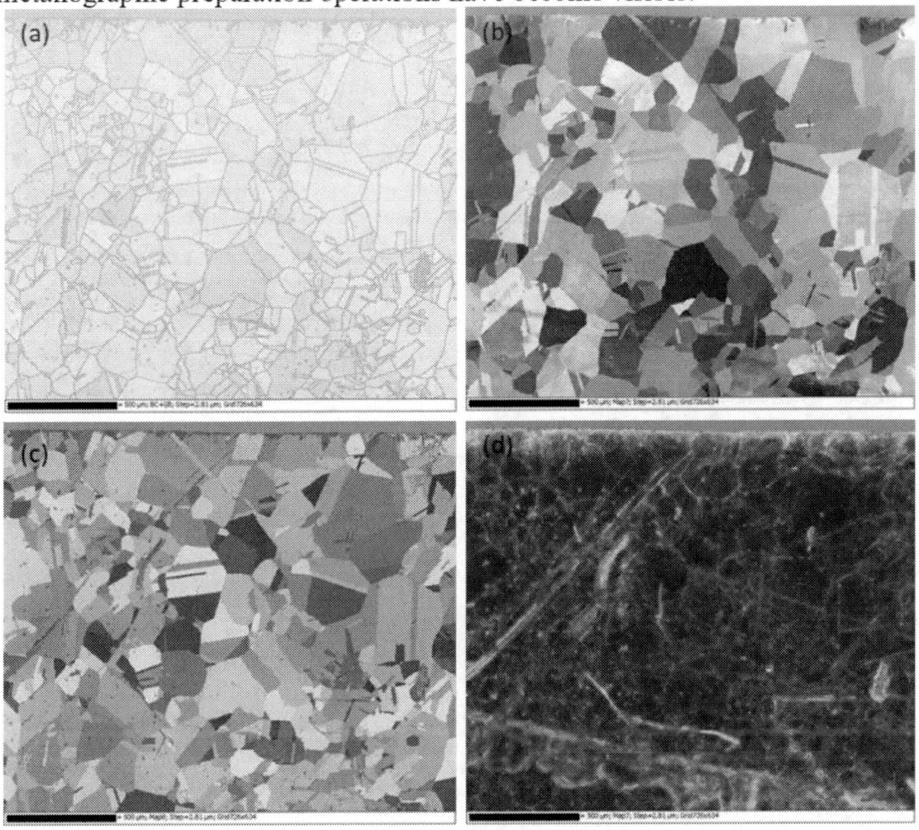

Fig. 8. (a) reconstruction of grain boundaries; (b) ipf map of the surface; (c) recrystallized fraction map, red colour shows plastically deformed grains, blue colour shows recrystallized grains; (d) local misorientation map shows areas with significant plastic deformation.

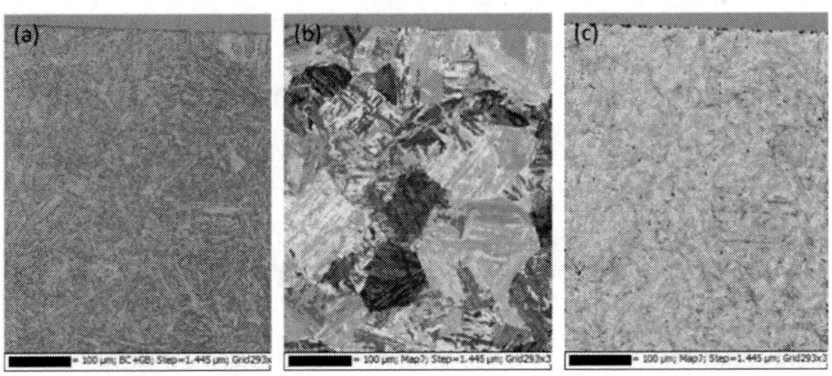

Fig. 9. (a) reconstruction of grain boundaries; (b) ipf map of the surface; (c) local misorientation map shows areas with significant plastic deformation

4 Conclusion

The aim of the study was to verify the quality of surface hardening of two types of steels, namely austenitic steel X6CrNiTi18-10 and martensitic steel X12CrNiMoV12-3 whose surface was shot peened by LSP. The results of the performed experiment confirmed that the surface layer of the substrate was influenced by its shot peening in the case of both analysed materials. The results obtained are summarized below:

- The more significant influence on the surface of the material occurred in the case of austenitic steel X6CrNiTi18-10. In this steel, a more pronounced plastic deformation occurred in the surface layer, which was manifested by a greater depth of the height profile.
- For both materials analysed, the value of the linear parameter of the mean depth of roughness Rz increased in both directions.
- Material hardness measurements showed an increase in surface hardness for both materials analysed. Higher depth of influence was achieved for austenitic steel X6CrNiTi18-10. The depth of influence on the surface of the material then varied for each measurement direction.
- The analysis of the structure of the samples by LM methods did not reveal any obvious differences between the LSP shot peened surface and the unshot peened surface for both the austenitic steel X6CrNiTi18-10 and the martensitic steel X12CrNiMoV12-3.
- EBSD analysis showed non-uniform plastic deformation in the surface layer of the X6CrNiTi18-10 austenitic steel samples to a depth of tens of micrometres. Other significant microstructural changes were not observed at the magnification used due to the low resolution of the method at this magnification.
- No significant change in the structure of the X12CrNiMoV12-3 martensitic steel samples was observed at the magnification used due to the low resolution of the method at this magnification.

From the above results, it is clear that the chosen shooting strategy did not bring significant changes in the microstructure of both analysed materials. One reason is probably the large coarseness of the grains of the analysed materials. Therefore, part of the follow-up experiments will deal with the optimization of the materials used and shooting parameters. The second part of the experiments will then deal with a more detailed description of the microstructure of existing samples at higher magnifications and use other metallographic methods such as TEM observation and local EBSD analyses.

This study was supported by the project SGS-2021-030 "Development of new materials, application of modern methods of their processing, ecological production, welding, and testing".

References

1. Kaufman, J., Špirit, Z., Vasudevan, V., Steiner, M., Mannava, S., Brajer, J., Pína, L. and Mocek, T., 2021. Effect of Laser Shock Peening Parameters on Residual Stresses and Corrosion Fatigue of AA5083. Metals, 11(10), p.1635.
2. Špirit, Z., Brajer, J., Kaufman, J., Chocholoušek, M., Bohm, M., Kott, J. and Strejcius, J., 2018. Effect of Laser Shock Peening on Fatigue life of Austenitic stainless steels. IOP Conference Series: Materials Science and Engineering, 461, p.012084.A. Mecke, I. Lee, J.R. Baker jr., M.M. Banaszak Holl, B.G. Orr, Eur. Phys. J. E **14**, 7 (2004)
3. R, S., P, G., Gupta, R., G, R., Pant, B., Kain, V., K, R., Kaul, R. and Bindra, K., 2019. Laser Shock Peening and its Applications: A Review. Lasers in Manufacturing and Materials Processing, 6(4), pp.424-463.

4. Spirit, Z., Kaufman, J., Strejcius, J., Chocolousek, M. and Kott, J., 2019. Increase of the Fatigue Life of Stainless Steel by Laser Shock Peening. DAAAM Proceedings, pp.0826-0831
5. Gujba, A. and Medraj, M., 2014. Laser Peening Process and Its Impact on Materials Properties in Comparison with Shot Peening and Ultrasonic Impact Peening. Materials, 7(12), pp.7925-7974
6. Montross, C., 2002. Laser shock processing and its effects on microstructure and properties of metal alloys: a review. International Journal of Fatigue, 24(10), pp.1021-1036
7. Qiao, H., Zhao, J. and Gao, Y., 2015. Experimental investigation of laser peening on TiAl alloy microstructure and properties. Chinese Journal of Aeronautics, 28(2), pp.609-616
8. Hu, Y., Yao, Z. and Hu, J., 2006. 3-D FEM simulation of laser shock processing. Surface and Coatings Technology, 201(3-4), pp.1426-1435
9. Keller, S., Chupakhin, S., Staron, P., Maawad, E., Kashaev, N. and Klusemann, B., 2018. Experimental and numerical investigation of residual stresses in laser shock peened AA2198. Journal of Materials Processing Technology, 255, pp.294-307
10. Zhou, L., He, W., Luo, S., Long, C., Wang, C., Nie, X., He, G., Shen, X. and Li, Y., 2016. Laser shock peening induced surface nanocrystallization and martensite transformation in austenitic stainless steel. Journal of Alloys and Compounds, 655, pp.66-70
11. Wang, Z., Sun, G., Lu, Y., Chen, M., Bi, K. and Ni, Z., 2020. Microstructural characterization and mechanical behavior of ultrasonic impact peened and laser shock peened AISI 316L stainless steel. Surface and Coatings Technology, 385, p.125403
12. Wang, C., Luo, K., Wang, J. and Lu, J., 2022. Carbide-facilitated nanocrystallization of martensitic laths and carbide deformation in AISI 420 stainless steel during laser shock peening. International Journal of Plasticity, 150, p.103191
13. Wang, C., Luo, K., Bu, X., Su, Y., Cai, J., Zhang, Q. and Lu, J., 2020. Laser shock peening-induced surface gradient stress distribution and extension mechanism in corrosion fatigue life of AISI 420 stainless steel. Corrosion Science, 177, p.109027

Heat Treatment Influence on Mechanical Properties of Nickel Based Cold Spray

Michal Chocholousek[1*], *David* Bricín[1] and *Zbynek* Spirit[1]

[1] Research Centre Rez (CVR), Material Testing Department (8720), Morseova 1245/6, 301 00 Plzen, Czech Republic

Abstract. The article deals with evaluation of Ni and NiCr Cold Spray layers as a potential method for repairing components in energy industry by replacing or refilling flawed material. Cold spray layers from two NiCr powders of different particle sizes sprayed with nozzle for inner tube application and from Ni powder of one particle size sprayed with different nozzle types – for inner and outer tube applications – were tested. The evaluation is aimed on mechanical properties and metallographic analysis of cold spray layers as received and after defined heat treatments. It summarizes tensile properties at room temperature and microstructure of used powders and final cold spray layers. Ultimate tensile strength, hardness and plastic elongation are evaluated from the mechanical point of view. Metallographic analysis is focused on ovality and size of particles in powders and of splats in final cold sprays. Porosity of sprayed layers is evaluated on tested specimens.

1 Introduction

Cold spray (CS) is a deposition process of material in form of powder, in which particles dimensions are typically of the order of micrometres. The powder is accelerated in the solid state by gas nozzle to velocities 200-1200 m/s [1]. As-sprayed CS coating has its advantageous properties, e.g. high density, hardness, increased strength and induced compressive residual stresses (potentially positive impact on fatigue and crack initiation), all with minimum thermal input to the substrate [2].

The above properties predetermine CS as a potential method for repairing components by replacing or refilling flawed material. However, it is necessary to determine the material properties of applied CS [3-5] and to prove the applicability. According to previous experience in Centrum Vyzkumu Rez (CVR), CS coatings (few millimetre thickness) exhibits sufficient strength and compactness to be tested separately without the substrate. Hence, standard tensile specimen shape is possible to use and evaluate according to appropriate standards [6].

As-sprayed CS coatings are normally characterized by high strength and low ductility [3], which lead to deterioration of fracture properties. Heat treatment (HT) can improve CS ductility for certain materials [4,5]. Hereafter, one of the determined goals is to obtain

* Corresponding author: michal.chocholousek@cvrez.cz

suitable CS coating, which can be used to repair certain pipe-components in power industry. The influence of heat treatment is thus examined in suitable temperature range considering the substrate material and for higher temperatures to evaluate the temperature influence trend.

Spraying system for industrial application should be adapted to limitations of component dimensions and work space limits. Spraying system adaptation includes non-orthogonal positioning of nozzle to substrate surface (Figure 1). Various angles of spraying can possibly have impact on CS quality. Thus the porosity and material properties are estimated for angle 45° of spraying nozzle.

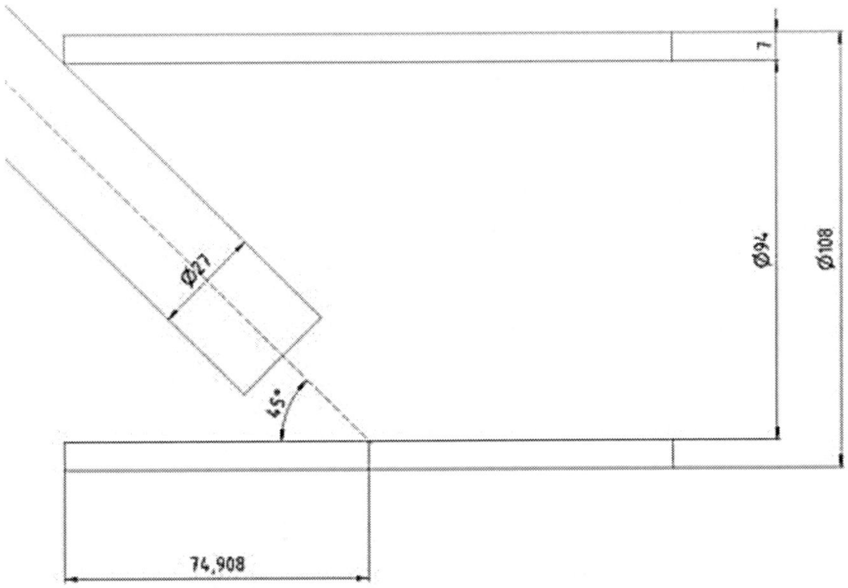

Fig. 1. Schema of CS deposition for tube 108×7 mm with nozzle angle at 45°.

The paper summarizes results of the subtask, which is aimed on verification of CS material properties. Two NiCr powders of different particle sizes sprayed with nozzle for inner tube application and Ni powder of one particle size sprayed with different nozzle types (for inner and outer tube applications) were tested.

2 Experimental procedure

As a candidate materials, nickel and NiCr (Ni 80 %, Cr 20 %) powders were used for cold spraying. Two different particle sizes (5-25 μm and 15-32 μm) of NiCr powder were used, sprayed with nozzle for inner tube application. The particle size 15-35 μm of Ni powder was sprayed with nozzles for inner and outer tube application. Non-optimal angle of the spraying nozzles was 45°, which is presumed to be used for in-situ applications. Other parameters of the spraying were guaranteed to be identical (Impact Innovations GmbH). A layer of 4.5 mm thickness was applied on aluminium substrate 65×280 mm. CS layers were cut off from substrate and used for electrical discharge machining of flat tensile specimens with 3.5×4 mm cross-section area, i.e. original 4.5 mm CS thickness was reduced to 3.5 mm.

Previous experience with NiCr CS [7] indicated low but not negligible influence of HT at 640 °C. However, potential application on tubes in energy industry allows temperature below 400 °C. Thus the HT temperature range was chosen with respect to the amount of available material from the temperature limit of considered application (390 °C) up to confirmed HT influence (600 °C). HT temperatures are summarized in Table 1. HT was performed at defined heating rate and at natural cooling rate in air. Hold time at lower temperatures was increased to support the HT influence on CS.

Table 1. Parameters of applied HT.

Designation	Heating rate (°C/h)	Max. T (°C)	Hold time at max. T (h)
HT1	80	390	120
HT2	80	500	48
HT3	80	600	24

Tensile tests were performed by means of ZwickRoell universal testing machine Z250. The tensile tests were conducted according to the standard [6] to determine ultimate tensile strength (UTS) and strain (A_{20}) of bone shape specimens with initial gage length 20 mm. Hardness measurement (HV1) was performed by means of Struers DuraScan hardness tester on as received CS layers and CS after heat treatment. The results are averages evaluated from at least four measurements. Metallographic analysis was performed by means of light microscopy (LM) – Carl Zeiss Observer Z1m – and scanning electron microscopy (SEM) – Tescan Mira3, focused on ovality and size of particles in powders (e.g. Figure 2) used for cold spraying and of splats (e.g. Figure 3) in final cold sprays. The analysis of powders was done from twenty measurements for each powder. The samples for analysing were prepared by floating in ethanol with aim to dissolve agglomerated structures. Porosity of sprayed layers was evaluated by means of LM on tested bone-shape specimens on the specimens' heads. The analysed area was in the middle of the CS layer due to perpendicular direction to substrate surface. Influence of HT on CS splat microstructure was not expected due to low temperatures, thus the analysis was performed only on as-received CS layers.

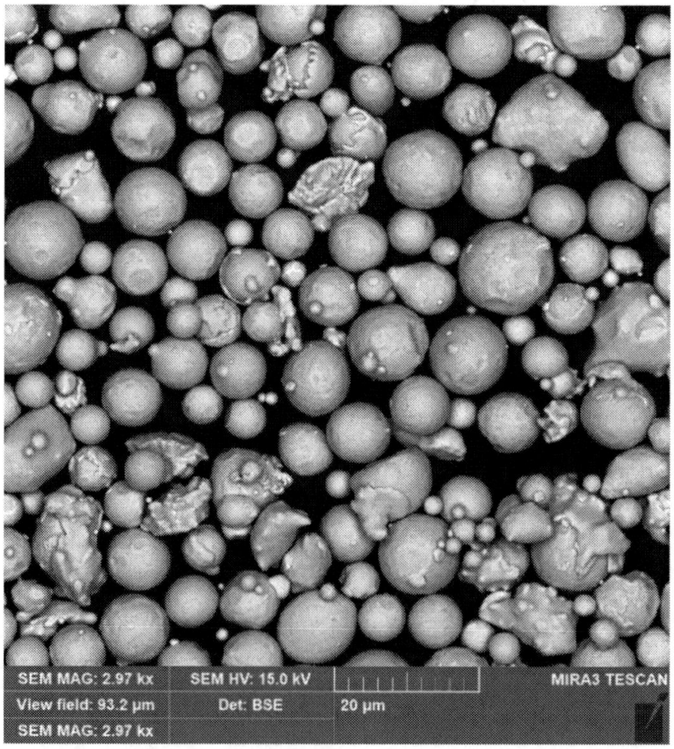

Fig. 2. SEM overview of NiCr powder with particle size distribution 5-25 μm.

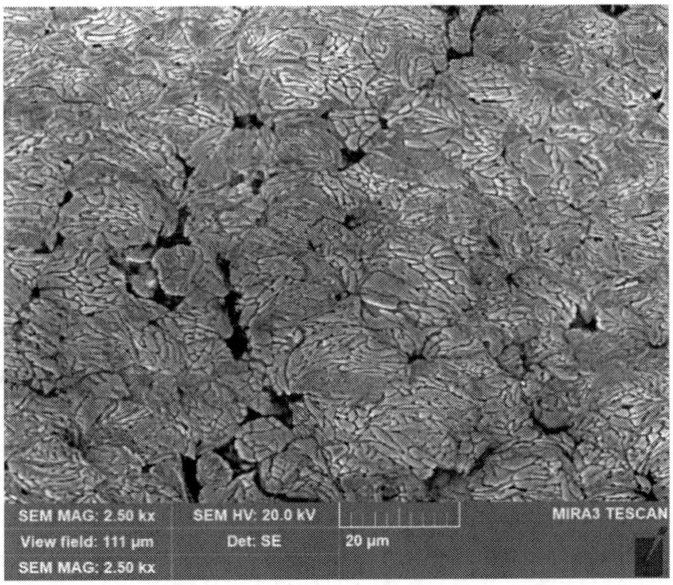

Fig. 3. SEM overview of CS layer of NiCr powder 5-25 μm sprayed by inner nozzle. CS is formed by splats with several grains. Dark areas indicate porosity of the layer.

3 Results and discussion

Particle size is generally presented by suppliers as a range of diameters (e.g. NiCr 5-25 μm), where the margins are given by sieve used for analysing. Major amount of powder, over 95 %, is in the proclaimed range. Laser diffraction analysis was not available in certificates for all powders, thus the particle size distribution was done for each powder using LM to obtain information about average particle size. Summary of particle sizes for each powder used for CS is presented in Table 2. However the proclaimed size ranges for both NiCr powders are similar, the more precise measurement reveals that the average size of one powder is significantly larger than the other one. The coarser NiCr powder is comparable with Ni powder due to the particle size. The average size of splats in CS layers is presented in Table 2. It can be noted that the size of splats should be equal to the size of powder particles independently on the used nozzle type. Comparing similar results from CS splats and powder particles gives general overview of the measurement accuracy. Table 3 contains splat size distribution in CS, which gives more precise information about sizes of used particles.

Table 2. Average particle size of powders used for cold spraying and average plat size of final CSs.

Designation	Average size ± standard dev. (μm)
NiCr powder, 5-25 μm	6 ± 3
NiCr powder, 10-32 μm	13 ± 6
Ni powder, 15-35 μm	13 ± 8
CS (NiCr 5-25 μm, inner nozzle)	10 ± 3
CS (NiCr 10-32 μm, inner nozzle)	17 ± 4
CS (Ni 15-35 μm, inner nozzle)	16 ± 7
CS (Ni 15-35 μm, outer nozzle)	17 ± 6

Table 3. Splat size distribution of CS layers.

Splat size (equivalent diameter) (%)							
µm	**0-5**	**5-10**	**10-20**	**20-30**	**30-40**	**40-50**	**50<**
CS (NiCr 5-25 µm, inner nozzle)	6	56	38	1	0	6	0
CS (NiCr 10-32 µm, inner nozzle)	0	0	81	19	0	0	0
CS (Ni 15-35 µm, inner nozzle)	7	3	61	27	1	7	0
CS (Ni 15-35 µm, outer nozzle)	2	2	69	27	0	2	0

The results from the ovality measurement of the powders and CS splats are shown in Table 4. The presented results of ovality representation are an arithmetical average obtained from three measurements from each powder and CS layer. It is obvious that the initial high ovality of the powders is deformed to lower ovality of CS splats, as expected. The ovality of splats in NiCr CS is similar for both layers made from powders of different particle sizes. Ni powder used with two different nozzles results in lower ovality of splats sprayed with outer nozzle, which is in agreement with the expectation of higher quality CS than CS sprayed with inner nozzle. Higher deformation of splats in Ni CS compared to NiCr CS is due to identical parameters of spraying caused by properties of material itself.

Table 4. Results of percentage ovality representation of powders particles and CS splats.

Designation	**Ovality representation (%)**				
	0-0.2	**0.2-0.4**	**0.4-0.6**	**0.6-0.8**	**0.8-1**
NiCr powder, 5-25 µm	1	5	14	15	66
CS (NiCr 5-25 µm, inner nozzle)	0	10	42	38	10
NiCr powder, 10-32 µm	0	3	8	13	75
CS (NiCr 10-32 µm, inner nozzle)	0	11	44	37	7
Ni powder, 15-35 µm	0	3	11	20	66
CS (Ni 15-35 µm, inner nozzle)	1	34	40	24	0
CS (Ni 15-35 µm, outer nozzle)	4	45	42	7	2

Mechanical properties together with porosity are summarized in Tables 5 - 8 for each CS layer as received and after HT (see Table 2). Note that elongation A_{20} was not evaluated for CS from NiCr powders (Tables 5 and 6). The elongation was negligible due to brittleness of the NiCr layers and it can be considered as zero. The porosity and pore sizes are evaluated for as received material and heat treated material too. There was no expectation of influence of HT on porosity, however, the influence of porosity have to be excluded in case of scattering results for specimen series after various treatments. The low scatter of porosity results of each CS layer and similar pore sizes are observed comparing one material with and without treatment, therefore each CS layer can be considered as a homogeneous material.

Table 5. Results of mechanical properties and porosity of CS (NiCr 5-25 μm, inner nozzle).

Designation	UTS (MPa)	A_{20} (%)	HV1	Porosity area (%)	Pore size (μm^2)
as received	124±16	N/A	334±21	1.6	676
HT1 (390°C, 120h)	157±23	N/A	400±23	2.2	434
HT2 (500°C, 24h)	182±32	N/A	386±25	1.9	344
HT3 (600°C, 12h)	198±29	N/A	279±12	1.8	354

Table 6. Results of mechanical properties and porosity of CS (NiCr 10-32 μm, inner nozzle).

Designation	UTS (MPa)	A_{20} (%)	HV1	Porosity area (%)	Pore size (μm^2)
as received	67±11	N/A	254±29	7.7	2064
HT1 (390°C, 120h)	92±21	N/A	304±30	7.6	2523
HT2 (500°C, 24h)	118±19	N/A	308±33	8.7	2538
HT3 (600°C, 12h)	125±11	N/A	228±16	7.4	2017

Table 7. Results of mechanical properties and porosity of CS (Ni 15-35 μm, inner nozzle).

Designation	UTS (MPa)	A_{20} (%)	HV1	Porosity area (%)	Pore size (μm^2)
as received	75±7	0	204±11	1.4	1146
HT1 (390°C, 120h)	204±7	1.9±0.7	105±8	0.9	917
HT2 (500°C, 24h)	199±4	2.5±0.5	101±6	1.1	769
HT3 (600°C, 12h)	213±5	5.3±1.7	95±3	0.8	955

Table 8. Results of mechanical properties and porosity of CS (Ni 15-35 μm, outer nozzle).

Designation	UTS (MPa)	A_{20} (%)	HV1	Porosity area (%)	Pore size (μm^2)
as received	508±45	0	220±10	0.2	410
HT1 (390°C, 120h)	378±8	17.0±2.0	106±7	0.1	269
HT2 (500°C, 24h)	384±6	21.6±2.6	114±12	0.2	486
HT3 (600°C, 12h)	396±2	28.6±3.2	104±4	0.1	389

Considering previous result [7] of NiCr layer sprayed by outer nozzle perpendicular to surface (i.e. UTS approx. 400-500 MPa without HT), the values of UTS were expected to be higher for NiCr layers. However, the high absolute value of porosity together with low ductility result in fracture in early stage of tensile loading. The influence of porosity can be seen by comparing the UTS results of NiCr CS (Tables 5 and 6), where the CS from the coarser powder shows a higher porosity and larger pores, which leads to lower UTS values. The influence of the nozzle type on porosity can be observed in Tables 7 and 8, where the outer nozzle results in significantly lower porosity. Generally, the UTS values scatter for as received material and the scatter of results decreases with ductility increase, which is observed on Ni CS (Tables 7 and 8). The scatter of UTS results persists for NiCr layers, the influence of HT on ductility was negligible (Tables 5 and 6).

Negligible ductility of as received CS layers is generally expected due to the high residual stress caused by deformation of powder particles during the spraying process. Brittleness removement due to HT depends on the material of used powder. NiCr layer is obviously more thermal stable, however, the influence of HT is still significant and it increases with higher temperatures. Small increase of ductility is expected, though it have to be confirmed with additional measurement with higher accuracy of deformation measurement and it is only hypothetical at the moment. On the other side, Ni layer is significantly loosing residual stress even for low temperature of HT, which leads to significant ductility increase at the expense of UTS (see Figure 4) and HV1. For quantifying residual stresses, additional analysis is recommended. Hardness decrease is significant especially between the heat treated and as received material. Dependance on HT temperatures cannot be affirmed due to scatter of similar results for Ni layer. Comparing influence of HT in Ni CS (Tables 7 and 8), it can be noted that the ductility improvement due to HT results in UTS increase in cases, where UTS is significantly low due to porosity (Table 7).

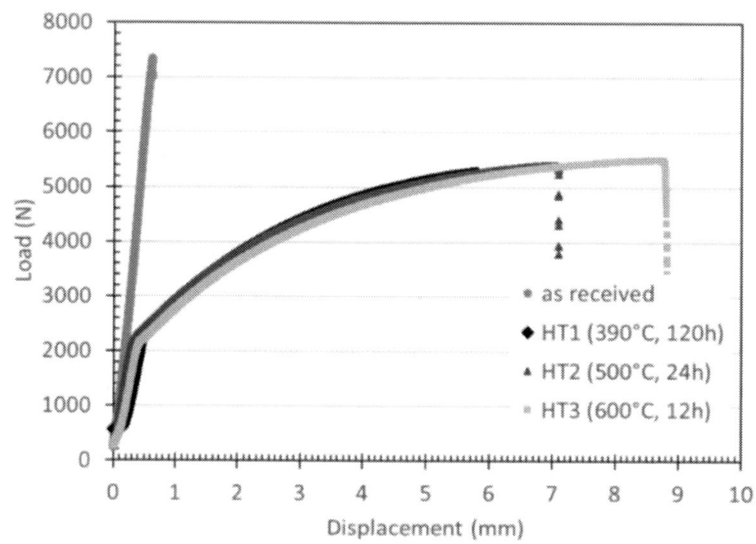

Fig. 4. Load-displacement tensile curves of Ni layers (outer nozzle). HT results in UTS decrease and in elongation increase with increasing HT temperature.

4 Conclusion

Tensile and hardness tests were performed on four different CS layers differing in used powder (Ni and NiCr) and nozzle type, supplemented by metallographic analysis of powders particles, CS porosity and CS splats. Three heat treatments in temperature range 390-600 °C were performed on each layer, on selected series of specimens.

The type of the nozzle used for cold spraying influences the final porosity of the layer. Optimizing spraying parameters for the nozzle for inner tube applications is recommended, however, that is not the aim of the study. Spraying powder with lower diameter of particles improves cold spray properties and results in lower porosity.

The porosity has significant influence on ultimate tensile strength values. The cold spray as received is generally brittle. The brittleness in combination with porosity leads to noteworthy decrease of ultimate strength.

The influence of heat treatment on tensile properties was observed on each cold spray. It differs with used material of powder. NiCr behaves as thermally stable material, heat treatment influence results in increase of ultimate strength, though the influence on ductility have to be confirmed by additional measurement. Heat treatment of Ni layers results in significant decrease of ultimate strength, however, it has positive impact on ductility. The ductility grows with increasing temperature of heat treatment. The ductility improvement due to heat treatment results in ultimate strength increase in cases, where ultimate strength is significantly low due to porosity.

The ovality of splats in cold spray layer appears to be material dependent parameter. The results of splat ovality representation for NiCr is equal for both powder particle sizes. NiCr is generally harder material and shows lower rate of deformation than Ni splats. Using outer and inner nozzle with Ni powder resulted in different values of splat ovality. Comparing the absolute sizes of powder particles and equivalent diameters of splats results in similar values.

Negligible ductility of as received CS layers is generally expected due to the high residual stress caused by deformation of powder particles during the spraying process. Removing the brittleness due to heat treatment depends on the material of used powder and it is expected to be connected to the loss of the residual stress. Measurement of residual stresses is recommended for further analysing of heat treated cold sprays.

The presented work has been realized within Institutional Support by Ministry of Industry and Trade (Czech Rep.). The submitted work was created thanks to a project supported by the TA CR No. TN01000007 within the Program of the National Center of Competence - 1st PC.

References

1. A. Papyrin, A. Kosarev, S. Klinkov, A. Alkimov, V. Fomin, Cold Spray Technology, pp 13-17, Elsevier (2007). ISBN 978-0-08-045155-8
2. J. Villafuerte (Ed.) Modern Cold Spray (Windsor, Ontario, Canada: Springer International Publishing, 2015). ISBN 978-3-319-16771-8
3. P. Cavaliere, A. Silvello, N. Cinca, H. Canales, S. Dosta, I. Garcia Cano, J.M. Guilemany, *Microstructural and fatigue behavior of cold sprayed Ni-based superalloys coatings*, Sur-face & Coatings Technology **324**, pp 390–402 (2017)
4. M.R. Rokni, C.A. Widener, O.C. Ozdemir, G.A. Crawford, *Microstructure and me-chanical properties of cold sprayed 6061 Al in As-sprayed and heat treated condition*, Surface & Coatings Technology **309**, pp 641–650 (2017).
5. J.W. Murray, M.V. Zuccoli, T. Hussain, *Heat Treatment of Cold-Sprayed C355 Al for Re-pair: Microstructure and Mechanical Properties*, J Therm Spray Tech, **27**:159–168, (2018). (https://doi.org/10.1007/s11666-017-0665-z)
6. EN ISO 6892-1: Metallic materials – Tensile testing – Part 1: Method of test at room temperature
7. M. Chocholousek, Z. Spirit, J. Brom, *Material Properties of Steel with NiCr-Cold Spray Coating,* IOP Conference Series. Materials Science and Engineering; Aug 2021, Bristol, **1178**, 1 (2021). DOI:10.1088/1757-899X/1178/1/012022

Numerically modelling temperature deformations and stresses in the exhaust system of a gas compressor unit

Maryna Demianenko [1*], *Michal* Volf [1], *Myron* Banias[2] and *Oleksandr* Starynskyi [3]

[1] University of West Bohemia, 22 Univerzitni, Pilsen, 32600, Czechia
[2] NAEK Energoatom, 22-24 Hoholivska, Kyiv, 01032, Ukraine
[3] Sumy State University, 2 Rymskogo-Korsakova, Sumy, 40007, Ukraine

Abstract. During the operation of a gas compressor unit, the temperature of the exhaust system can cause failure or burnout of the inner facing sheets of the unit. The article describes a technique for determining the maximum temperature deformations of these elements. The technique enables more accurate calculation of the temperature deformations by determining the influence of the contacts between the inner facing and the power frame on the temperature distribution and mobility of the exhaust system. The temperature deformations and stresses of the exhaust system elements are determined, and the level of the permissible stresses is evaluated. The results of the work can be used to prevent possible ruptures and burning of the inner facing, as well as to calculate the design dimensions of the lens expansion joint.

1 Introduction

The operational reliability and efficiency of gas-turbine stations and gas compressor units (GCUs) depend on their main elements, which comprise the axial compressor, combustion chamber, and high and low-pressure turbines. But no less important for ensuring uninterrupted energy-efficient operation of the installation are the inlet air preparation device and the exhaust system of the unit [1]. The GCU exhaust system removes the exhaust gases from the gas turbine engine, reduces the concentration of pollutants into the atmosphere, lowers the noise levels and keeps the hydraulic resistance within the limits necessary for the normal operation of the gas turbine engine. The GCU exhaust tract is a steel casing (power frame) placed on movable and fixed supports, consisting of a gas duct, a diffuser, an adapter, a rotary device, and a vertical exhaust pipe.

Designing an exhaust system includes the calculation of hydraulic resistance and the degree of purification of gases, and strength calculations. The strength calculations consider the main loads acting on the elements of the exhaust system, i.e. thermal and dynamic loads. As indicated in [2], thermal loads are calculated at the design stage and checked in the launch process, which allows some design solutions to be specified (e.g. design of lens expansion

* Corresponding author: demianen@fst.zcu.cz

joint). The existing engineering methods for calculating thermal loads do not allow the precise determination of the deformations of the GCU exhaust system elements. This is primarily due to the complexity of the exhaust system, the presence of many detachable and non-detachable joints, as well as basalt and sound-absorbing plates, which contribute to the temperature distribution, overall mobility of the power frame, and consequently, the deformations of the exhaust system's structural elements.

The dynamic processes in the exhaust system, namely the vibroacoustic influence of the flow on the inner facing, have not been taken into account by researchers. In [2] it was found that the gas flow is highly unstable, which is associated with the formation of vortices and their passage through the GCU exhaust system. The gas pressure fluctuations reach an amplitude of 110 kPa with a frequency of 10-15 Hz. These factors can lead to failures during operation, where the inner facing sheets collapse and there are blowouts of the lens expansion joints.

Therefore, the research goal is to conduct refined calculations of thermal loads on the GCU exhaust system to determine its deformations and stresses. In the future, it will allow us to determine the optimal sizes of the lens expansion joints and avoid the collapse of the inner facing. It will therefore ensure the reliable operation of a GCU and its exhaust system in various microclimatic zones. The research will be divided into two stages, which will allow us to separately assess the effects of thermal and dynamic loads on the elements of the GCU exhaust system. This article describes the calculation of the effects of the thermal loads.

2 Literature Review

The influence of the thermal load on a GCU exhaust system can be determined using numerical methods. This is confirmed in many works devoted to the numerical research of various elements of a GCU, which show that numerical methods allow us to reliably determine the operating parameters of a unit [3-6]. One method which can be used is finite element analysis (FEA), which is implemented in many software packages, including the Static Structural module of ANSYS. This module has also proved to be applicable for estimating the strength of complex industrial structures [7-10].

In [11] FEA and computational fluid dynamics (CFD) are used to evaluate the thermomechanical fatigue life of a cylinder head. The cylinder head has a complex internal design (including water and air ducts, etc.), and a full simulation would be quite complex and time-consuming. Therefore a simplified model of the cylinder head was developed for the simulation to find the deformations and stresses.

In [12], FEA of the thermal load of a small piston engine cylinder was carried out in ANSYS. The complex design of the cylinder head was simplified at the geometry preparation stage. The geometry was divided into separate parts, which allows a mesh of higher quality to be built. In such cases, a certain type of contact is specified between the separate bodies, which determines their interaction.

In [13], the thermal load characteristics of a double dry clutch were studied using ABAQUS software. Because of the focus on the temperature field distribution of the friction plate, the pressure plate, and the intermediate master plate, these parts were divided into separate bodies. The "standard" contact type was set between these plates, which determines the interaction of the surfaces and the possibility of common temperature deformations.

It is necessary to develop a simulation technique for each design on the basis of the reviewed works, which will consider features such as geometry, types of mountings, and materials. CFD and the methods described in papers [14, 15] can be used to estimate the dynamic loads on the GCU exhaust system elements caused by the vibroacoustic effects of the gas flow.

3 Research Methodology

Simulating the temperature deformations and stresses of the GCU exhaust system elements requires data on the maximum possible temperature gradients during operation in a microclimatic region with a cold climate. The temperature of the internal facing of the exhaust system is +514°C, the minimum temperature of the outside air is -50°C, the maximum temperature of the outside air is +32°C. The material configuration of the exhaust pipe walls is shown in Fig. 1. The temperature deformations and stresses will be modelled using FEA using the Static Structural module of ANSYS 2019 R2.

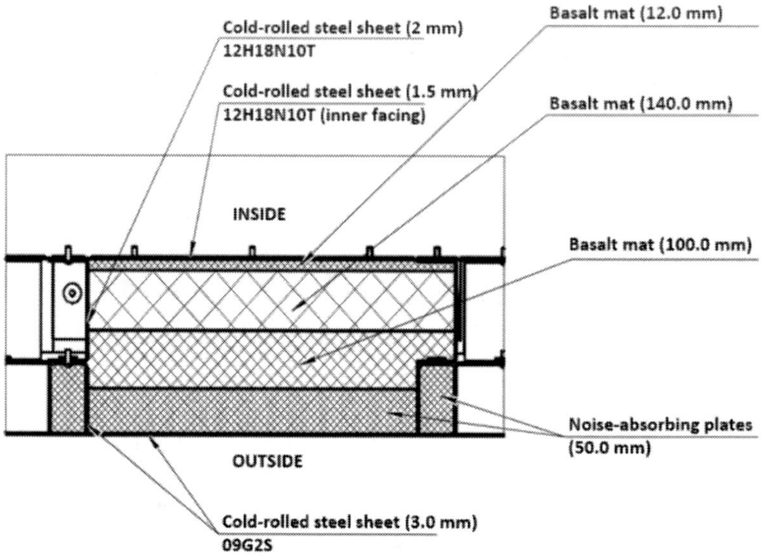

Fig. 1 – Design of the horizontal part of the GCU exhaust system

Due to the absence of fixed connections between the inner facing and the exhaust system power frame, as well as a large number of contacting surfaces, the numerical simulation of the temperature-stress state of the exhaust system horizontal section is extremely resource-consuming. Therefore, an analysis of the thermal stress state for each element was performed, and the total displacements were determined analytically according to these results.

During heating, the inner facing sheets expand, and the forces occurring in them can be divided into two types: forces resulting from constrained temperature deformations, and forces from outer loads (forces from the set-back reactions of the studs). The impact of the inner facing thermal deformations into the stress state of the power frame was evaluated by solving a test problem. The test section is shown in Fig. 2. The temperature of the inner facing equal to +514°C was set as the load, and the load from the weight of the power frame was set by including gravitational force in the calculation. The boundary condition "convection" was set on the outer walls of the outside facing of the exhaust system.

The contact type 'Bonded' is set between washer 1 and stud 2, meaning that the contact surfaces cannot move relative to each other, but joint temperature deformation is possible. The contact type 'Frictional' was set between washer 1 and facing 3, with a friction coefficient equal to 0.15, in which case slippage of the surfaces and their separation is possible. Stud 2, frame 4, and the outer facing 5 are fused into one body, as they are welded together. 'Joint General' joints were specified between stud 2 and facing 3, with the ability to move in all three coordinate axes and the ability to rotate about the axis of stud 2. Figure 3 shows the mesh, and the results of strain calculations are shown in Figure 4.

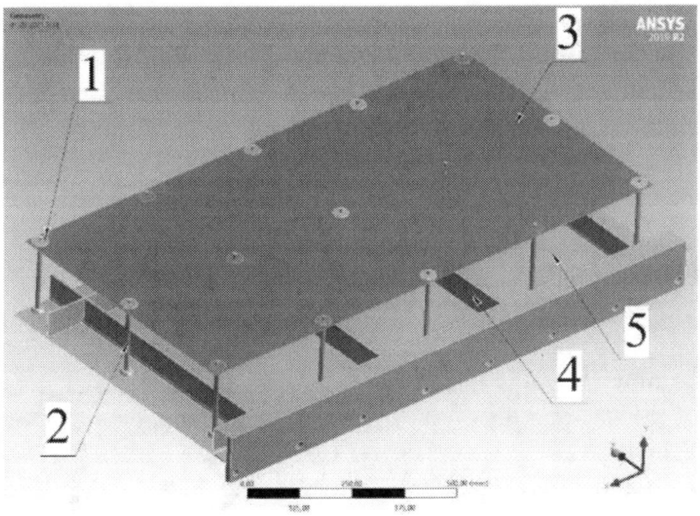

Fig. 2 – Design of the GCU exhaust system section for the test case: 1 – washer, 2 – stud, 3 – inner facing; 4 – power frame; 5 – outside facing

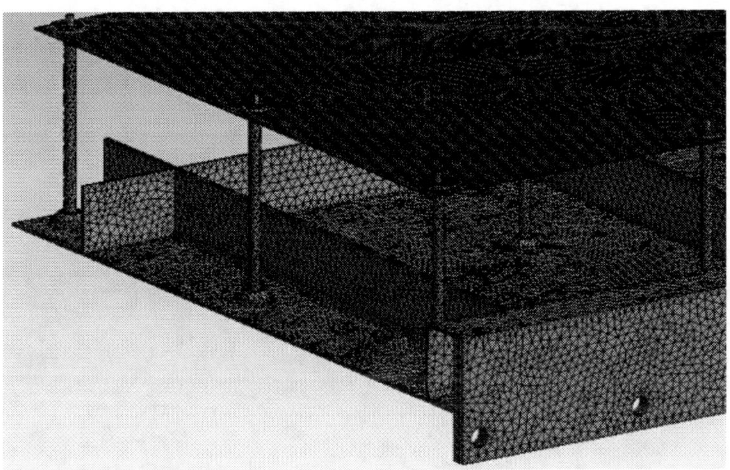

Fig. 3 – Computational mesh (Number of elements – 289 860)

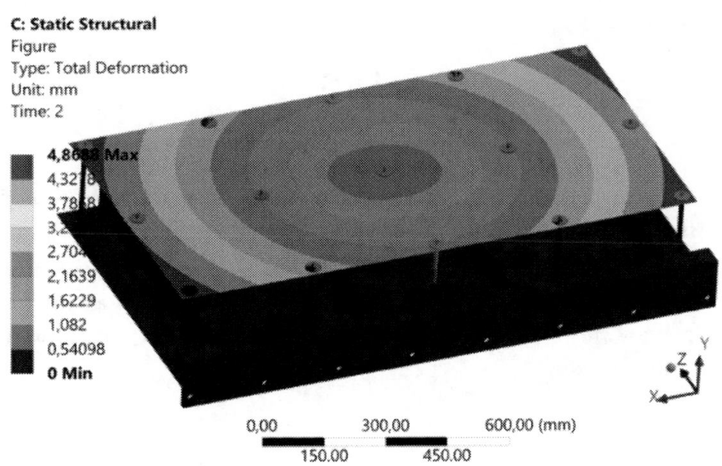

Fig. 4 – Temperature deformations of the section for the test case

The maximum calculated thermal deformations are detected in the corners of the facings and do not exceed 5 mm. The radial gap in the holes is 7 mm, so the temperature deformations of the inner facing are compensated for by the radial gap. Since the deformations of the inner facing are not constrained, it can be excluded from the calculation model of the whole exhaust system and only its weight is taken into account.

4 Results

Considering the placement and direction of the gaps in the sliding supports of the exhaust system, the outermost supports from the scroll housing side are fixed in the axial direction (Fig. 8). In this case, the adapter, diffuser, and the part of the gas duct from the left side of the fixed supports move towards the rotary device (to the left) under the action of thermal loads, and part of the gas duct from the right side of the support moves towards the scroll housing (to the right).

The maximum temperature differentials inside and outside the exhaust system were selected for the computations. The maximum temperature of the internal facing is +514°C and the minimum temperature of the outside air is -50°C. How was mentioned above, deformation of the inner lining does not affect to the deformations and stresses of the force frame. Hence, only weigh of the each plate of the inner lining was attached to this frame. This allows us to determine the maximum deformations and stresses of the exhaust system elements which may occur during operation.

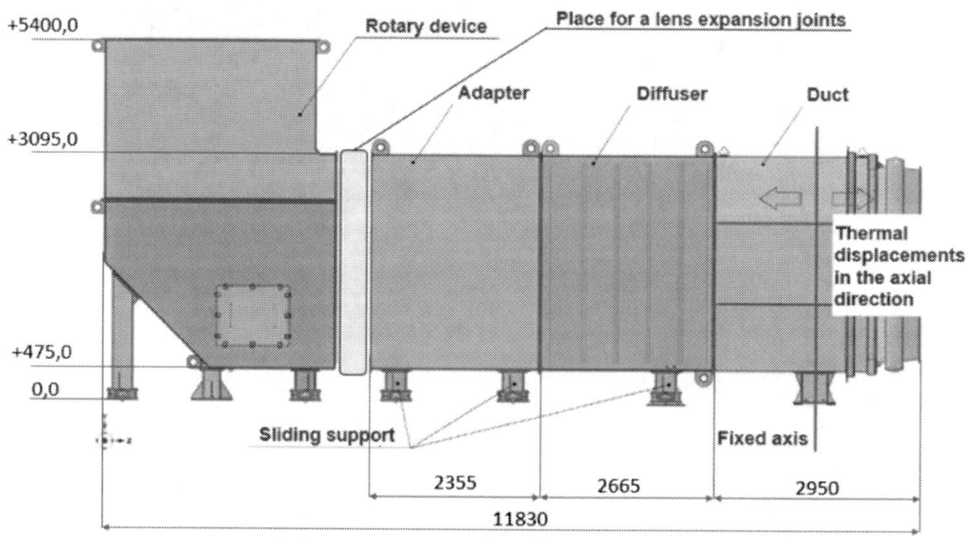

Fig. 5 – Design of the GCU exhaust system and location of lens expansion joint (dimensions in mm)

Figures 6a, 6b and 6c show the deformations of the gas duct, diffuser, and adapter. The thin lines show the initial state. The maximum total displacements are: gas duct flange 3 mm, diffuser flange 5.5 mm, and the adapter 3 mm. Table 1 shows the results of the calculation of thermal displacements of the diffuser and adapter flanges and the gas duct adjacent to the compensators under the action of thermal loads. Figures 7a, 7b and 7c show the locations of the control points on the flanges.

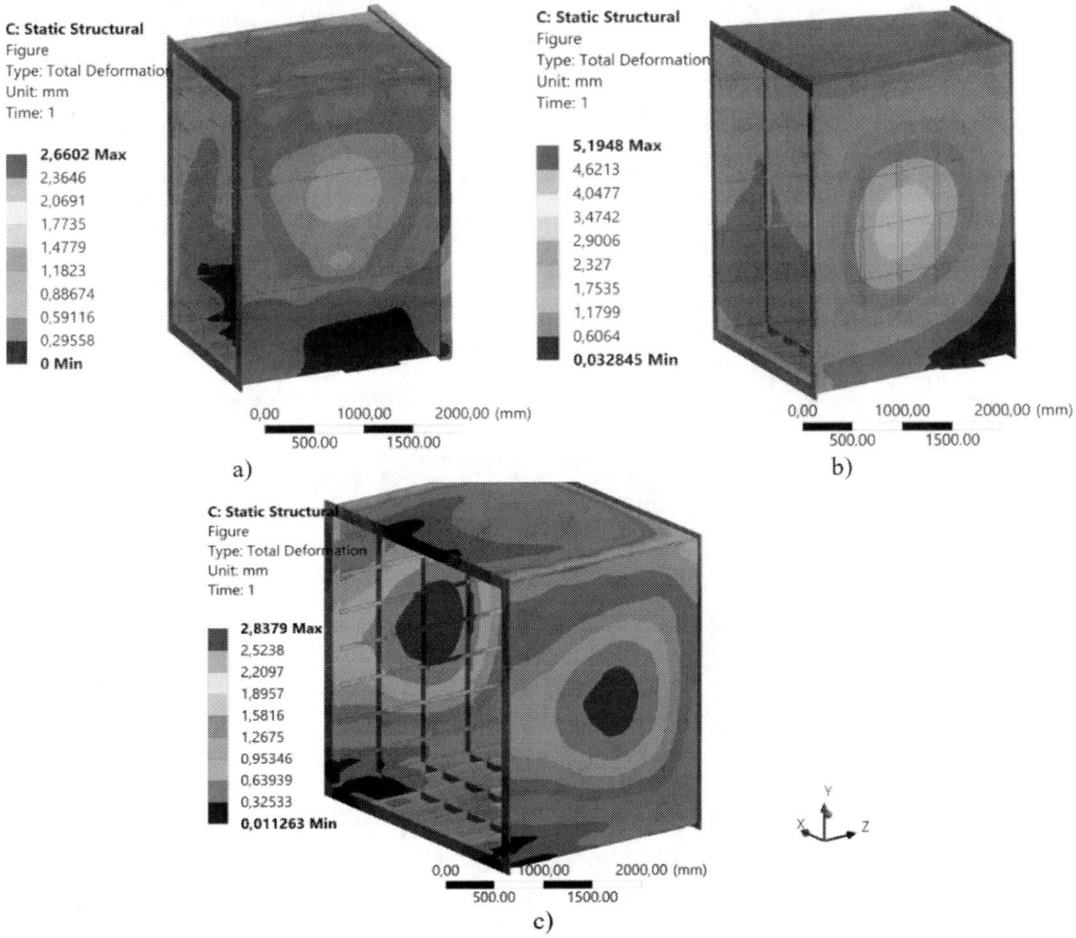

Fig. 6 – Total deformations of the GCU exhaust system elements (a – gas duct; b – diffuser; c – adapter)

Table 1. Thermal displacements of the adapter flanges relative to the initial position.

	Flange 1 (from the scroll housing side)			Flange 2 (form the rotary device side)			Flange 3 (form the rotary device side)		
Direction	x	y	z	x	y	z	x	y	z
T. 1	1.0	0.9	0.3	-2.5	1.0	-1.8	-1.1	2.0	-3.0
T. 2	1.1	1.3	0.3	-1.7	0.4	-1.7	-0.1	2.0	-3.2
T. 3	1.3	0.8	0.2	-1.0	2.0	-1.5	1.0	2.0	-2.7
T. 4	0.3	0.5	0.5	-0.7	1.0	-1.6	-0.2	1.0	-3.1
T. 5	0.3	0.3	0.4	1.0	-0.4	-1.5	0.5	-0.2	-2.7
T. 6	0.1	-0.2	0.4	0.1	-0.4	-1.9	-0.1	-0.1	-3.4
T. 7	-0.3	0.4	0.4	-1.2	-0.5	-1.6	-0.5	-0.15	-2.9
T. 8	1.0	0.6	0.4	-2.1	0.5	-4.0	0.1	1.0	-5.5
Maximum displacement	1.1	1.3	0.5	-2.5	2.0	-4.0	-1.1	2.0	-5.5

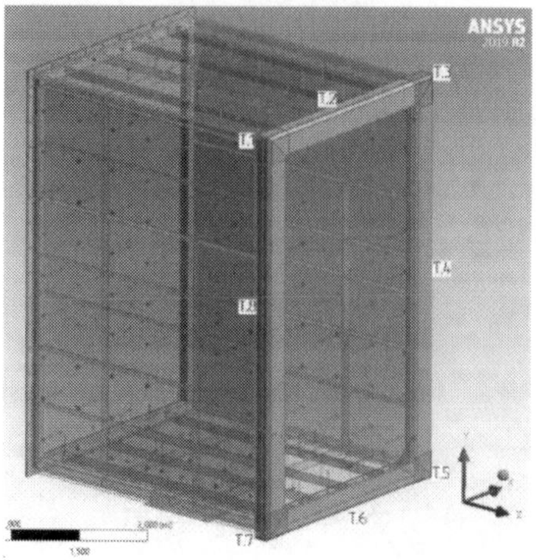

a)

b)

c)

Fig. 7 – The control points located on the flanges of the GCU exhaust system elements (a – duct from the scroll housing side; b – diffuser form the rotary device side; c – adapter form the rotary device side)

Figures 8a, 8b and 8c show the stress fields for the gas duct, diffuser, and adapter. It must be noted, that is no standards, which are determined allowable stress exactly for exhaust system of a gas compressor unit. In this case, was desisted to use for this purpose GOST 34233.1-2017 [16]. In its application area are written that, if the properties of materials, requirements for the design, manufacture and control of vessels and apparatus match the requirements of GOST 34347, them stress can be calculated according to GOST 34233.1-2017. This exhaust system of a gas compressor unit exact this case, it design, material properties manufacturing method are matched to requirements.

Hence, according to GOST 34233.1-2017 allowable stresses for carbon steel (09G2S) must be determined like:

$$[\sigma] = \eta \cdot min\left\{\frac{R_{elt}}{n_t} or \frac{R_{p0,2/t}}{n_t}; \frac{R_{m/t}}{n_b}; \frac{R_{m/10^n/t}}{n_d}; \frac{R_{p1,0/10^n/t}}{n_p}\right\},$$

where $R_{elt}, R_{p0,2/t}, R_{m/t}, R_{m/10^n/t}, R_{p1,0/10^n/t}$ - minimum yield strength at design temperature and minimum conditional yield strength at residual elongation of 0.2% at design temperature; minimum value of tensile strength (tensile strength) at design temperature; the average value of the limit of long-term tensile strength for a resource of 10n at design temperature; average 1.0% tensile creep strength in 10n at design temperature, respectively, MPa;

n_t, n_b, n_d, n_p - safety factor for yield strength; tensile strength (margin for tensile strength); limit of long-term strength; creep limit.

Allowable stresses for austenitic chrome nickel steel (12H18N10T) must be determined by GOST 34233.1-2017 like:

$$[\sigma] = \eta \cdot min\left\{\frac{R_{p0,2/t}}{n_t}; \frac{R_{m/t}}{n_b}; \frac{R_{m/10^n/t}}{n_d}; \frac{R_{p1,0/10^n/t}}{n_p}\right\}.$$

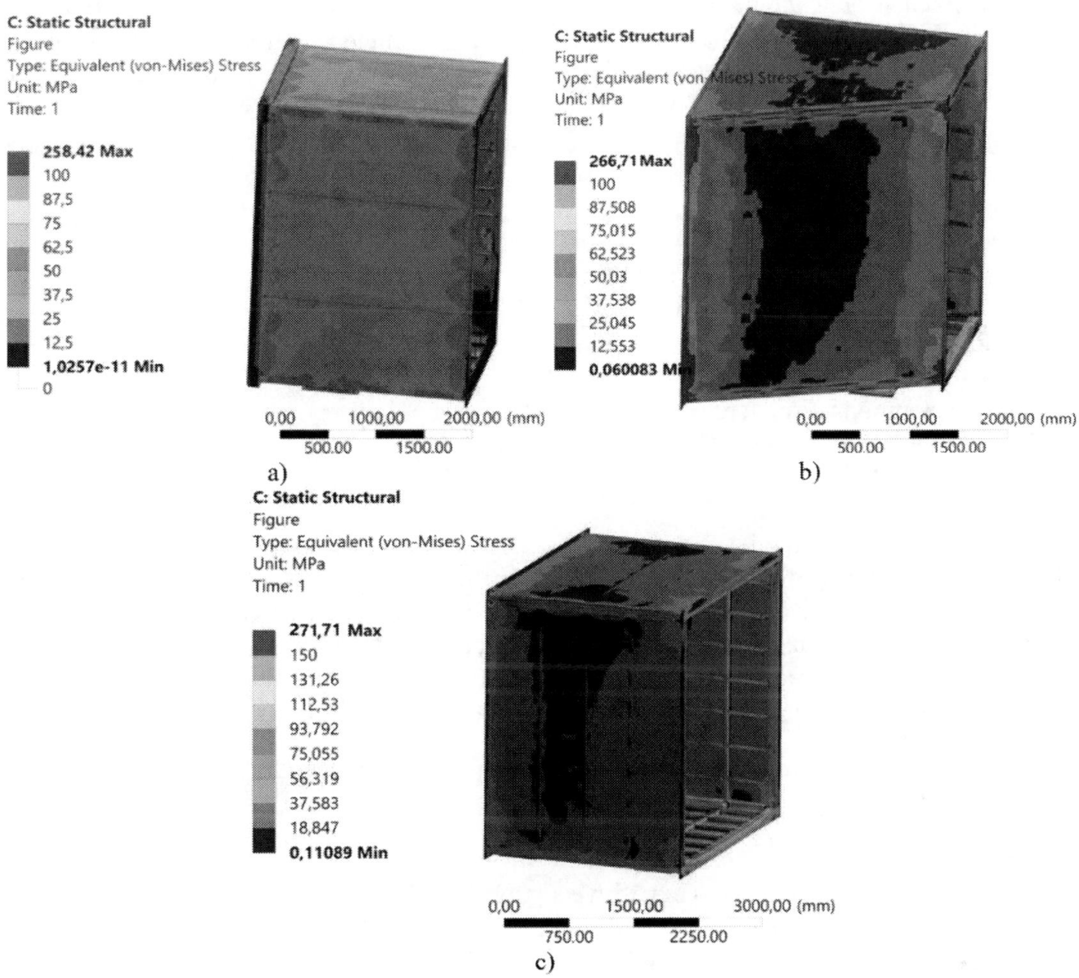

Fig. 8 – Equivalent Von-Mises stresses of the GCU exhaust system elements (a – gas duct; b – diffuser; c – adapter)

Determined values was written in table 2. These stresses were compared with those given in dangerous sections, determined by linearization according to third stress theory.

The linearized stress of the gas duct is 90 MPa, while the maximum stress in the flanges of the gas duct is 120 MPa. The stress is 50 MPa for the exhaust system diffuser, and the maximum stress in the diffuser flanges is 150 MPa. The linearized stress for the adapter is 70 MPa, and the maximum stress in the adapter flanges is 130 MPa.

Table 2. Stress analysis results according to GOST 34233.1-2017

Maximum stress location	Maximum stress	Temperature	Accepted allowable stress
gas duct flanges (12H18N10T)	120 MPa	76℃	180 MPa
diffuser flanges (09G2S)	150 MPa	82℃	177 MPa
adapter flanges (09G2S)	130 MPa	82℃	177 MPa

5 Conclusions

A technique for determining the temperature deformations and stresses in GCU exhaust system elements using numerical methods is designed. This includes carrying out a test problem to determine the influence of the mounting methods of the inner facing on the deformations of the power frame, and the subsequent selection of the calculation method for the elements.

The maximum temperature displacements of gas duct flange 1 adjacent to the lens expansion joint are determined to be:
- axial direction: 0.5 mm
- vertical direction: 1.3 mm
- exhaust system axial direction: ±1.1 mm

Maximum temperature displacements of the diffuser flange 2:
- axial direction: 4.0 mm
- vertical direction: 2.0 mm
- exhaust system axis direction: ±2.5 mm

Maximum thermal displacements of the adapter flange 3:
- axial direction: 5.5 mm
- vertical direction: 2.0 mm
- exhaust system axis direction: ±1.1 mm.

These results can be used to calculate the stress-strain state of the elements of a GCU exhaust system and to predict their further performance. These displacements make it possible to determine the dimensions of the lens expansion joint more accurately.

The strength of the flanges of the gas duct, diffuser and adapter is also evaluated and it can be confirmed that the strength conditions in the flanges of the exhaust system elements are satisfied.

References

1. A. Poletaev and A. Nedoshivina, "Investigation of characteristics and modernization of the design of exhaust tract gas turbine and gas compressors units", Young scientists conference – 2017, UralENIN, «UrFU», pp. 123 – 124, 2017.
2. A. Lobanov, Evaluation of the exhaust system dynamic loads of a gas compressor unit by vibroacoustic characteristics: PhD Thesis. Perm: PNPU, 135 p., 2008.
3. O. Baturin, A. Krivcov, D. Kolmakova and V. Matveev, "Modeling, Analysis and Improvement of the Input Cleaning Device of a Gas Turbine Plant for Pumping Gas," in Proceedings - 2019 21st International Conference; Complex Systems: Control and Modeling Problems; CSCMP 2019, vol. 2019–September, pp. 365–369, 2019.
4. E. Mekhonoshina and V. Modorskii, "Numerical modeling of aeroelastic behavior of GTU centrifugal compressor rotor," in Procedia Engineering, vol. 201, pp. 655–665, 2017.

5. O. Kyrylash, V. Kostiuk, A. Smirnov, D. Tkachenko, and I. Loboda, Mathematical Simulation of the Gas Turbine Packages Thermal State, vol. Volume 5C: Heat Transfer. 2018.
6. A.A. Mesropyan and I.I. Mukhametzyanova, "Numerical Modeling of Gas Dynamics and Heat and Mass Transfer in the Cooling System of the GTE Box." USATU Bulletin Vol. 14 No. 1: pp. 25-31, 2010.
7. V. Anisimov, I. Zubrilin and M. Orlov, "Investigation of thermal and stress states of the annular combustion chamber flame tube walls," in Proceedings of the ASME Turbo Expo, vol. 4B, 2016.
8. A. Pugachev, A. Sheremetyev, V. Tykhomirov and A. Petrov, "Structural and dynamic optimization of a single axial compressor blade using the gradient-based method," in Proceedings of the ASME Turbo Expo, vol. 7B, 2014.
9. Micha Premkumar T. et al., "Optimizing the Design and Material used for Connecting Rod through Finite Element Simulation," International Journal of Vehicle Structures and Systems, vol. 13, no. 5, pp. 548–552, 2021.
10. A. Belhocine, D. Shinde and R. Patil, "Thermo-Mechanical Coupled Analysis-Based Design of Ventilated Brake Disc Using Genetic Algorithm and Particle Swarm Optimization," SAE International Journal of Passenger Cars - Mechanical Systems, vol. 14, no. 2, pp. 137–150, 2021.
11. W. Yang et al., "Thermo-mechanical fatigue life prediction based on the simulated component of cylinder head," Engineering Failure Analysis, vol. 135, p. 106105, 2022.
12. L. Lei and C. Bing-Wei, Finite element analysis of thermal load on cylinder head of small-sized piston engine, U.P.B. Sci. Bull., Series D, Vol. 81, Iss. 2, 2019.
13. C. Xie, G. Wu, S. He, Y. Zhang, and J. Tian, "Finite Element Analysis of Thermal Load Characteristics of Dry Dual Clutch," Journal of Physics: Conference Series, vol. 1748, no. 5, p. 52047, Jan. 2021.
14. O. Liaposhchenko, I. Pavlenko, K. Monkova, M. Demianenko and O. Starynskyi, "Numerical simulation of aeroelastic interaction between gas-liquid flow and deformable elements in modular separation devices," Lecture Notes in Mechanical Engineering, pp. 765–774, 2020.
15. M. Demianenko, M. Volf, V. Pavlenko, O. Liaposhchenko, I. Pavlenko, The solution of the stationary aeroelasticity problem for a separation channel with deformable sinusoidal walls. Journal of Engineering Sciences, Vol. 7(1), pp. D5–D10, 2020.
16. GOST 34233.1-2017, Vessels and apparatus. Norms and methods of strength calculation. General requirements, 35 p., 2018.

Preliminary results of PIV measurement past a stator wheel inside the VT-400 test turbine

Daniel Duda[1], *Vít* Horáček[1], *Marek* Klimko[1], *Petr* Milčák[3], *Václav* Uruba[1,2], *Vitalii* Yanovych[1,2] and *Pavel* Žitek[1*]

[1]Faculty of Mechanical Engineering, University of West Bohemia in Pilsen, Univerzitní 22, 306 14, Pilsen, Czech Republic
[2]Institute of Thermomechanics, Czech Academy of Sciences, Dolejškova 5, 180 00, Prague, Czech Republic
[3]Doosan Škoda Power, Tylova 16, 301 00, Pilsen, Czech Republic

Abstract. The feasibility study of Particle Image Velocimetry (PIV) measurements inside a test turbine at the University of West Bohemia. The current VT-400 turbine is not prepared for optical measurement with seeding particles, thus several technical issues had to be addressed until low-quality data were obtained only at low speed of 2000 RPM (rounds per minute). Even the low quality data are able to show the fluctuation anisotropy or the size of fluctuation structures, which are quantities not measurable by classical pressure methods.

1 Introduction

This year, 2022, we "celebrate" two centuries since phenomenological definition of turbines in 1822 [1] by Claude Burdin. The phenomenological definition is mentioned, because such type of machines is much elder, however, it has been constructed in a form of windmill (6th century, Persia), windpump or waterwheel (for powering watermill (3rd century BC, Byzantinum [2]) or waterpump (5th century BC, China [3])). Probably, it has to be much elder: although there is no any direct archeologic or written evidence, it seems, that Sumerians had to have this technology [4] as well.

Since 18th century, there was added the *steam* as the working medium and the turbines working in closed channel were constructed. The phenomenological definition as a *machine, which rotates similarly as the vortices* – from the Greek word *τυρβη = vortex* [1], opens the way to exact scientific exploration, engineering optimization and to systematic education.

There is a lot of insight obtained by experiments and numerical simulations about what happens inside the turbine. Let us mention, among others, the Ilieva's review [5], Langston's review [6], Sieverding's review [7], analysis of Denton and Pullan [8] or our own previous measurements [9], [10].

* Corresponding author: zitek@kke.zcu.cz

2 Experimental setup

2.1 VT-400 test turbine

The experimental test rig VT-400 [11] is a 1:2 scale model of a high-pressure steam turbine part. The working medium is air sucked form the atmosphere by a compressor. It is a high reaction axial single turbine stage [12]. The VT-400 test rig consists of the basic parts (see *Fig. 1*).

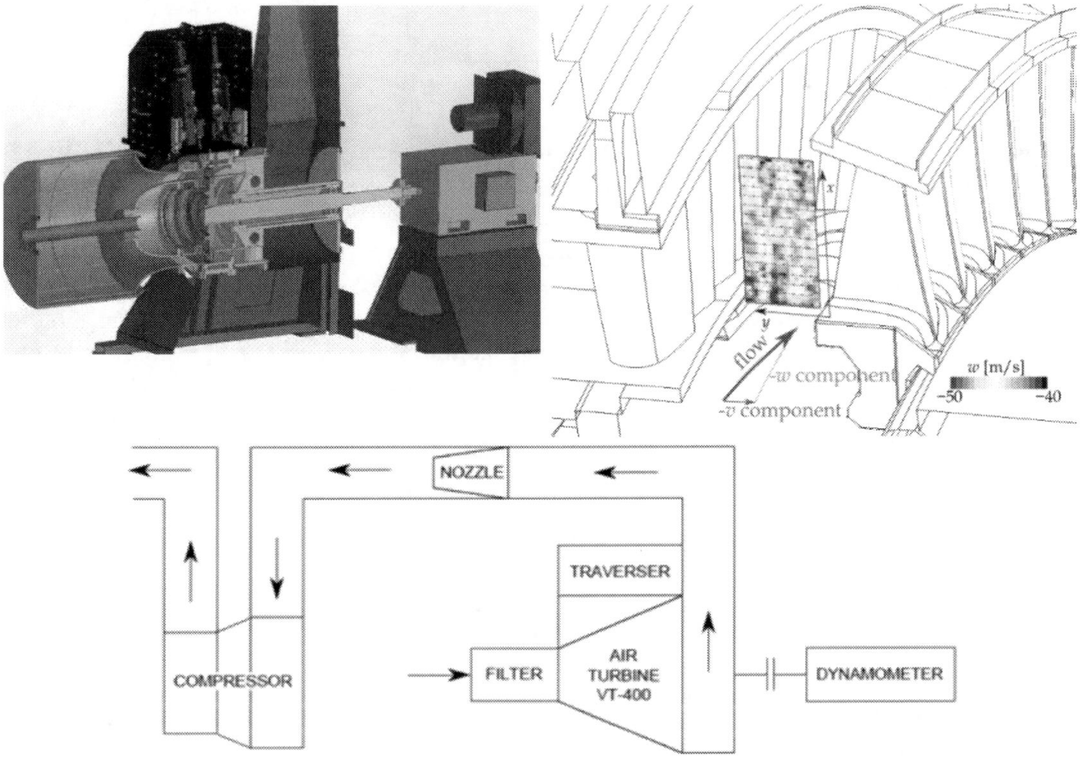

Fig. 1. Experimental VT-400 test rig (upper left) with corresponding 3D sketch of the measured PIV plane (upper right). Scheme of the entire test stand (bottom)

Some basic measuring tasks can be performed on this experimental device. The first task is the determination of the basic turbine stage characteristics, i.e. dependence of the stage efficiency on the velocity ratio $u_m/c_{02,\text{is}}$ from torque measurement and static pressure taps. Parameter u_m is circumferential velocity at mid radius, and $c_{02,\text{is}}$ is the isentropic velocity (see *eq. 1*).

$$c_{02,\text{is}} = \sqrt{2H_{\text{is}}^{ST}} \tag{1}$$

where H_{is}^{ST} is the isentropic turbine stage enthalpy drop.

The second measurement method is a detailed wake traverse measurement behind the nozzle and bucket using two 5-hole pneumatic probes. The basic parameters of the tested turbine stage are summarized in *Tab. 1*. [13]

More detailed information about the experimental device and evaluation methodology can be found in the publications [12], [11] and [13].

Tab. 1 Basic geometric parameters of tested turbine stages

Nozzle			Bucket		
Root diameter [mm]	D_P^{NZL}	400	Root diameter [mm]	D_P^{BCK}	400
Blade length [mm]	L^{NZL}	45.5	Blade length [mm]	L^{BCK}	47
Chord [mm]	c^{NZL}	22.5	Output rel. velocity angle [°]	β_2	14.5°
Output absolute angle [°]	α_1	14.5°			

2.2 PIV measurement system

Particle Image Velocimetry (PIV) [14] system at the University of West Bohemia (UWB) is a standard one supplied by the Dantec company. It consists of pair of FlowSense MkII cameras with resolution of 4 MPix, solid-state double pulse laser New Wave Solo with emitting energy of 0.5 J during 5 ns pulse (i.e. power during pulse is around 100 MW). The optical access for both (laser and camera) is realised through a gap for pressure probes covered by 4 mm thick desk of transparent polystyrene. The observed plane in axial × radial direction is crossed by the flow at sharp angle, thus the Stereo [14] configuration has to be used to capture the out-of-plane velocity component (see right top panel of Fig. 1). However, the Stereo configuration is much more sensitive to the calibration quality and to the homogeneity of fog as well as to the vibrations, which at the end stopped our attempts to measure even faster rotations than the mentioned 2000 RPM.

3 Results

3.1 Mean quantities

Figure 2 shows the mean vector field in the observable area. The air flows from top in the image and from behind the paper-plane. First column shows the mean vector field, where the tangential component perpendicular to the measurement plane is displayed by color. It is needed to mention, that the data are very noisy especially close to the boundary of Field of View (FoV) and at the largest probed pressure drop (bottom line in Fig. 2). It is even more apparent in the plot of turbulence intensity calculated as

$$I_T = \sqrt{\frac{\langle u^2\rangle + \langle v^2\rangle + \langle w^2\rangle - \langle u\rangle^2 - \langle v\rangle^2 - \langle w\rangle^2}{\langle u\rangle^2 + \langle v\rangle^2 + \langle w\rangle^2}} \quad (2)$$

Where $\langle\cdot\rangle$ means ensemble averaging. The large values close to the FoV boundaries are probably caused by the spurious vectors in those areas. In the central region we can observe a pair of horizontal (radial) strips of slightly higher turbulence intensity, which can be identified with the wakes past stator blades. The data at highest pressure drop are not significant.

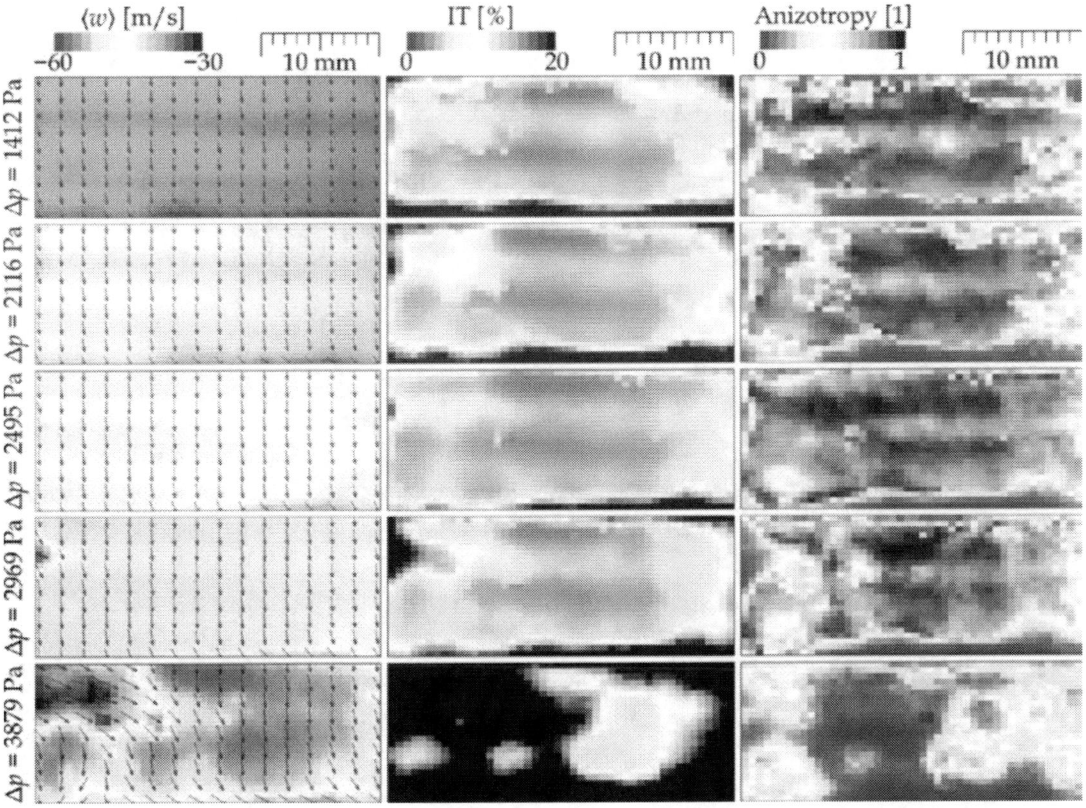

Fig. 2. Ensemble averaged field in the radial × axial plane. First column shows the mean velocity vectors only every third vector is displayed for sake of clearance; color represents velocity component (w) perpendicular to the measurement plane. Second column is the turbulence intensity showing the large amount of noise at the area borders and at the largest pressure drop. Third column shows the anisotropy of fluctuations.

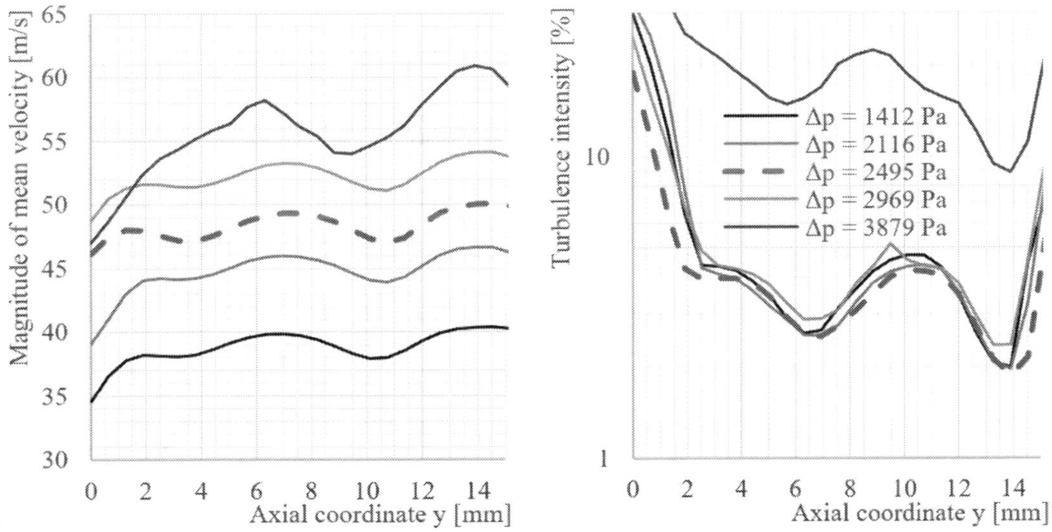

Fig. 3. Axial cut of mean velocity magnitude (left) and turbulence intensity (right). The color corresponds to pressure drop: blue for smallest, red for largest.

3.2 Anisotropy

The last column of Fig. 2 shows the anisotropy coefficient of 3D velocity fluctuations according to Lumley [15], Choi [16] and Simonsen [17], (an overview about the interpretation of the invariant of fluctuation matrix can be found in [18]). The analysis is based on the Reynolds stress tensor τ_{ij}

$$\tau_{ij} = -\rho\langle u_i' \cdot u_j' \rangle, \text{ where } u_i' = u_i - \langle u_i \rangle \tag{3}$$

Whose non-dimensionalised anisotropic part $b_{ij} = \tau_{ij}/\tau_{kk} - \delta_{ij}/3$ is eigen-decomposed getting three invariants, whose combination describes the level of 3D anisotropy:

$$F = 1 - \frac{9}{2} b_{ij} b_{ji} + 9 b_{ij} b_{jk} b_{ki} \tag{4}$$

A pure isotropic turbulence of Kolmogorov style [19] would display $F = 0$, however even the grid-generated turbulence [20] [21] displays a significant anisotropy [22] [23]. On the other hand, a noise might be isotropic as well (there is no reason for anisotropic behaviour). In this case, we believe, that the isotropy signal ($F \approx 0$ in Fig. 2) comes from the noise, not being a sign of ideal isotropic turbulence. The flow past an airfoil (such as turbine blade) is strongly anisotropic, as observed by other authors [24], [25] and by us as well [26] [27].

3.3 The size of fluctuation structures

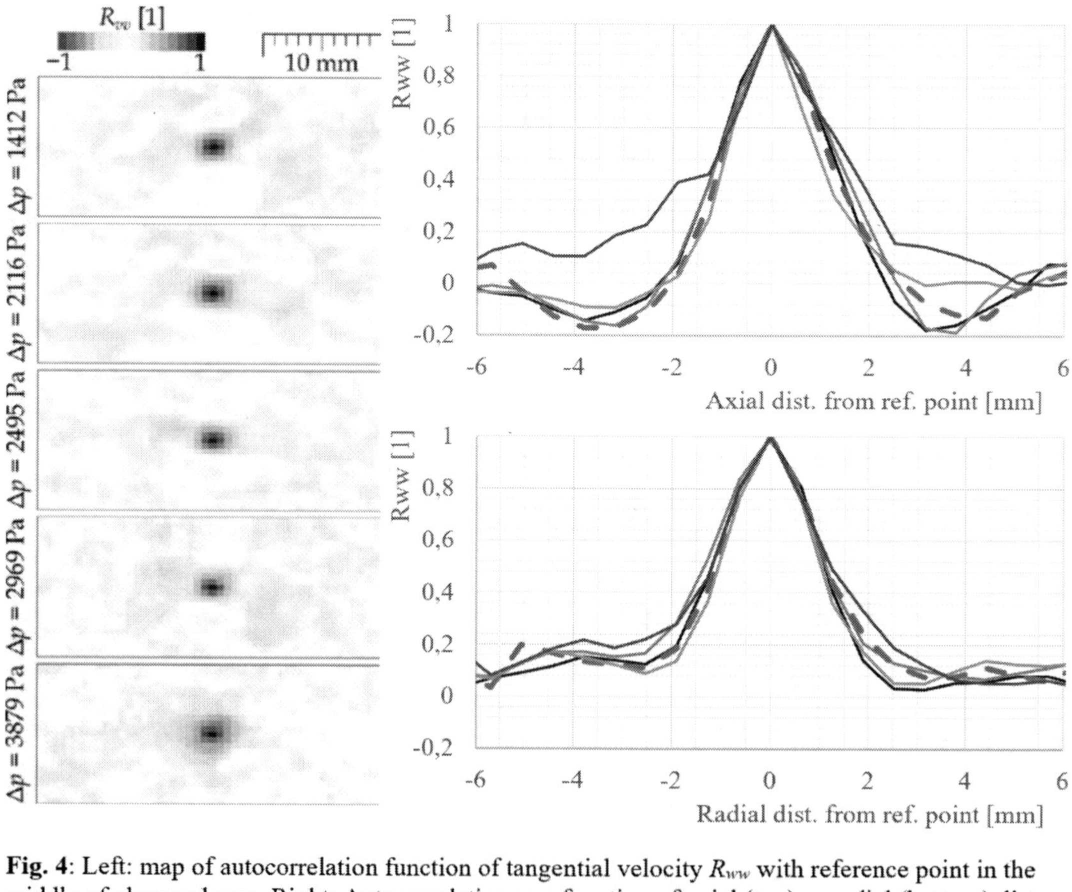

Fig. 4: Left: map of autocorrelation function of tangential velocity R_{ww} with reference point in the middle of observed area. Right: Autocorrelation as a function of axial (top) or radial (bottom) distance from reference point. The color represents the pressure drop as in Fig. 3 does.

The size of coherent structures is quantified in terms of autocorrelation function in Fig. 4. In the present case, it does not depend on the pressure drop (and thus velocity, see Fig. 3). Additionally, there is an evidence of a "wave" in axial direction caused by the structure of alternate jet and wake past the stator wheel. On the other hand, the comparable range in radial direction suggests no two-dimensionality of the turbulence (2D turbulence would display fully correlated fluctuations along the radial coordinate).

Alternative way [19] is in the turbulence spectrum, which quantifies, how much energy is at which length-scale, standardly represented by the wave number of corresponding oscillations. We use our algorithm described in more details in [28]. The data at largest pressure drop are not valid as discussed before and again proved by the non-Kolmogorov spectrum in Fig. 5. The other data show more-or-less Kolmogorov scaling with some depletion at $k \approx 0.3\ \mathrm{mm}^{-1}$. For deeper exploration of the spectrum we would need cameras with higher resolution and better optical access into the turbine VT-400.

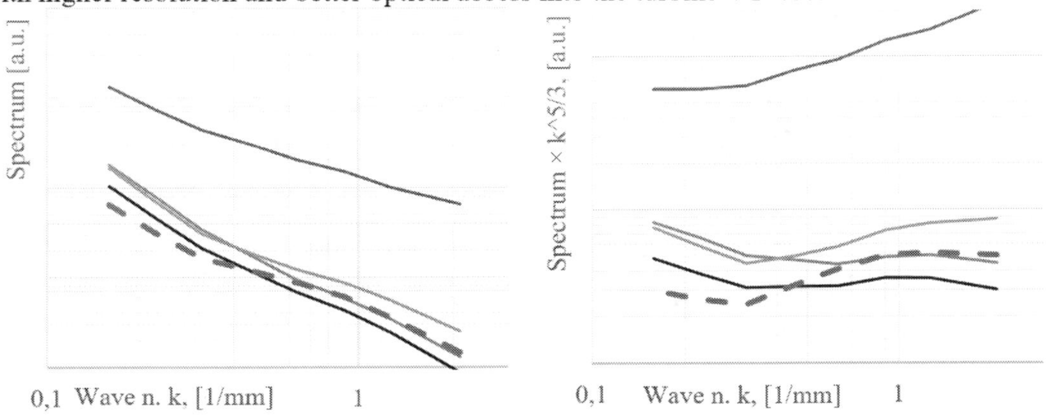

Fig. 5: Left: Power spectral density of fluctuations as a function of wave number k. Right: the same data as in left but multiplied by $k^{5/3}$ to highlight the differences from Kolmogorov scaling (ideal turbulence would display a horizontal line in this plot). The color represents the pressure drop same as in Fig. 3.

4 Conclusion

The possibility of PIV measurement inside the test turbine VT-400 at UWB has been proven. However, here is so many technical issues decreasing the quality and credibility of the data, that it is needed to improve our experimental setup and equipment significantly before the next measurement.

Acknowledgments

The presented work was financially supported by the TA ČR project TN01000007 and the student project SGS-2022-023 (Research and development of power machines and equipment).

References

1. C. Burdin, Annales de chimie et de physique **24**, (1824)
2. Ö. Wikander, in *Handbook of Ancient Water Technology* (Brill, Leiden, 2000), pp. 371-400
3. P. Jahren and T. Sui, *How Water Influences Our Lives* (Springer, Singapore, 2016)

4. I. Vargas and H. Gallegos, Journal of Professional Issues in Engineering **116**, (1990)
5. G. Ilieva, Defect and Diffusion Forum **379**, (2017)
6. L. Langston, Annals New Yourk Academy of Sciences **934**, (2001)
7. C. Sieverding and M. Manna, International Journal of Turbomachinery Propulsion and Power **5**, (2020)
8. J. Denton and G. Pullan, Proceedings of ASME Turbo Expo **GT2012**, (2012)
9. D. Duda, T. Jelínek, P. Milčák, M. Němec, V. Uruba, V. Yanovych and P. Žitek, International Journal of Turbomachinery Propulsion and Power **6**, (2021)
10. D. Duda, T. Jelínek, M. Němec, V. Uruba, V. Yanovych and P. Žitek, AIP Conference Proceedings **2323**, (2021)
11. M. Klimko and D. Okresa, AIP Conference Proceedings **1745**, (2016)
12. M. Klimko and D. Okresa, Acta Polytechnica **56**, (2016)
13. M. Klimko, R. Lenhard, P. Žitek and K. Kaduchová, Processes **9**, (2021)
14. C. Tropea, A. Yarin and J. Foss, *Springer Handbook of Experimental Fluid Mechanics* (Springer, Berlin, 2007)
15. J. Lumley and G. Newman, J. Fluid Mech. **82**, (1977)
16. K. Choi and J. Lumley, J. Fluid Mech. **436**, (2001)
17. A. Simonsen and P. Krogstad, Phys. Fluids **17**, (2005)
18. V. Uruba, Proceedings Topical Problems of Fluid Mechanics **2015**, (2015)
19. A. Alexakis and L. Biferale, Physics Reports **767-769**, (2018)
20. T. Kurian and J. Fransson, Fluid Dynamics Research **41**, (2009)
21. M. Mohamed and J. LaRue, J. Fluid Mech. **219**, (1990)
22. P. Roach, International Journal of Heat and Fluid Flow **8**, (1987)
23. D. Duda, V. Yanovych and V. Uruba, Processes **8**, (2021)
24. I. Solís-Gallego, A. Meana-Fernández, J. Fernández Oro, K. Argüelles Díaz and S. Velarde-Suárez, Journal of Applied Fluid Mechanics **10**, (2017)
25. G. Ozkan and H. Egitmen, Experimental Thermal and Fluid Science **134**, (2022)
26. V. Yanovych, D. Duda, V. Uruba and T. Tomášková, Processes **10**, (2022)
27. D. Duda, V. Yanovych, V. Tsymbalyuk and V. Uruba, Energies **15**, (2022)
28. D. Duda and V. Uruba, Journal of Nuclear Engineering and Radiation Science **5**, (2019)

Impact of dust erosion on the reduction of axial compressor efficiency of a turboshaft engine and on the stability of its operation

Marián Hocko[1*], *Samer* Al-Rabeei[2]

[1]Technical University of Kosice, Faculty of Aeronautics, Department of Aerospace Engineering, Rampová 7, 041 21 Kosice, Slovakia
[2] Technical University of Kosice, Faculty of Aeronautics, Department of Aerospace Engineering, Rampová 7, 041 21 Kosice, Slovakia

Abstract. This paper aimed to solve the impact of operational abrasive wear on the rotor blades of the axial compressor of the turboshaft engine on the decrease in its total compression efficiency η_{Ct} and its transition into unstable work mode (surge). This process is analyzed based on the obtained data by operating TV3-117 helicopter turboshaft engines in high dust atmosphere conditions. Abrasive wear of the rotor blades of axial compressors causes mechanical damage to the blades, reducing their strength, changing their geometry, and aerodynamic properties, reducing the life of the whole compressor and thus the entire engine. Destruction of the compressor of a turboshaft engine may occur suddenly as a result of unstable compressor operation caused by damaged axial compressor blades due to their damage by the abrasive effect of dust.

1 Introduction

The first research on the dust erosion impacts was carried out in Germany already in 1930. These studies were based on a theoretical analysis of the influence of erosion on various materials based on elaborated models for registration of the volume of worn material, deformation and abrasive wear, particle motion, etc. A great scientific contribution to the creation of erosive wear models was brought by I. Finnie, J. G. Bitter, J. Neilson, A. Gilchrist, W. Tabakoff, G. Grant, and others. In the Russian Federation, scientists from many universities and organizations (CIAM, UGATU, SGAU, IAIA, VIVA, and others) have made a major contribution to experimental and numerical research of the influence of operational factors on the characteristics of aircraft turbo-compressor engine (ATCE) (CIAM, UGATU, CGAU, MAI, VVIA, and others). [2, 3, 4].

The basic wear mechanism is a complex function of the physical properties of the damaged material, a composition of abrasive particles, their dimensions, the angle of a collision (interaction), and the collision velocity. This mechanism has been investigated by many scientists and it has resulted in a definition of an experimental correlation between the

* Corresponding author: marian.hocko@tuke.sk

material wear level as a function of erosion parameters. [5] J. Goodwin and others have found that the erosion level practically directly depends on the percentage of quartz in the dust. [6] This is due to the fact that quartz is one of the most present dust components. At the same time, they have shown that the degree of erosion at the given interaction velocity increases with a growing dust particle diameter. The dust particle size is one of the important factors determining the level and nature of erosive wear. Particles with dimensions of less than 70 micrometres are classified as dust.
The results of further research illustrate that the level of wear increases with growing particle size. [7] In the case of high dust levels, the turboshaft turbo compressor engine (TsTCE) shows equal wear and tear on the first stage blades and increased wear on the peripheral parts of the last stage blades. With a moderate dust content in the air, the blade's wear is local in nature and depends on the specificity of the flow, which causes a local increase in particle concentration and their velocity with regard to the compressor blades. The wear of the compressor blades results in a change in the shape of the blade's aerodynamic profile and an increase in the blade surface roughness. Due to decreasing the total efficiency of the compressor η_{Ct}, the total compression rate π_{Ct}, the mass flow rate Q_{Air}, the gas-dynamic stability of each stage of the compressor, and the compressor as a whole of K_y decreases, resulting in unstable operation of the TsTCE and combustion turbine (CT) compressors. [8]

2 TV3-117 turboshaft helicopter engine

To analyze the influence of the erosive wear of the rotor blades of the axial compressor of a turboshaft engine on the unstable work of the axial compressor, one of the most widely used turboshaft helicopter engines at present, the TV3-117, was selected. Its selection was also influenced by the fact that the mentioned turboshaft engine was and still is massively used in the conditions of countries with atmosphere with considerable dust content (Afghanistan, Iraq, Iran, Syria, Egypt, Algeria and others). The advantage of this selection is that statistical data of individual parameters are available and usable from the operation of these turboshaft engines in the above conditions.

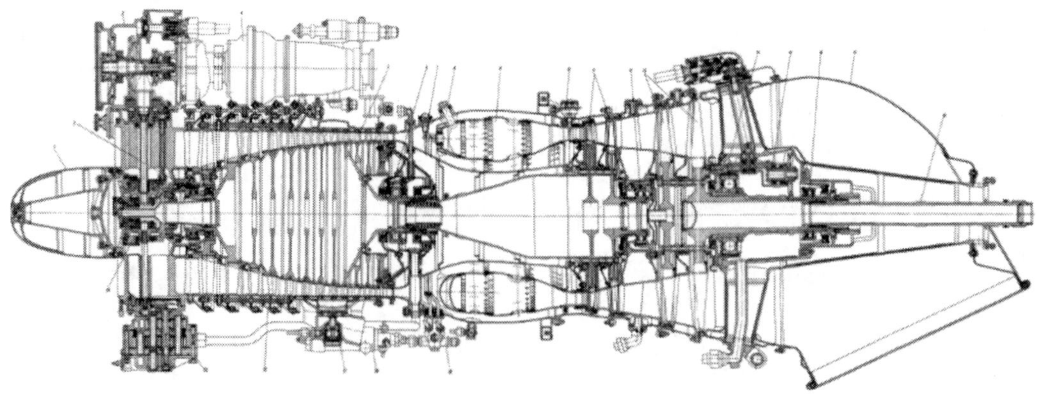

Fig. 1. TV3-117 turboshaft helicopter engine section. [6]

The TV3-117 is a turboshaft engine with an axial inlet, a twelve-stage axial compressor with rotary vane blades and two air bleed valves, an annular combustion chamber, a two-stage axial, cooled gas turbine compressor, a two-stage axial, free gas turbine, and an outlet

system with the outlet tube rotated at an angle of 60° to the right or left with respect to the engine [11].
Two TV3-117 3rd series turboshaft engines are used to drive the Mi-24D and Mi-24DU helicopters, two TV3-117V engines of the Mi-24V drive the helicopter, and two TV3-117MT engines drive the medium transport Mi-17 helicopter [11].

Table 1. Modes of the turboshaft engine TV3-117, 3rd series.

Mode	P_e	c_e	T_{3t}	n_{TC}	n_{MR}	Operating time (min.)	
	kW	$kg.kW^{-1}.h^{-1}$	K	%	%	Continuously	3x for TL
Take off	1628	0,314	1248	97,6	98±1	6	≤ 5 %
Nominal	1427	0,342	1208	94,7	98±1	60	≤ 40 %
1. Travel	1244	0,369	1173	93,5	100±2	Unlimited	Unlimited
2. Travel	1098	0,396	1143	91,5	100±2	Unlimited	Unlimited
Idle	879	≤ 155 $kg.h^{-1}$	1103	73+1	65	20	-

The TV3-117 turboshaft engine compressor is an axial, single-shaft, twelve-stage compressor with adjustable guide vanes on the first four stages of the compressor stator. The adjustment of the guide vanes is carried out automatically in the range from $\alpha_{GV} = 27°$ to $\alpha_{GV} = 0°$, depending on the calculated rotor speed of the turbo-compressor. To ensure the stable operation of the compressor, two air release valves are also used downstream of the seventh compressor stage.

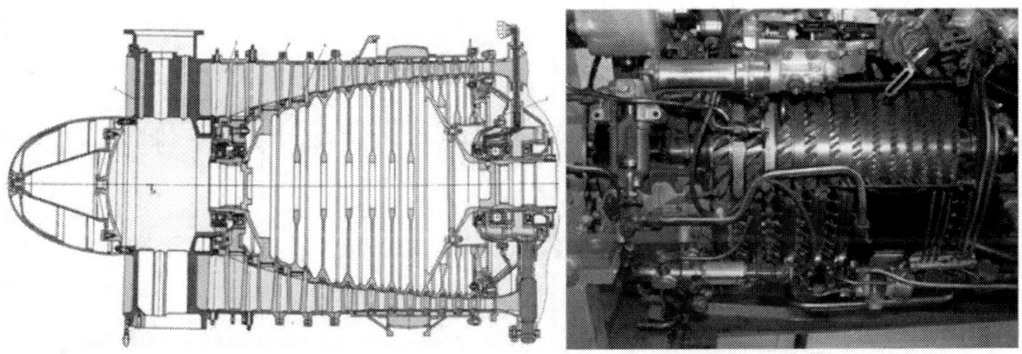

Fig. 2. Compressor section of the turboshaft helicopter engine TV3-117.

To protect against the occurrence of unstable compressor operation, the compressor uses adjustable stator vanes (variable stator vanes) of the first four stages and the inlet rectifier, which are set to a given position according to the specified program $\alpha_{GV} = f(n_{C,calc.})$. A deviation from the program of 1,5 ° to the opening side results in a reduction of the K_y compressor's steady-state operating margin of 1,5 % to 2,0 %. A deviation of 1° to the closing side results in a reduction of 66 kW in engine power. Control of the compressor vane rotation mechanism is provided by two hydraulic slave cylinders, which are supplied with pressurised fuel from the NR-3A pump-regulator.
Two blow-off valves, which discharge part of the air outside the compressor depending on the recalculated speed of the turbo-compressor rotor according to a program controlled by the pump-controller NR-3A, also provide an increase in the supply of stable operation of the engine compressor on the transient modes.

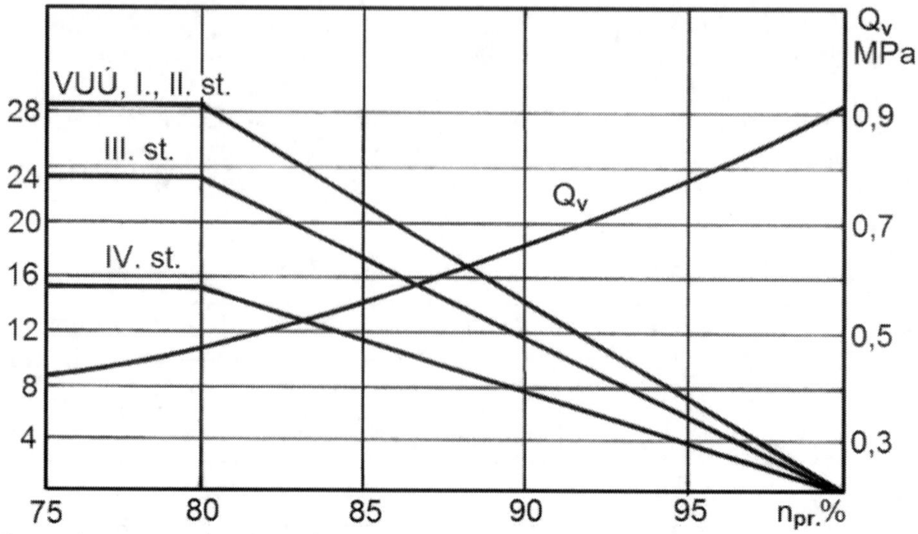

Fig. 3. Control program for the adjustable guide vanes of the inlet rectifier and the first four stages of the compressor stator of the TV3-117 engine.

Table 2. Gas-dynamic parameters of the compressor of the turboshaft engine TV3-117 in take-off mode

Parameter	Compressor stage											
	1	**2**	**3**	**4**	**5**	**6**	**7**	**8**	**9**	**10**	**11**	**12**
T_{2t} (K)	312	340	370	398	427	456	484	511	533	564	589	613
ΔT_t (K)	24	28	30	29	29	29	28	27	27	26	25	24
π_{Cc} (1)	1,289	1,314	1,298	1,271	1,246	1,228	1,205	1,192	1,175	1,159	1,146	1,28
η_{Cc} (1)	0,889	0,899	0,904	0,907	0,902	0,904	0,904	0,904	0,901	0,896	0,823	0,857
z_{stator}	60	60	60	60	80	84	84	86	88	90	90	114
z_{rotor}	37	43	59	67	73	81	89	89	89	89	89	89

Where:
T_{2t} – temperature at the outlet of the stage;
ΔT_t – heating the air to v degree;
π_{Ct} – the overall degree of compression of the grade;
η_{Ct} – overall degree efficiency;
The total efficiency of the compressor is $\eta_{Ct} = 0{,}86$.

3 Effect of dust erosion on the characteristics of a turboshaft engine axial compressor

The disturbance of the stable work of the axial compressor of a turboshaft helicopter engine, caused by the loss of gas-dynamic stability, is one of its most dangerous failures. For this reason, operation on modes, where the working point in the compressor characteristic is close to the limit of unstable work, i.e. where the stock of stable work is small, is inadmissible.

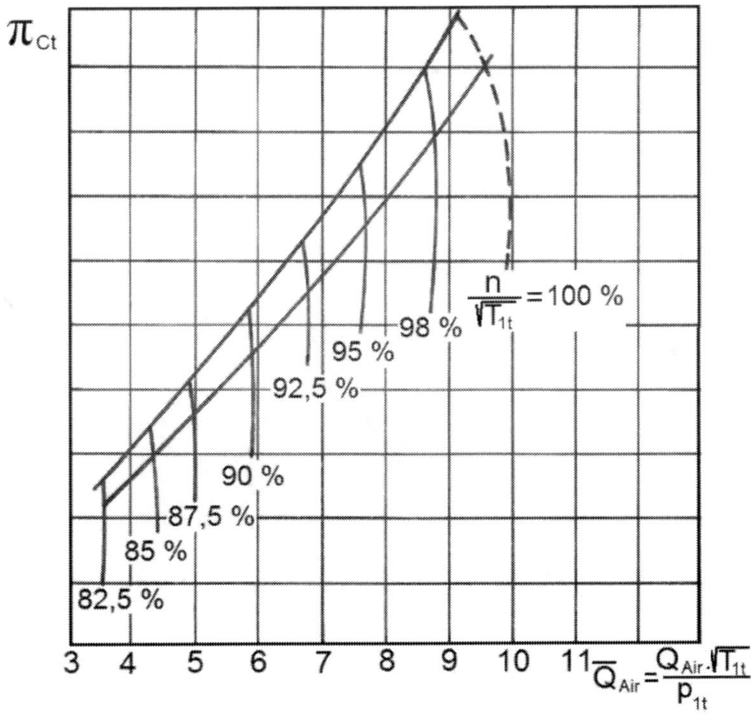

Fig. 4. Characteristics of TV3-117 turboshaft engine axial compressor.

The steady-state workload of an axial compressor can be expressed by the equation

$$\Delta K_y = \left[\frac{\left(\frac{\pi_{Ct}}{Q_{Air}}\right)_{SL}}{\left(\frac{\pi_{Ct}}{Q_{Air}}\right)_{WP}} - 1\right].100\ \% \tag{1}$$

Where:
$\pi_{Ct,SL}$ – the overall degree of compression at the limit of unstable work (on surge line),
$Q_{Air,SL}$ – air mass flow rate at the limit of unstable work (on surge line),
$\pi_{Ct,WP}$ – the total degree of air compression at the working point,
$Q_{Air,WP}$ – air mass flow rate at the working point.

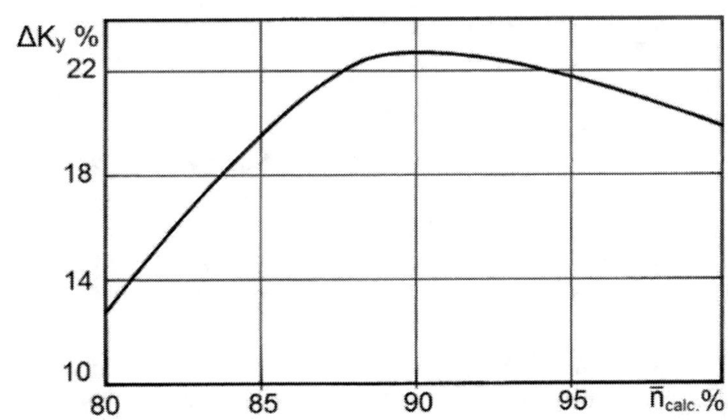

Fig. 5. Dependence of the steady-state work reserve of the axial compressor of the TV3-117 turboshaft engine on the engine operation mode.

When the compressor is tested above the calculation mode, the compressor's steady-state operating margin reaches $K_y = 2 - 5$ %.

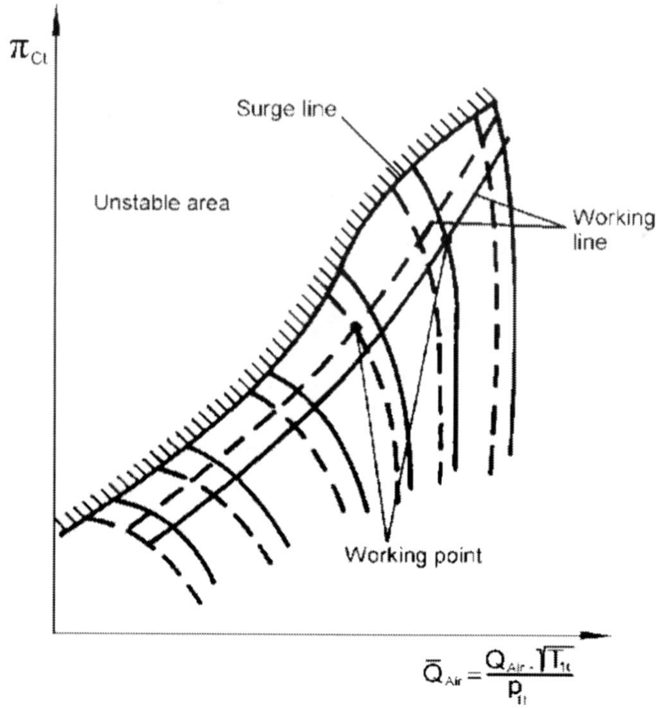

Fig. 6. Characteristics of a compressor with intact rotor blades and blades damaged by dust erosion.

Statistical analysis of the measurement results of the blades of different stages of the axial compressor of the TV3-117 engine, which were subjected to abrasive action of dust in operation, showed that the shortening of the blade bowstring was manifested only in the peripheral part of the blades, at the level of the upper third of the blade (Fig. 5). The analysis of the blade leaf wear values in the individual stages of the axial compressor allowed to detect the regularities caused by the design peculiarities of the compressor of the TV3-117 engine. It was confirmed that there are close correlations between the value of blade wear of the second to the twelfth stage of the compressor, which is due to the homogeneity of the processes that lead to blade wear. Maximum wear occurs in the first and sixth stage blades of an axial compressor. However, the nature of the wear is different. The rotor blades of the first stage show signs of damage mainly on the leading edge and on the front face, which is the result of contact with the largest abrasive particles that are fragmented in other parts of the axial compressor. The rotor blades of stages one to four had basic wear at the leading edges and in the blade trough. Rotor blades of the fifth to twelfth stages had wear at the inlet and outlet edges and the trailing edge of the blade. The above wear patterns can be explained by the fact that large dispersive abrasive particles encounter the leading edges of the rotor blades of the first stage of the axial compressor, fragment and move with the air stream towards the next stages. As the rotor blades of the first stage encounter relatively large particles, these blades are the most damaged. Already in the second stage, the shortening of the bowstring due to wear is considerably less, but it is of an increasing character up to the sixth (seventh) stage of the compressor. This effect is due to the fact that the rotor blades of the subsequent stage have a smaller profile thickness compared to the previous blade, but the volume of abrasive particles acting on it remains constant. At the same time, the rotor blades of each successive stage operate at a higher temperature (Table 2) resulting in a reduced ability of the material to resist wear.

The blades from the eighth to the twelfth stage have considerably less, approximately equal wear when compared to the sixth and seventh stages. This is due to the air intake chamber to the blow-off valves located after the seventh stage (Fig. 6).

When the engine is running at low speed (idle, right correction) on the ground, under dusty conditions, some of the dust (sand) along with the air escapes through the blow-off valves to the atmosphere. As a result, less abrasive particles reach the next stages, resulting in less blade wear.

Change in overall adiabatic compressor efficiency $\eta_{Ct,ad.}$ as a function of the engine operating mode is shown in Fig. 7.

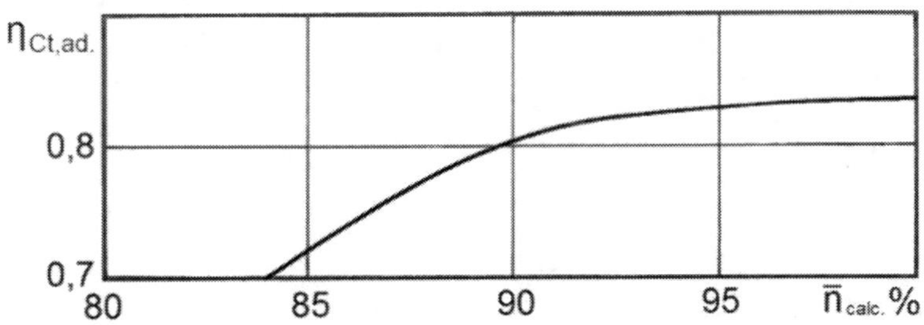

Fig. 7. Change in the total adiabatic efficiency of the axial compressor of the TV3-117 turboshaft engine $\eta_{Ct,ad.}$ as a function of the engine operating mode.

The research results in the change of the TV3-117 compressor condition as a function of the number of hours worked in dusty atmosphere conditions, which with the increase of the leaf erosion intensity leads to a shift of the whole compressor characteristic (Fig. 8), as a result of which the overall compressor compression ratio π_{Ct} and the overall compression efficiency η_{Ct} decreases.

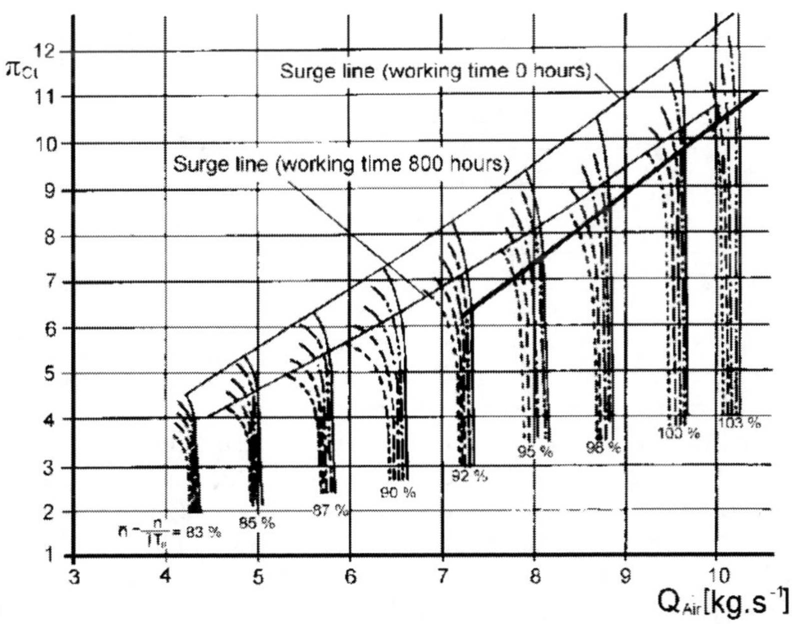

Fig. 8. Characteristics of TV3-117 turboshaft engine compressor with marked changes after working different number of hours in dusty environment.

_____ 0 working hours, _ _ _ _ 200 working hours, _ . _ . _ . _ 400 working hours, _ .. _ .. _ 600 working hours, - - - - - - - 800 working hours

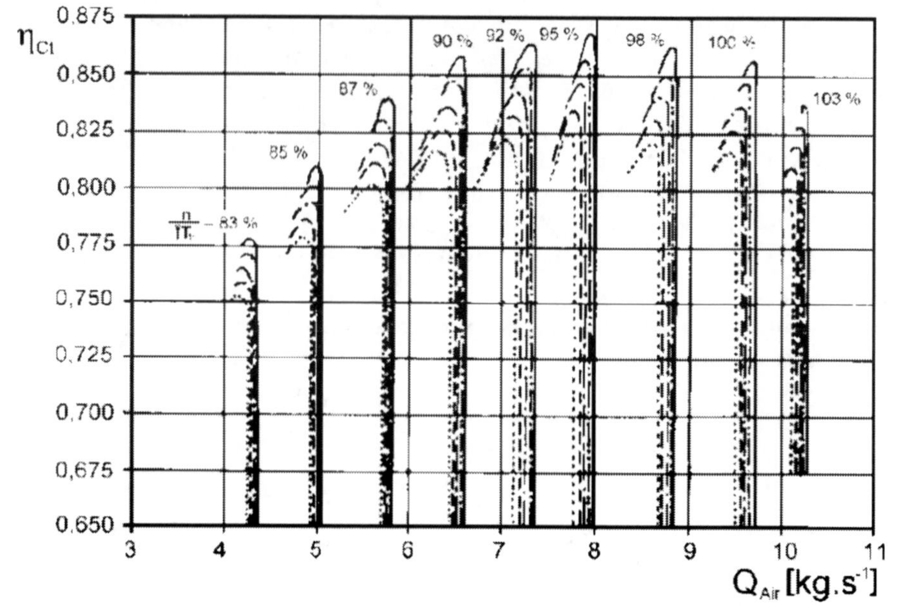

Fig. 9. Characteristics of TV3-117 turboshaft engine compressor with marked changes total compressor efficiencies after working different number of hours in dusty environment. ______ 0 working hours, _ _ _ _ 200 working hours, _ . _ . _ . _ 400 working hours, _ .. _ .. _ 600 working hours, - - - - - - - 800 working hours

It is known that a 15 % reduction in the gasodynamic stability margin of an axial compressor induces the development of unstable compressor operation in bench tests of the TV3-117 turboshaft engine.

Table 3. Change in the gasodynamic stability reserve of the compressor after working different periods of operation

Earned time (hours)	$\frac{n}{\sqrt{T_{1c}}} = 95\ \%$	$\frac{n}{\sqrt{T_{1c}}} = 98\ \%$	$\frac{n}{\sqrt{T_{1c}}} = 100\ \%$	$\frac{n}{\sqrt{T_{1c}}} = 103\ \%$
0	25,25 %	22,14 %	20,58 %	19,10 %
200	21,30 %	18,32 %	16,52 %	15,13 %
400	17,35 %	14,51 %	12,46 %	11,16 %
600	13,39 %	10,69 %	8,39 %	7,18 %
800	9,44 %	6,87 %	4,33 %	3,21 %

In addition to the influence of flow section erosion, damage, erosion, pollution, frost and others, the characteristic of the TV3-117 engine compressor is affected by the angle of adjustment of the engine compressor guide gear. The compressor motor alignment gear enables better coordination of the operation of the individual stages and the achievement of high values of the overall compression efficiency η_{Ct} and the stock of stable work of the axial compressor. Such research requires technical personnel to periodically check the characteristics of the directing gear and achieve their nominal values through control.
For a helicopter turboshaft helicopter turbo-compressor engine operated in dusty atmosphere conditions, one of the current challenges is to evaluate the effect of compressor blade erosion on its gasodynamic characteristics. Its solution, simultaneously with the evaluation of the influence of blade erosion on the vibro frequency characteristics, allows to elaborate recommendations in terms of the compressor blade boundary condition.

To evaluate the influence of dust erosion on the change of the compressor's gasodynamic characteristics, calculations of the air velocity and pressure in the flow section for the nominal (initial) blade geometry, as well as the geometry corresponding to the operating time of the engine in operation during 200, 400, 600 and 800 hours, were carried out. [1]
The results of calculations of the compressor characteristics (Fig. 8., 9.) π_{Kc}, $\eta_{Kc} = f(Q_v)$ from the operating time in dusty atmosphere showed that with the increase of the blade erosion intensity, the overall compression ratio in the compressor and its overall compression efficiency decreases. This implies that the limit of stable operation decreases, while at the same time the stable work margin of turboshaft engine ΔK_y decreases. The steady-state work reserve of the turboshaft engine was calculated according to the relation (1).
The results of the calculation are shown in Table 3.
It is known that a reduction of the compressor's gasodynamic stability margin below 15 % causes the compressor to develop unstable compressor operation (pumping) during bench tests of turboshaft engine TV3-117.
The analysis of the obtained results shows that the investigated compressor reaches its limit state when working in conditions of dusty atmosphere for 730 to 750 hours. The analysis of the absolute current velocity in the peripheral region of the flow area of the compressor with blades having different degree of wear, as well as the absolute current velocity fields at the height of 90 % from the compressor blade root, showed that due to the erosive wear in the peripheral region of the blades 6. Up to the 9th stage of the compressor develops air stream tear-off, which induces the development of unstable compressor operation. The cause of the phenomenon was the reduction of the rotor blade chord and the increase of the radial gap above the blades.

4 Conclusion

Maximum wear of the rotor blades of the axial compressor of the TV3-117 turboshaft engine due to dust erosion occurs at the first and sixth rotor blade stages.
From the point of view of the gas-dynamic stability of the axial compressor of the TV3-117 turboshaft engine, the erosive wear of the rotor blades of the sixth stage of the axial compressor is limiting, which can cause the jet to break off and cause the axial compressor to work unstably. When the bowstring in the peripheral part of the rotor blades is shortened by 6.19 mm, the stock of stable work of the axial compressor decreases by ΔK_y = 15 % to 17 %.
Determination of the limit value of permissible erosive wear of rotor blades of all stages of the TV3-117 turboshaft engine is possible on the basis of the required level of gas-dynamic stability.

References

1. Павленко, Д. В.; Двирник, Я. В.: Закономерности изнашивания рабочих лопаток компрессора вертолётных двигателей, эксплуатирующихся в условиях запыленной атмосферы. Вестник двигателелестроения Но1/2016, ISSN 1727-0219.
2. Suzuki, M.: *Numerical simulation of sand erosion phenomena in rotor/stator interaction of compressor / M. Suzuki, K. Inaba and M. Yamamoto //* Proceedings of

the 8th International Symposium on Experimental and Computational Aerothermodynamics of Internal Flows. – Lyon. – July 2007.

3. Neilson, J.: *Erosion by a stream of solid particle wear* / J. Neilson and A. Gilchrist // 1968. – Vol. 11, No 2. – pp. 111–122.

4 Tabakoff, W.: *Blade deterioration in a gas turbine engine* / W. Tabakoff, A. Hamed, V. Shanov // International journal of rotating machinery. – 1998. – Vol. 4, No. 4. – pp. 233– 241.

5 Grant, G.: *Erosion prediction in turbomacninery resulting from environmental solid particles* / G. Grant, W. Tabakoff // Journal of Aircraft. – May 1975. – Vol. 12, No. 5. – pp. 471–478.

6 Goodwin, J.: *Study of erosion by solid particles* / J. Goodwin, W. Sage, G. Tilly // Proceedings of the Institution of Mechanical Engineers. – 1970. – Vol. 184, Pt. 1, No. 15. – pp. 279–292.

7 Montgomery, J. E.: *Dust erosion parameters for a gas turbines* / J. E. Montgomery, J. M. Clark // SAE Summer Meeting Preprint 538A, Sos. Automative Engrs. – New York, 1962.

8 Биксаев, А. Ш.: *Анализ методов защиты авиационных газотурбинных двигателей от вредных факторов среды эксплуатации* / А. Ш. Биксаев и др. // Молодой ученый. – 2013. – №8. – С. 75.

9 Камаева, Р. Ф., Стругвец, С., Кривошеев, И. А., Галиуллн, Р.: *Использование закономерностей протекания характеристик компрессора при пармерической диагностике техического сотояния ГТД.* Молодой ученый. 2010. Н°1-2. С. 59 – 66.

10 Rapp, G. C., Rosenthal, S. H.: *Problems and Solutiuns for Sand Environment Operation on Helictoper Gas Turbines.* ASME 1968. P . 16.

11 Hocko, M.: *Turboshaft helicopter engines used in Czechoslovak, Czech and Slovak military aviation*, LF TUKE Košice 2019, ISBN 978-80-553-3493-6, p. 220. (SK).

MATEC Web of Conferences **367**, 00009 (2022) https://doi.org/10.1051/matecconf/202236700009
Power System Engineering 2022

Measuring deformations of the fan blade by optical methods

Vít Horáček[1,2*], *Vitalij* Yanovych[1,3], *Filip* Bastl[1]

[1]University of West Bohemia, Faculty of Mechanical Engineering, Department of Power System Engineering, Univerzitní 2732/8, 301 00 Pilsen, Czech Republic

[2]Department of Noise and Vibration, Research and Testing Institute Plzen Ltd., Tylova 1581/46, 301 00 Pilsen, Czech Republic

[3]Department of Fluid Dynamics, Institute of Thermomechanics, Czech Academy of Sciences, Dolejškova 5, 182 00, Prague, Czech Republic.

Abstract. Across industries, the demand for the ability to analyze components in operation and make decisions on repair based on the data obtained is growing rapidly. There is the potential to ensure safer operations while saving the cost of unnecessary new parts. A quick and relatively inexpensive method is a 3D scanner - the device that can scan a real component and create an accurate 3D model. For stationary cases - inspecting parts or assemblies - these are already established methods. In contrast, measurement of vibrations and deformations in operation is still not widespread and for turbomachinery, it can be a new direction of development.

1 Introduction

Currently, there are not many ways to easily observe deformations on turbomachines in operation. It is possible to simulate the task, but a simulation requires a largely idealized model, which may not correspond to reality at all. There are some necessary simplifications in mesh and the model does not include operational wear. Currently, there are various 3D scanning methods have been used to find out the real condition of equipment, which are able to provide a fairly detailed model of the real component. The output can be various comparisons with the original geometry or with scans from previous years. Based on these scans is possible to make numerical analysis with "real" blades. There is still the major limitation that the scanned parts have to be decomposed and it is required to create the mesh. But this step necessarily introduces inaccuracies into the analysis due to the need for the mesh appropriate to computational possibilities. One possibility to solve the problem of blade deformation in real operation is to use a dynamic 3d scanner - a device that is still not widely used compared to "classical" scanners, but theoretically meets all the prerequisites to give good results in certain situations. The advantages of modern 3D scanners include actively

* Corresponding author: vhoracek@kke.zcu.cz

monitoring the studied parameters of scanning models. This feature is most critical for analyzing the behavior of various machine parts in the dynamics and development of their automation systems [1].

It is clear that turbomachines are an integral part of today's energy industry. Therefore, many scientific institutions and researchers have devoted their attention to the research and improvement of these constructions. It should be noted that the behavior of turbulent flow is one of the key issues in these studies. A specialized laboratory was based at the Department of Power System Engineering of the University of West Bohemia to study the topology of turbulent flow. Its work provides by modern measuring equipment and technics. Among them are various high-precision pressure sensors [2], optical measuring systems (Particle Image Velocimetry) [3], and a technique for the measurement of turbulence (Constant Temperature Anemometry) [4]. The laboratory also includes a wind tunnel characterized by high-laminar flow [5].

This tunnel is popularly used to study various aerodynamic phenomena. Their design significantly influences the quality of the created airflow. During the flow generation, the construction parts of the wind tunnel get deformed due to pressure variations. This phenomenon significantly complicates the creation of high-level laminar flow and worsens the results of experimental studies. Therefore, we used the ARAMIS optical measuring system to study this phenomenon [6, 7]. At the University of West Bohemia, we have a dynamic scanner GOM ARAMIS. It is a device made mainly for industrial use, i.e. for deformation checks of structures and constructions [8-10], beams, or in the field of biomechanics. Our desired use for turbomachines, where deformations are mainly due to the combined action of the flow medium and centrifugal forces, is not typical and has not been applied frequently. One experiment carried out with this apparatus on turbomachinery was the measurement of the vibration of a wind turbine tower [11], but here, according to the available record, they only monitor the motion of points located on a "stable" tower. Measurements on turbine blades are also mentioned by the manufacturer of the ARAMIS system on its website, but this is a measurement of vibrations on a removed blade, excited by hammer blows [12].

2 Methods and equipment

2.1 GOM ARAMIS stereo system

The GOM ARAMIS is a smart solution for determinating deformations by measuring displacements of the measured object. These measured values are observed on the prepared surface. There are two options, for how to prepare the surface for measuring – use reference points or create a stochastic pattern. Stochastic patterns are created by spraying drops on the surface and are suitable for vast objects. Reference points are easier and require only glueing reference points in the area of interest. Both methods depend on the contrast between observed objects (points or drops) and surroundings, the choice between them is based on the specific properties of the measured object. Usually depends on a shape, size, or color of the object – for a variegated object is better to use reference points.

The main components of the GOM ARAMIS system are a 3D camera with a controller, situated on a moveable stand. The ARAMIS Camera is built as a stereo system that can take two pictures from cameras deployed on a frame. There are several sizes of frames, whose choice depends on the volume of the measured object. Based on photoshoots from the camera is possible to determine very accurate coordinates of reference points or stochastic patterns. The acquisition of measured data is managed by the controller, which is responsible for controlling lightning the scene too. Ideal illumination of the measured object is one of the most important conditions and at the same time for this experiment extremely complicated problem. Originally as the accessories, there are two kinds of lights – tracking spots for bigger frames and light projector for the smaller ones. Both lights use narrow-band blue LED light.

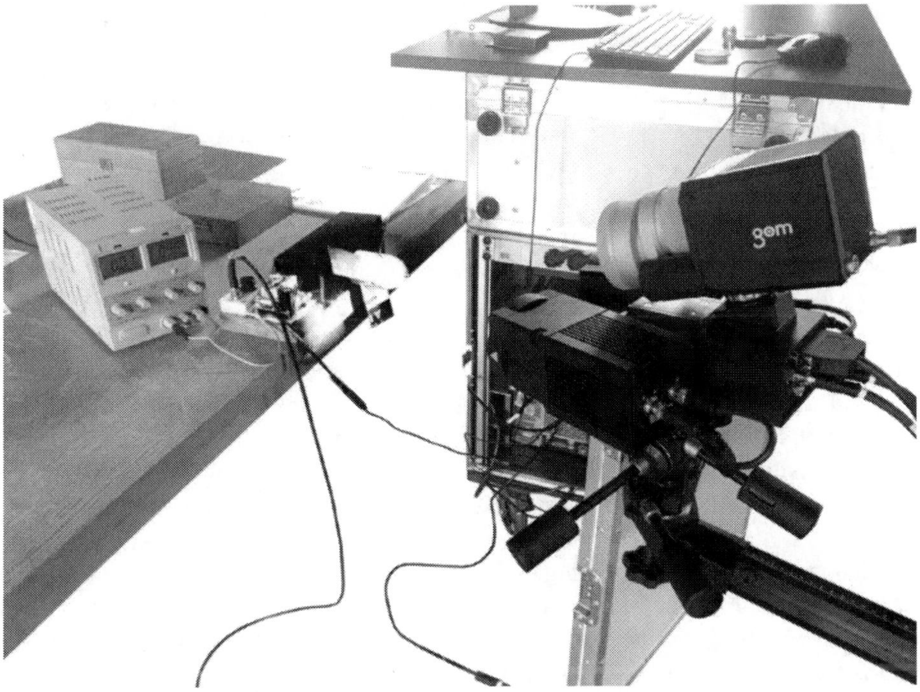

Fig. 1. The ARAMIS camera with light projector focused on the measured fan.

2.2 Method and equipment

This experiment is performed as an initial before measuring deformations of blades on real turbines, so there were used some assumptions and specific properties. In the experiment was use a homemade fan with paper blades. For this choice, there were several reasons, but the main was the quantity of deformation. Factory fan had blades from plastic with high rigidity, as was tried, deformations of the blade from classic office fan were almost negligible. This property is important for use but not very suitable for the initial experiment. The goal of our effort was to create a small and simple fan where We need only low speed however with easily observed deformations of the blades. The emphasis was focused on simplicity. Most components were designed in CAD and printed from the PLA filament, electric parts are purchased as pre-made modules and assembled into a simple adjustable fan with very elastic blades.

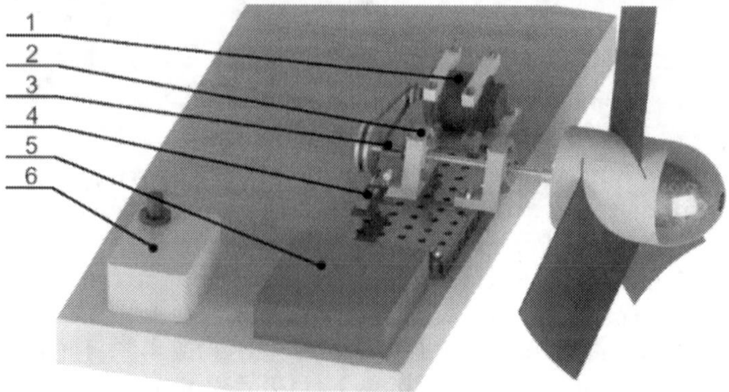

Fig. 2. 3D model with marked main parts: 1 – DC motor, 2 – Bearing hose, 3 – contrast element, 4 – phototransisor sensor, 5 – pulse regulator, 6 – control panel

The final design (on Fig. 2) is based on several tries and it contains six main parts assembled on the base. Electric DC motor with pulse regulator powered by 10 volts from a stabilized laboratory source. Torque from the motor is transferred to the shaft with a bladed wheel via a rubber belt (transmission ratio 0,3). To eliminate tension forces from the belt, which became a significant source of vibration, the shaft was mounted on two ball-bearings (SKF CSN 02 4630). The speed is set by a potentiometer allowing continuous regulation from zero to maximum. For monitoring the speed, there is a sensor module based on a phototransistor placed nearby the driven shaft. On the shaft is one special cylindrical element with very contrasting segments. The number and position of elements determines the frequency and the time of photoshoots. This sensor has a lot of advantages – for operation needs only 5 volts DC, so it is very easy to use every USB as a power source. It includes regulation of sensitivity, so it can be easily tuned to optimal settings. The output is a signal useable for the ARAMIS software as a trigger for creating a photoshoot. Thanks to this sensor, photos are taken synchronized in the required position, automatically sorted, and only the demanded are saved for processing. This is very important for use in the future, now the homemade fan has only three blades, so it is not hard to recognize which is on the shot. But the ambitions are higher, and the option to focus only on a few precisely selected blades from the whole is very useful. In service, the sensor, DC motor, and shaft must be undercover, because light from the projector affects the sensor, and photos are not created at the correct time.

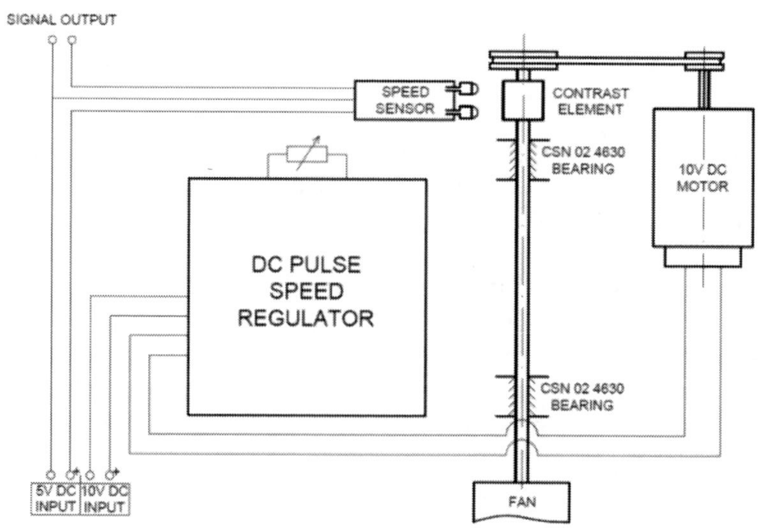

Fig. 3. The electric scheme of measured fan.

From experiments with a factory office fan, there was some data showing that a bigger blade is worse at measuring deformation than a tiny one. So it was decided to use cameras on the smallest frame (180 mm) and a direct light projector. This assembly is suitable for blades smaller than 120 mm (finally it was used 95 x 35 mm). The measured blade is relatively small, so it was used reference points glued to the surface of the blade. The used size of the points is 0,8 mm. These points were placed on the blade as positions where will be the deformation determined, and on the hub as a point to measure the displacement of the (theoretically) rigid body. Differences between the rigid hub and deformative blades are values of the measured deformation. Thanks to this difference, there were pure data of deformation independent of the angle of rotation.

3 Results and discussion

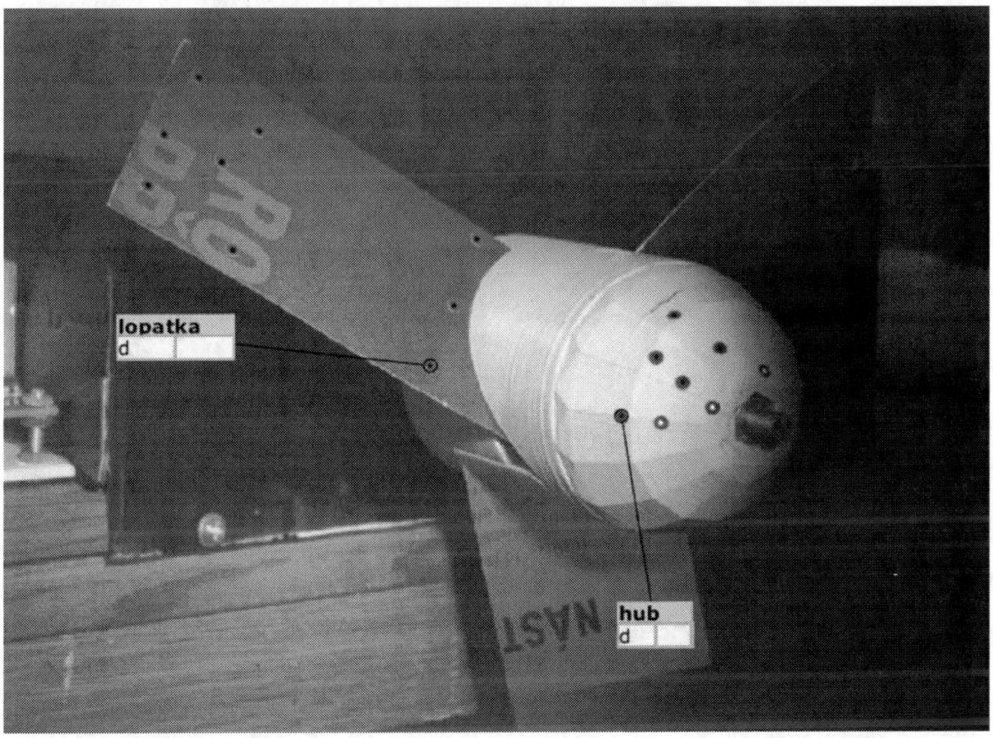

Fig. 4. Reference image.

The figure 4 was created as a reference shot. It shows the observed part of the assembly with reference points on the hub and blade. This shoot shows the maximum quality from the camera because it was photographed when the fan stayed.

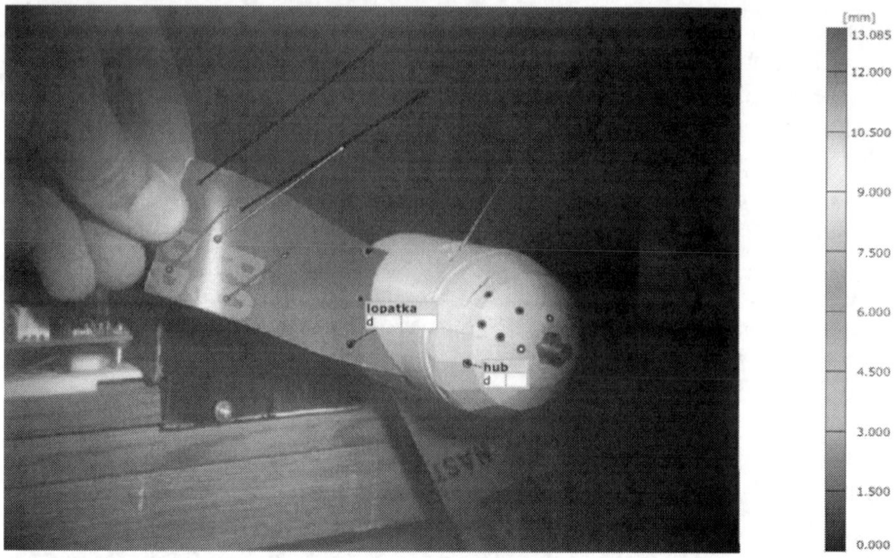

Fig. 5. The measured blade slowly deformed with a finger.

The figure 5 was created when the blade was deformed manually with a finger. The data is clear, changes were slow. So there is no problem in determining the deformations because the ARAMIS software can reliably recognize the coordinates of reference points.

Fig. 6. Unusuable data from measuring the fan in service.

Picture taken while the fan was running, fully showing problems with using the ARAMIS dynamic scanner for measuring on turbomachinery. Points on the hub are useable, but with growing distance from the center of rotation, the velocity is linearly growing. For this experiment, we used small velocities (frequency 4 Hz). Despite this, reference points on the blade are not sharp and not useable to determine a deformation. It is clear, that with bigger velocities, these problems will grow. At now, there is no useable solution to determine deformations by ARAMIS dynamic scanner.

4 Conclusion

It is clear that the current state is not satisfying. This initial experiment shows the way, but there must be added some breakthrough solution leading to sharpening photoshoots. The biggest problem is about the illumination of a scene. So first, what we want to do is move the equipment outside the laboratory and try to measure deformations in the sunshine. This can lead to solving the problem for small velocities. But it means making it all measurement outside, and the majority of interesting objects are somewhere inside, usually in bad light conditions. So we believe, that the solution can be based on a combination of particular improvements current status. One of these solutions can be using stronger contrast by using reflexive reference points. It can be a possibility for bigger measured objects. But unfortunately, there are no 0,8 mm reflexive points to purchase. Another direction of the consideration is the type of illumination – there is a requirement for a very powerful lighting system, based on LED or laser. This system can be based on very short flashes directed by signals from a phototransistor sensor. Solution of the problems with illumination is the first and most important goal in the future.

References

1. V. Yanovych, T. Honcharuk, I. Honcharuk and K. Kovalova. Design of the system to control a vibratory machine for mixing loose materials. Eastern-European Journal of Enterprise Technologies, 6 (3-90), pp. 4 - 13, (2017)
2. V. Yanovych, D. Duda, V. Horáček and V. Uruba. *Creation of recombination corrective algorithm for research of a wind tunnel parameters* in Proceedings of the Power System Engineering, AIP conference 2118, Pilsen, Czech Republic (2019).
3. D. Duda, V. Yanovych, V. Tsymbalyuk, V. Uruba. Effect of Manufacturing Inaccuracies on the Wake Past Asymmetric Airfoil by PIV. *Energies* **2022**, *15*, 1227.
4. V. Yanovych, D. Duda, V. Uruba and T. Tomášková. Hot-Wire Investigation of Turbulence Topology behind Blades at Different Shape Qualities, *Processes* **10(3)**, 522 (2022).
5. V. Yanovych, D. Duda, V. Horáček and V. Uruba. *Research of a wind tunnel parameters by means of cross-section analysis of air flow profiles*, in Proceedings of the Power System Engineering, AIP conference 2189, Pilsen, Czech Republic (2019).
6. V. Yanovych, D. Duda. Using ARAMIS system for measurement of structural stability of running wind tunnel. MATEC Web Conf. 328 01003 (2020)
7. V. Yanovych, D. Duda. Structural deformation of a running wind tunnel measured by optical scanning. (2020) Strojnicky Casopis, 70 (2), pp. 181 – 196
8. Peterkova, Eva & Srefl, Milan. (2016). Use of 3D measuring system aramis for analysis of tube flaring process. MM Science Journal. 2016. 1392-1397.
9. Walotek, Konrad & Bzówka, Joanna & Ciołczyk, Adrian. (2021). Examples of the Use of the ARAMIS 3D Measurement System for the Susceptibility to Deformation Tests for the Selected Mixtures of Coal Mining Wastes. Sensors. 21. 4600. 10.3390/s21134600.
10. Urbańska, Krystyna. (2017). Application of aramis system for measurement of deformations of masonry. Zeszyty Naukowe Uniwersytetu Zielonogórskiego / Inżynieria Środowiska. 166. 48-59. 10.5604/01.3001.0010.6034.
11. CIMCo – Especialistas en CAD/CAM y Metrología 3D. (2017, May 5). *GOM ARAMIS - Vibración en turbina* [Video]. YouTube. https://www.youtube.com/watch?v=RpOl-XEvNwQ
12. GOM a ZEISS company. (n.d.). *Speeding Up Fan Blade Integrity Testing with ARAMIS.* https://www.gom.com/fr/industries/aeronautique/vibration-analysis-fan-blade.html

Steam turbine flow path drainage system under different operating conditions

Michal Hoznedl[1*] and *Jindřich* Bém[1]

[1]Doosan Škoda Power, 301 28 Tylova 1/57, Plzeň, Czech Republic

Abstract. One of the important elements of anti-erosion protection of the steam turbine last blades is the use of hollow stator blades. There are grooves on the blade surface that lead the steam mixture inside the blades. This mixture is blown to the diffuser area. However, the drain principle fails for low outputs of the turbine, during the so-called ventilation regimes. In the part of blade section the steam flow is reversed and the flow part drain stops working. It was confirmed using experiments of steam turbine carried out in a wide range of power outputs.

1 Introduction

Steam turbine operation in long-term low outputs raises a number of problems. These are caused mainly by a significant output lowering of last stages when the expansion turns into compression. It is accompanied with the occurrence of compression heat and the necessity to cool last rotor blades. Cooling is achieved by injecting water through the nozzles and must be sufficient. At the same time there must be no excessive erosion of the trailing edge of rotor blades near the heavily loaded root section. Furthermore, the compressor operation of the steam turbine blades can cause undesired excessive blade vibrations.

When researching the behaviour of the turbine in low operation, it is always necessary to address the flow and vibration in the area of last rotor blades and the whole stage as a common phenomenon.[1-5]. Many authors are engaged in experimental research of the steam mixture flow near the suction grooves. The dimensions, shapes and locations of the grooves on the surface of stator blades are verified with regard to maximum efficiency of the water film suction with minimal amount of steam sucked from the flow part [6, 7].

The results of experimental and analytical studies are valuable which assume that the water film on the surface of stator blades is not sucked, but on the contrary, that the inside of the hollow blade is heated by warmer steam and the water film evaporates [8, 9]. However, these studies do not address the overall energetic balance of the steam in the turbine. The energy required to evaporate the water film can thus be greater than the energy escaped from the flow part due to the suction of part of the steam.

However, none of the above mentioned works deals with the ability of the drain system to suck effectively the steam mixture direction at low steam turbine output during the so-

* Corresponding author: michal.hoznedl@doosan.com

called ventilation. The submitted paper aims to compensate for this shortcoming. The behaviour of this suction tract at nominal and a number of low turbine operations is determined on the basis of a number of experimentally obtained data.

2 Experiment description

Experiments were carried out on the steam turbine of the 200 MW output in the combined cycle power plant. The last stage of the turbine had the rotor root diameter of 1416 mm and the last blade length of 729 mm. Design pressure at the rotor outlet is 11.4 kPa. Root reaction reaches the value of 0.58. Stator blades are hollow. The schematic section of the flow and the last stage with the designation of important elements is in Fig. 1. It is obvious that after the last stage there is the axial diffuser.

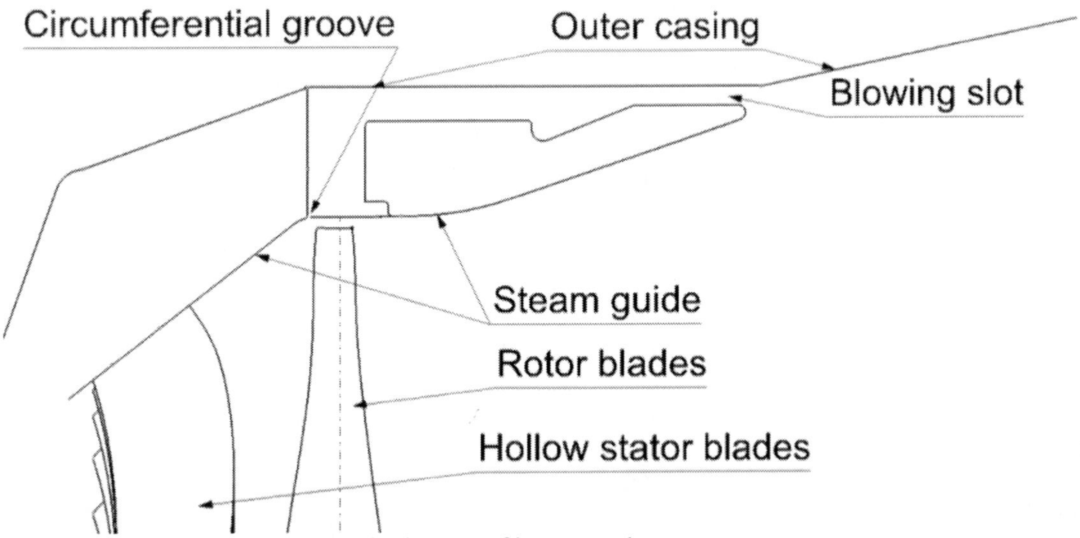

Fig. 1 Steam turbine cross section in the area of last stage tip

In Fig. 2 the points with static pressure samples are labelled using circles. Pressures were measured on the tip limiting wall before (p_{0t}) and behind (p_{1t}) the stator blades, on the tip limiting wall also behind the rotor blades (p_{2t}) and behind the place where the steam sucked from the stator blades is blown back to the main flow (p_{d3t}). The static pressure in the mixing chamber of the diffuser is also measured. All given pressures were measured in pairs at two circumferentially different places of the turbine. Pressure on the blade surface at the suction groove is not measured directly, but is based on a number of CFD calculations:

$$p_{gr} = p_{0t} - 0.45 \cdot (p_{0t} - p_{1t}) \quad (1)$$

For low outputs it is often valid that $p_{0t} \leq p_{1t}$. As the flow in blading increases, all three pressures, namely p_{gr}, p_{0t} and p_{1t}, increase linearly.
Static pressure in the blowing slot was not measured directly. Due to low velocities of the steam in this place (Ma < 0.2) it is supposed that the pressure in this place will be the same as the pressure p_{d3t} that is measured about 100 mm axially behind the blowing slot. Steam mass flow through stator blades is labelled as G_{sb}. The amount of steam G_{hb} is sucked to hollow blades. The volume flow of water through suction grooves is negligible, it makes about 2 % of the total flow through the grooves. Therefore for further calculations it can be assumed that the construction area of the grooves is the same as the area for steam flow rate.

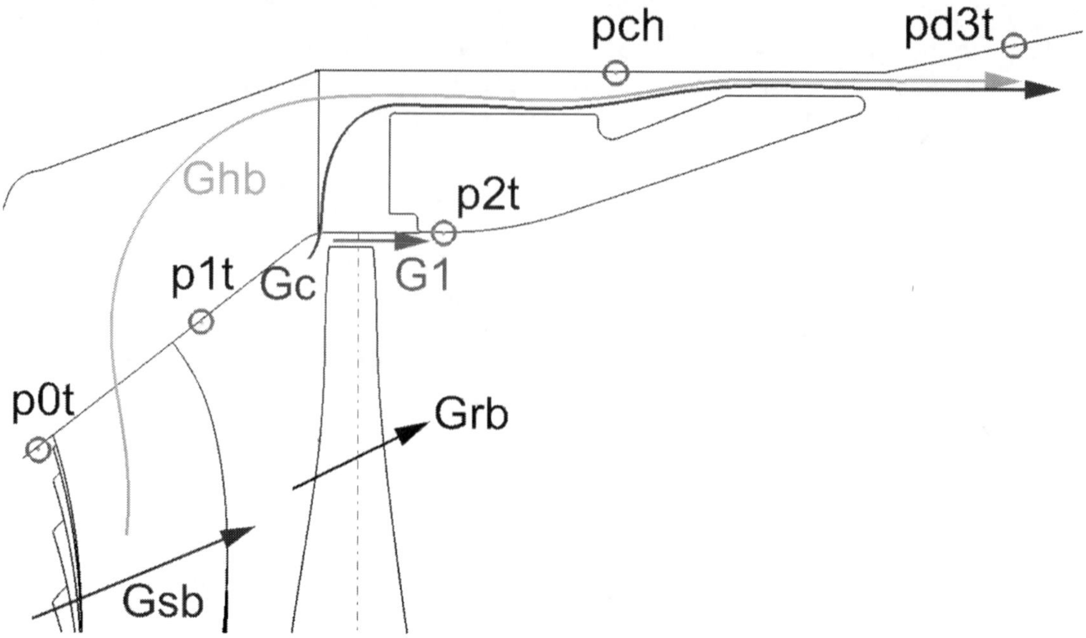

Fig. 2 Measured pressured and calculated mass flows

In equation (2) G_1 is the flow through rotor blades radial clearance and flow G_2 is then:

$$G_2 = G_{hb} + G_c \tag{2}$$

Here G_c is the steam flow through circumferential groove. It captures a part of water films flowing along the outer circumferential wall of the flow part.
The total steam flow that does not work because it bypasses the last blade is:

$$G = G_1 + G_2 \tag{3}$$

Analyses were carried out for 8 operating regimes of the turbine differing in turbine output, the flow through the last stage stator blades, outlet axial velocity from the last stage c_{2x} and possibly the so-called ventilation coefficient c_{2x} / u. For values roughly $c_{2x} / u \leq 0.17$ the turbine is in the ventilation regime, i.e. the regime when the last stage has zero or negative output. The „u“ value is the circumferential velocity at the mean blade diameter. The condenser pressure was in the range of 0.1 to 0.2 bar. The overview of basic parameters for all measured variants is given in Table 1. The measured regimes cover the whole range of the last stage and the exhaust hood work from the zero output to aerodynamic throttling.

Table 1. Overview of measured variants.

Nr.	**1**	**2**	**3**	**4**	**5**	**6**	**7**	**8**
c_{2x} [m/s]	4	40	61	123	209	249	266	360
c_{2x}/u [-]	0.01	0.10	0.15	0.29	0.51	0.61	0.65	0.88
p_{2t} [bar]	0.112	0.164	0.208	0.158	0.121	0.138	0.132	0.100
Output [MW]	0	20	40	122	135	183	198	200
G_{rb} [kg/s]	-	20.5	42.0	63.8	90.3	119.9	125.6	131.0

3 Results

First of all it is necessary to define the pressure ratios over the individual throttling elements of the inner bypass. Specifically these are:

- p_{ch} / p_{gr} (pressure ratio in the diffuser chamber and on the blade surface in the groove area, the pressure loss between the hollow blade and the diffused chamber is neglected),
- p_{ch} / p_{1t} (pressure ratio in the diffuser chamber and on the outer limiting wall between the last stage stator and rotor blades),
- p_{d3t} / p_{ch} (pressure ratio on the outer circumferential wall of the diffuser and the diffuser chamber),
- p_{2t} / p_{1t} (pressure ratio over the last stage rotor blade tip).

For wet steam the critical pressure ratio is 0.583. The decision whether the ratio is critical or not is important for correct calculation of the steam flow. The overview of pressure ratios is in Table 2. Regimes with pressure ratios higher than 1 are marked in red. In these cases there is a back stream flow, in Fig. 2. In this case the suction tract loses its function. Specifically these are regimes 1, 2 and marginally also 3. The output of regime 3 is at the level of 20% of the nominal turbine output. The green fields represent the states when the pressure ratios are critical.

Table 2. Pressure ratios

Nr.	1	2	3	4	5	6	7	8
p_{ch} / p_{gr}	1.070	1.001	0.935	0.625	0.748	0.376	0.363	0.329
p_{ch} / p_{1t}	1.017	1.033	1.001	0.745	0.868	0.450	0.436	0.394
p_{d3t} / p_{ch}	1.023	1.005	0.992	0.953	0.963	0.848	0.825	0.741
p_{2t} / p_{1t}	1.055	1.063	0.986	0.697	0.834	0.364	0.339	0.266

The relation for critical flow e.g. through the grooves in hollow blades is:

$$G_{hb} = \mu_{hb} \cdot S_{hb} \cdot \chi \cdot \sqrt{\frac{p_{hb}}{v_{hb}}} \tag{4}$$

It is valid that the coefficient χ = 0.63045, flow coefficient of grooves μ_{hb} = 0.7, flow coefficient of radial clearance leakage μ_{Gl} = 0.67, flow coefficient of circumferential slot μ_{Gc} = 0.82, flow coefficient of blowing slot μ_{G2} = 0.96.
In the case of subcritical flow it is possible to calculate the flow (e.g. by blowing to the diffuser wall) as follows:

$$G_2 = \mu_{G2} \cdot S_{G2} \cdot w_{G2} \cdot \rho_{G2} \tag{5}$$

Density ρ_{G2} is calculated from pressure p_{d3t} and wetness y = 0. Velocity w_{G2} is calculated based on standard Saint-Venant-Wanzel equation, or using relation for critical velocity. The overview of velocities in all parts of the drain system and for all regimes is in Table 3. Negative velocities correspond with places and regimes with pressure ratios higher than 1.

Table 3. Velocities in individual components of the drain system

Nr.	1	2	3	4	5	6	7	8
w_c [m/s]	-77.6	-98.9	-18.4	206.3	291.3	410.6	411.2	411.7
w_{G1} [m/s]	-136.0	-131.5	0.2	233.0	323.7	410.6	411.2	411.7
w_{hb} [m/s]	-136.1	9.5	148.5	294.9	371.3	412.0	412.6	413.1
w_{G2} [m/s]	-41.2	-19.5	25.8	53.7	59.8	114.0	122.9	152.4

Important parameters are relative flow rates, i.e. the values of the flows through the individual parts of the drains related to the flow rate through the stator blade G_{sb}. The summary of flows is in Table 4. Due to low flows through the stator blades and pressure ratios higher than 1, the values of relative flow for regimes 1 and 2 are significantly negative. It is caused by the nature of the flow in the ventilation regime, which is indicated by the static pressure contours in Fig. 3, determined by CFD calculation, for regime 2. Significant compression over the rotor blade tips and higher pressure in the axial outlet area than the pressure at the suction grooves p_{gr} is evident.

Table 4. Relative mass flows in the drain system

Nr.	1	2	3	4	5	6	7	8
G_C / G_{sb} [%]	-7.1	-1.5	-0.2	2.0	0.8	1.6	1.5	1.5
G_{hb} / G_{sb} [%]	-14.9	-0.2	1.4	2.8	1.2	1.7	1.7	1.7
G_1 / G_{sb} [%]	-9.1	-1.5	0.5	1.5	0.6	1.1	1.1	1.1
G_2 / G_{sb} [%]	-22.0	-1.7	1.2	4.8	2.0	3.3	3.3	3.3
G / G_{sb} [%]	-31.1	-3.2	1.6	6.4	2.6	4.4	4.3	4.3

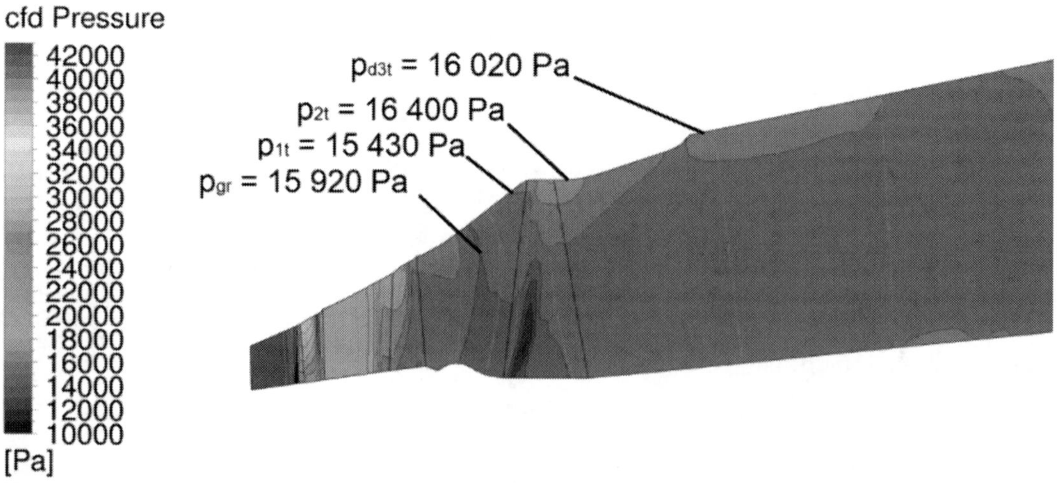

Fig. 3 Contours of static pressure for regime 2

The course of differences of measured pressures p_{gr} - p_{ch} and p_{1t} - p_{ch} depending on c_{2x} / u is shown in Fig. 4. For c_{2x} / u < 0.1 the flow direction in sucking grooves of hollow blades is reversed. However, for the ratio c_{2x} / u < 0.1 the last stage operates in a significant ventilation regime. Long term low output turbine operation is dangerous from the viewpoint of anti-erosion protection of the last rotor blades. In the range 0.1 < c_{2x} / u < 0.17 the turbine

operation is possible, but it is still the area of compression heat generation. It is related with the necessity of cooling rotor blades using water jets and with possible erosion of trailing edges at the root. For values $c_{2x} / u > 0.17$ it is possible to operate the turbine for a long time without the need for time constraints of operations due to erosion. In the interval $0.17 < c_{2x} / u < 0.6$ the turbine is loaded and the flow through the grooves in hollow blades and also through the circumferential groove increases. For $c_{2x} / u > 0.6$ the drain system is aerodynamically clogged and operates according to design assumptions.
In Fig. 4 also the courses of relative flofws into hollow blades calculated from pressure conditions are added. Sucking from hollow blades works for $c_{2x} / u > 0.1$, sucking by circumferential groove for $c_{2x} / u > 0.16$. This value is given by the intersection of the curve $p_{1t} - p_{ch}$ and $\Delta p = 0$.

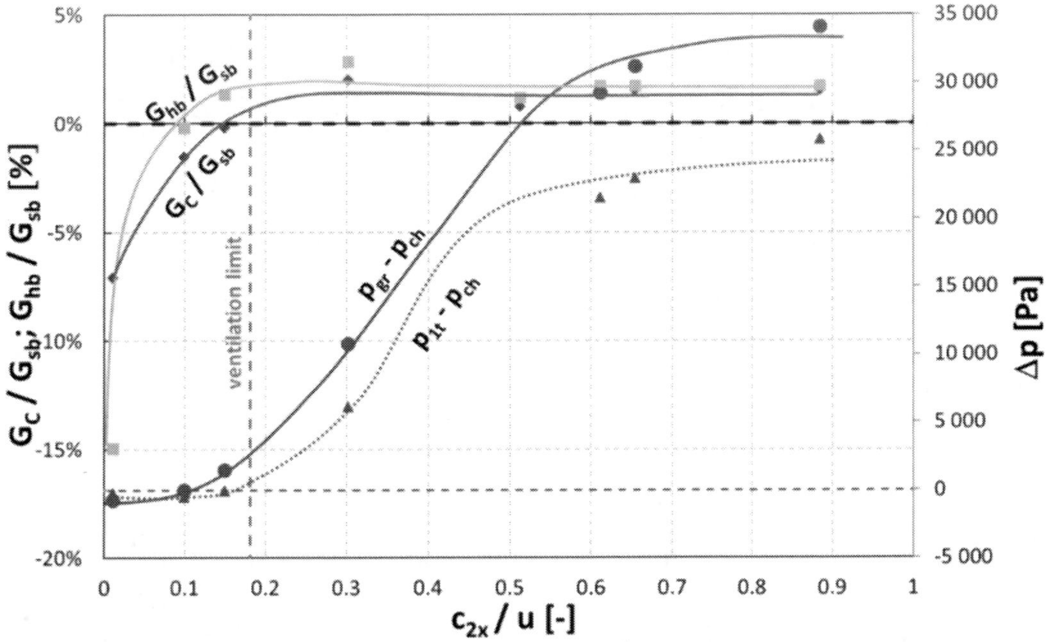

Fig. 4 Courses of pressure differences and relative flows depending on c_{2x} / u for the monitored turbine

4 Conclusions

Extensive experimental research of the last stage flow and the outlet area of the 200 MW steam turbine was performed focused on the functionality of the drain system in a wide range of operations, from zero output to the overload regime.
Based on 10 measured static pressures and geometry knowledge, pressure drops and relative flows in individual parts of the steam inner bypass were defined. For $c_{2x} / u < 0.1$ the steam flow from the outlet goes back to the stator blades and the drain system does not work. There is a very significant erosion of rotor blades on the trailing edge and when using water jets for outlet cooling even the trailing edge of stator blades can erode at the root. Border limit for long-term turbine operation is $c_{2x} / u = 0.17$; however, the temperature in the outlet area is also monitored. Now the suction to hollow blades already works effectively and the suction through the circumferential grooves on the tip limiting wall between the stator and rotor blades also starts working.
If long-term operation is required for $c_{2x} / u < 0.1$, which for the monitored turbine corresponds with 10 % of nominal output, it is necessary to use other elements of anti-erosion protection than just suction using hollow blades. It can be e.g. laser hardening of trailing edges or using hydrophobic surfaces on stator blades.

The work was financially supported by the Technology Agency of the Czech Republic; project FW03010515 Ultra flexible operation regimes of steam turbines and by European Regional Development Fund-Project "LoStr" (No. CZ.02.1.01/0.0/0.0/16_026/0008389).

References

1. Haller, B. *Alleviation of rotating pressure oscillations in the last LP turbine stage during low-load conditions.* Proc. ASME TurboExpo (2016). GT2016-56088 (2016), https://doi.org/10.1115/GT2018-76498 .
2. SHIBUKAWA, N. et al. *A Correlation Between Vibration Stresses and Flow Features of Steam Turbine Long Blades in Low Load Conditions*. Proc. ASME TurboExpo (2011), GT2011-46368, https://doi.org/10.1115/GT2011-46368 .
3. Sigg, R. et al., 2009, *Numerical and experimental investigation of a low-pressure steam turbine during windage,* In: Proceedings of the Institution of Mechanical Engineers, Part A: Journal of Power and Energy, Volume: 223 issue: 6, page(s): 697-708, https://doi.org/10.1243/09576509JPE826.
4. Megerle, B., 2014, Unsteady Aerodynamics of Low-Pressure Steam Turbines Operating under Low Volume Flow Conditions, In: Dissertation Theses, THÈSE NO 6096 (2014), Lausanne, Switzerland.
5. Tanuma, T. et al. *Unsteady flow effects on steam turbine last stage blades at very low load operating conditions*. Proc. ASME TurboExpo (2018). GT2018-76498, https://doi.org/10.1115/GT2018-76498 .
6. Xiaoming, W. et al. *Study on water extraction of hollow stationary blade under wet steam flow conditions.* Proc. 12th European Conference on Turbomachinery, ETC2017-349 (2017), Stockholm, Sweden.
7. Wang X et al. *Water Droplets Dispersion and Optimization of Water Removal Suction Slot Locations on Stationary Blades in Last-Stage of a 600 MW Steam Turbine*, Proc. ASME TurboExpo (2009), GT2009-59490, https://doi.org/10.1115/GT2009-59490
8. Akhtar, M.S. Prevention of Steam Turbine Blade Erosion Using Stator Blade Heating, Proceedings of the Institution of Mechanical Engineers, Vol 191, Issue 1 (1977), https://doi.org/10.1243/PIME_PROC_1977_191_042_02
9. Gribin, V. et al: *Performance of a wet steam turbine stator blade with heating steam injection,* Proc. 12th European Conference on Turbomachinery, ETC2017-349 (2017), Stockholm, Sweden.

CFD benchmark of flat plate fuel assembly

Štěpán Hrouda[1,2*], *Jan Syblík*[1,2], *Miroslav Gleitz*[1] and *Michal Cihlář*[1,3]

[1] Department of Energy Engineering, Czech Technical University, Technická 4, 160 00 Prague, Czech Republic
[2] National Radiation Protection Institute, Bartoškova 1450/28, 140 00 Prague 4, Czech Republic
[3] Research Center Řež, Hlavní 130, Husinec 250 68, Czech Republic

Abstract. Diversification of the nuclear fuel suppliers is an inseparable component of nuclear safety of power and research nuclear reactors. So far, the Czech research reactor LVR-15 has been operated with the Russian fuel IRT-4M, because there was no alternative to this fuel. Recently, the French company CERCA has developed an alternative to this fuel Flat Plate Fuel Assembly (FPFA). Hence, the influence of the different fuel characteristics (especially neutronic, thermomechanic and hydraulic) on nuclear safety has to be provided. This work involves a comparison of the flow characteristic of these two types of fuel assemblies using the ANSYS Fluent 2021 R1 and STAR CCM+ 2022 CFD programs. Moreover, these computational models of FPFA fuel are validated in an experimental facility.

1 Introduction

The determination of a neutron-physical, thermohydraulic and thermomechanical properties of any nuclear fuel are crucial for nuclear safety.

To diversify the research nuclear fuel market, Cerca has developed a new Flat Plate Fuel Assembly (FPFA), which should replace the current fuel IRT-4M for the LVR-15 reactor.

It was necessary to simulate the flow characteristics of both fuel types (mainly pressure losses) due to the different design, construction and material. Simulated data of FPFA model was verified on mockup model of FPFA in experimental loop build in CTU laboratories. The simulation was done in CFD codes ANSYS Fluent and STAR-CCM+.

* Corresponding author: stepan.hrouda@fs.cvut.cz

1.1 Research reactor LVR-15

The reactor LVR-15 in research centre Řež has been in operation without major design changes since 1989. It is a light-water tank-type reactor with a nominal thermal output of 10 MW. The core is cooled by forced downward circulation of coolant. The reactor vessel is a pool-type with a volume of 22 m^3. The core is located about 3-4 m below the surface (pressure in core 130 to 140 kPa). The reactor is cooled by a coolant flow up to 2000 $m^3 \cdot h^{-1}$ and maximal temperature 56°C. So far, the reactor has been operated with Russian fuel IRT-4M with a UO_2 matrix. New fuel FPFA has U_3Si_2 matrix. [1, 2]

An interesting feature of the LVR-15 reactor is the downward flow, i.e. against the natural circulation. The reactor was designed with a downward flow because fuel elements are sitting on a grid. In the axial direction they are secured only by their own weight.

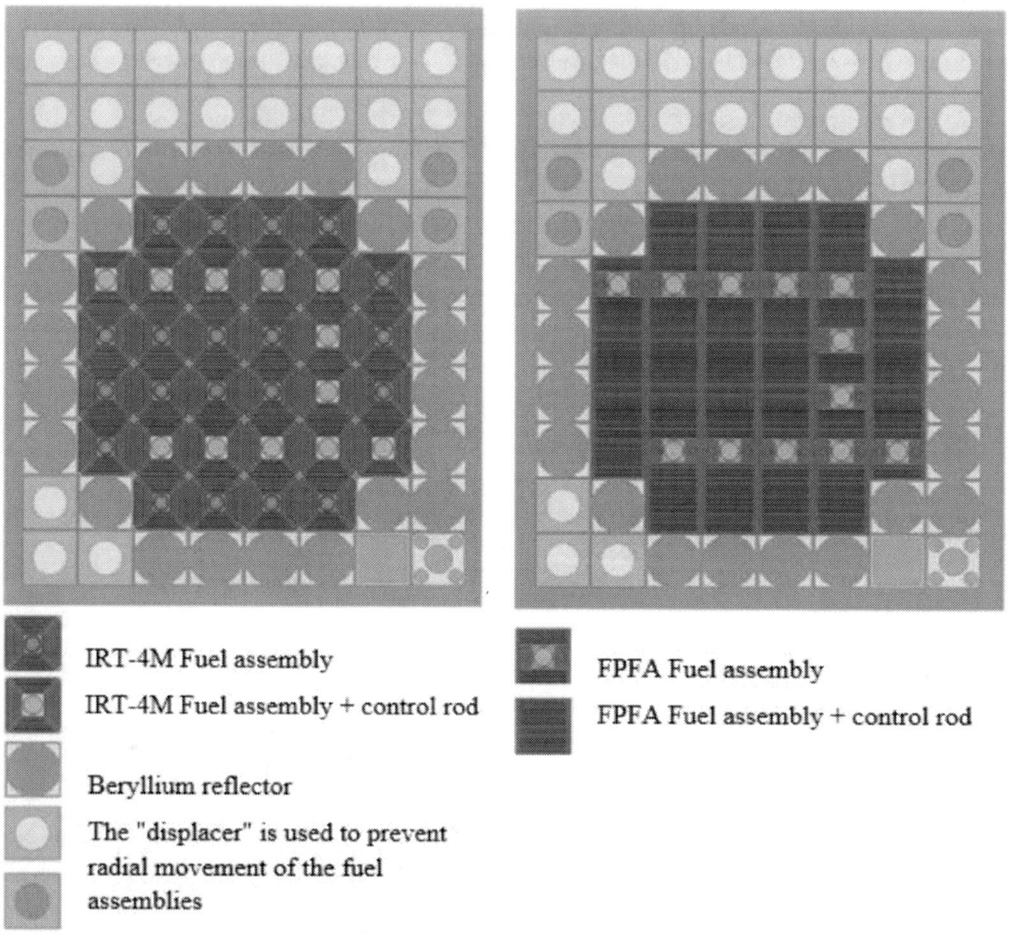

Figure 1. LVR-15 reactor theoretical core composition [2]

1.2 Project LEU-FOREvER

LEU-FOREvER stands for Low Enriched Uranium Fuels fOR REsEarch Reactors. The project brings together major European institutions (CV-Řež. ÚJV, CEA, FRAMATOME – CERCA, ILL, NCBJ, SCK-CEN, TechnicAtome and TUM) in the field of nuclear research. The project has two goals. The first goal is the diversification of nuclear fuel suppliers for Soviet-type research reactors, e.g., VR-1 (Vrabec. FNSPE CTU in Prague) or LVR-15 (CV-Řež. ÚJV). Until now, the company ROSATOM had a monopol on this fuel and the reactors were only compatible with IRT-4M fuel. To offer different choices, FRAMATOME - CERCA developed FPFA fuel assemblies that can replace IRT-4M fuel. The second goal is the conversion of HPRRs (High Performance Research Reactors) from highly enriched to low-enriched fuel. [1, 2]

2 Comparison of IRT-4M and FPFA fuel

The main difference between fuel assemblies is in geometry, material and manufacturing process of uranium plates. In IRT-4M is 8 plates bend in a rectangle shape and set concentrically. Developers of FPFA fuel have problems with bending the plates, in corners appears cracks, for this reason, they decide for 22 straight plates. The matrix UO_2 was replaced by U_3Si_2 for better accident resistance

Table 1. IRT-4M and FPFA parameters comparison [2]

	IRT-4M	FPFA
Fuel assembly cross-section [mm]	71.5x71.5	
Length FA [mm]	882	
Core (uranium layer) length [mm]	600	
Matrix type	UO_2	U_3Si_2
Fuel layer thickness [mm]	0.7	0.51
Surface layer thickness Al 5754 [mm]	0.45	0.38
Plate total thickness [mm]	1.6	1.27
Uranium alloy density in fuel layer [$kg \cdot m^{-3}$]	2700	4800
Number of fuel plates [-]	8	22
Amount of Uranium in FA [g]	300	384
FA power [kW]	328	

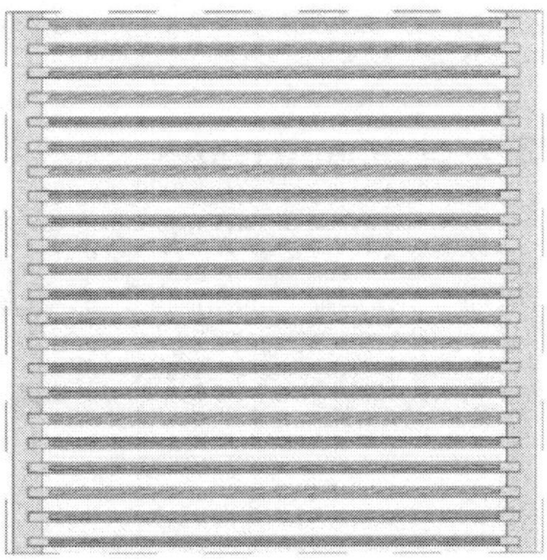

Figure 2. Cross-section of nuclear fuel. On the left IRT-4M and right FPFA [2].

2.1 Advantages of U_3Si_2 UO_2

Following the accident at the Fukushima nuclear power plant in 2011, the development of ATF (Accident Tolerant Fuel) intensified. This involves developing a new U_3Si_2 matrix and developing a new or upgrading existing zirconium fuel cladding.
Advantages of U_3Si_2 against UO_2:

- Higher thermal conductivity
- Lower specific heat
- Higher corrosion resistance
- Lower neutron absorption
- Extension of the reactor operating cycle time

In case of interruption of the coolant circulation through the reactor core, passive cooling of the fuel is necessary. U_3Si_2 has a higher passive cooling capability, and it achieves up to 10 times higher thermal conductivity compared to UO_2 at temperatures above 1400 °C, which reduces the probability and severity of a possible nuclear accident. [4]

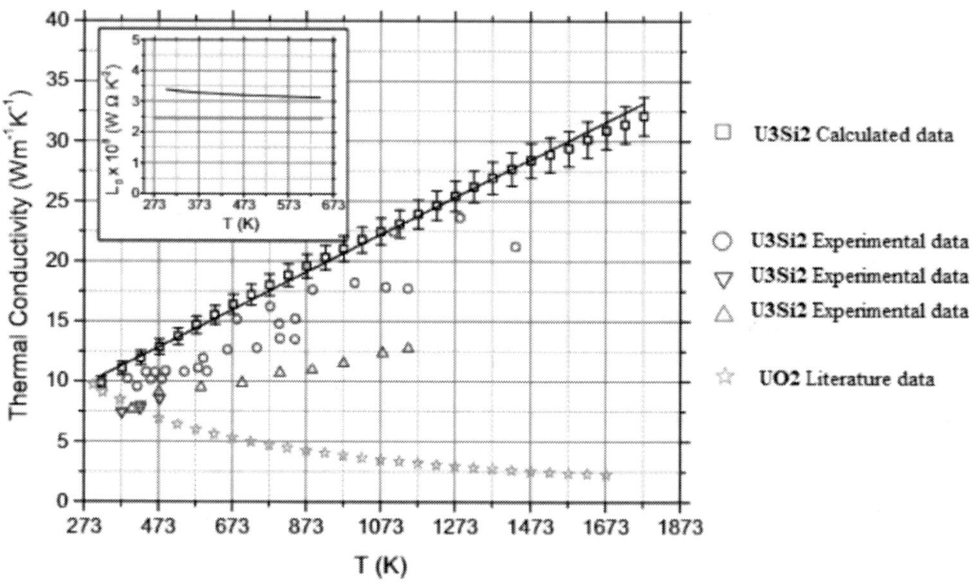

Figure 3. Calculated thermal conductivity data for U_3Si_2 as a function of temperature and literature values for UO_2 [4].

3 Experimental loop

The structure and the main parts of the experimental loop can be seen in figure 4 with detailed the FPFA model in figure 5. The experimental facility contains an open pool tank. During measurement, the experimental test section was approximately 0.3 m below the coolant level.

Experimental facility consists of the following parts:

- PVC pipes
- Water tank
- Valves regulation system
- Electrical pump 2.6 kW
- Electrical frequency changer for regulation pump power
- Test section + FPFA mockup model
- Coolant = tap water + additives against corrosion
- Ultrasonic flow meter
- Difference pressure gauge
- Thermocouples for temperature control

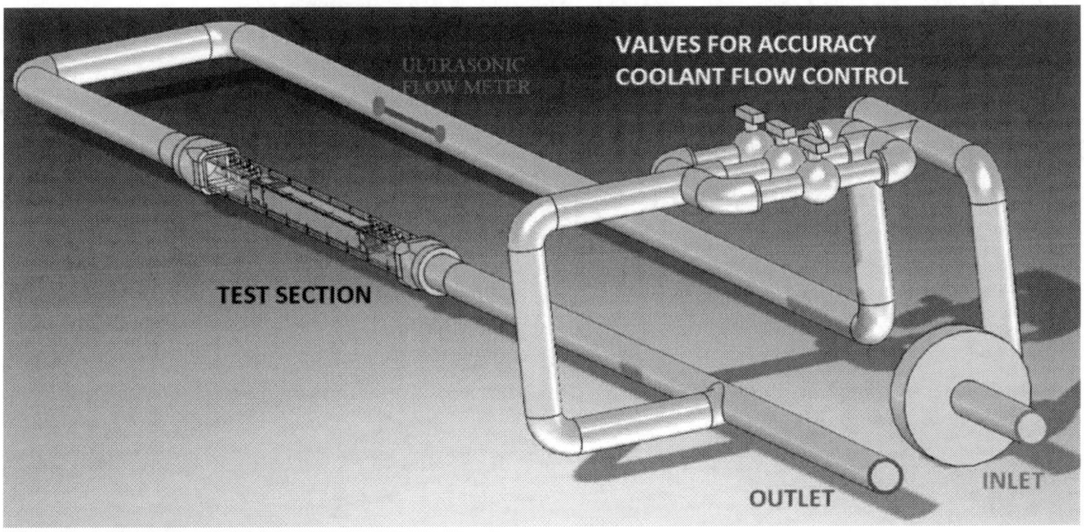

Figure 4. Experimental loop

The experimental data in Figure 6 shows pressure drop, mass flow and temperature are directly dependent.

Figure 7 describes the dependency of pressure drop on mass flow for experimental and simulation results. The yellow curve was obtained from the experiment with a coolant temperature of 50 °C. The other two curves are data from simulation (ANSYS Fluent or Star CCM+). Properties of coolant were set constant, temperature 50 °C, density 998 kg·m^{-3} and viscosity 546.858 μPa·s.

Figure 5. Model FPFA inserted in the test section

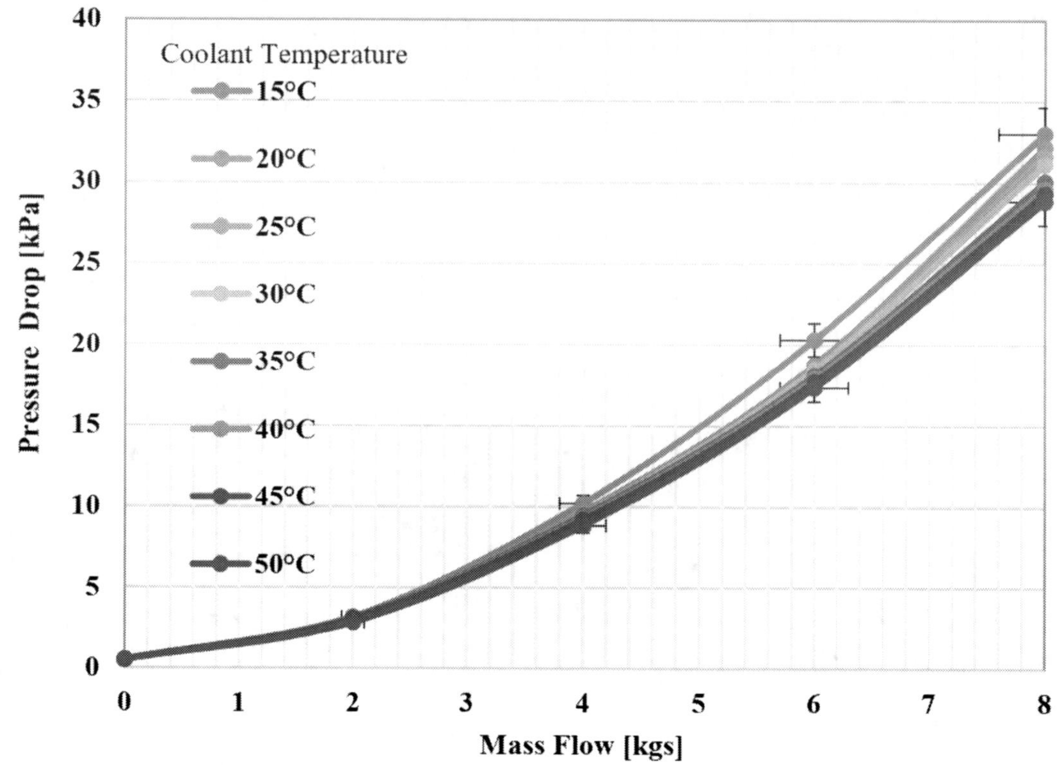

Figure 6. Data obtained from the experiment - dependence of pressure drop on coolant temperature and mass flow through FPFA.

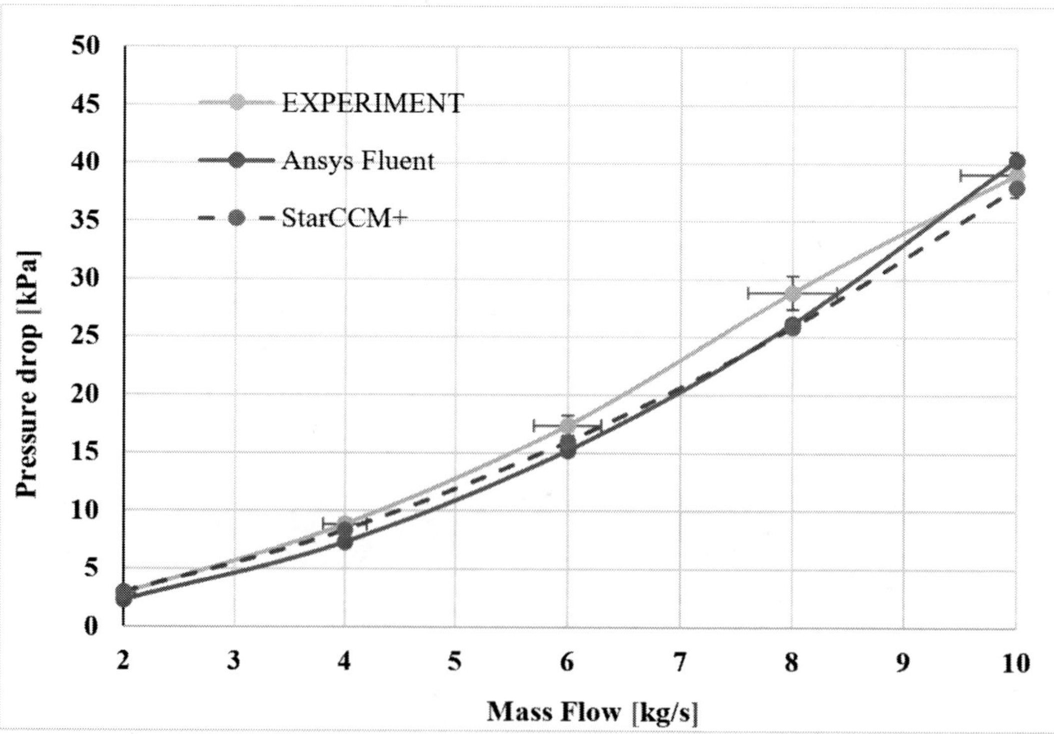

Figure 7. Comparison of experimental and simulation data – coolant temperature 50°C

3.1 Measurement uncertainty

This experiment was dealing with the uncertainty of the ultrasonic flow meter GE AT600, Rosemount 1154 pressure transmitter, and thermocouples type K.
In the following measurements have been established uncertainties type A (U_A) based on statistical means expressed by standard deviation and type B (U_B) depends on the error of the measuring device, surrounding environment, used method or approach etc.

Transmitter supplier states accuracy ±0.25 % of range 0.6 bars includes combined effects of linearity, hysteresis, and repeatability. Error is valid only at a temperature 20±2 °C.
Thermocouples have an accuracy of ±0.75 %. [7,8]
The ultrasonic flow meter according to the GE measuring and calibration list has accuracy depending on mass flow. For mass flow in the experiment was considered accuracy ±0.46 %.

$$U_B = \sum_{i=1}^{n} \frac{i}{n\sqrt{3}} \quad (1)$$

Combined uncertainty is the square-root of the linear sum of square uncertainty components.

$$U_{COMBINED} = \sqrt{\sum_{i=1}^{n} U_i^2} \quad (2)$$

Expanded uncertainty is the last calculation when estimating uncertainty in measurement based on 95% Gauss law.

$$U_{EXPANDED} = 2 \cdot U_{COMBINED} \quad (3)$$

Table 2. Combined uncertainties all measure devices

Mass flow [kg·s⁻¹]	Flow [kg·s⁻¹]	Pressure [kPa]	Temperature [°C]
2	±0.09	±0.77	±1.23
4	±0.14	±0.78	±1.23
6	±0.09	±0.78	±1.23
8	±0.22	±0.79	±1.23

Uncertainties of flow and pressure measurement depend on mass flow, temperature did not.

4 CFD simulation

The CFD simulation was using the FPFA CAD model described in Figure 8. The fluid volume model is in Figure 9.

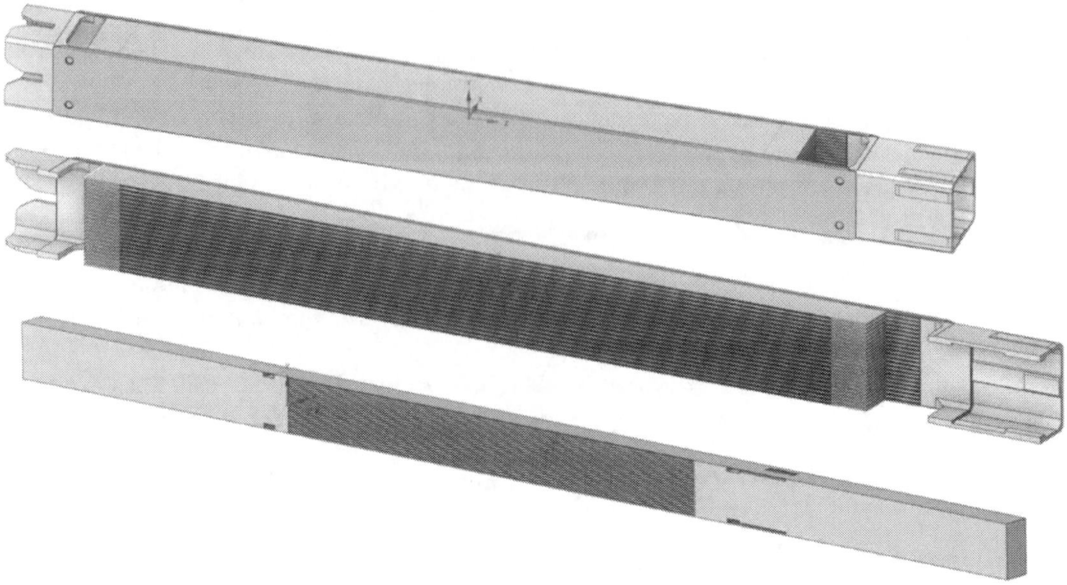

Figure 8. Cross section of the FPFA CAD model and the computational domain

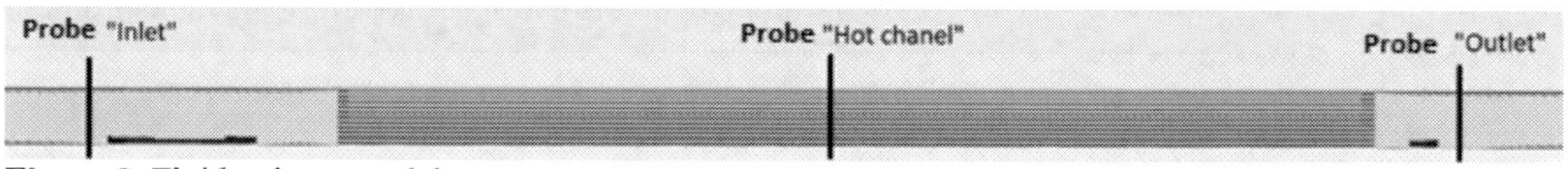

Figure 9. Fluid volume model

4.1 Fluent simulation and mesh independent study

The calculations were performed in the ANSYS Fluent 2021 R1 and Star CCM+ softwares. ANSYS independent mesh study compared 8 IRT-4M and 7 FPFA meshes. The geometry of the calculation model is an approximation of the inverse volume of the fuel assembly. For flow stabilization is empty space before and behind solid parts 300mm. The "Inlet" and "Outlet" parameters were measured 10mm before and behind solid parts. The "hot channel" parameters were measured in the middle (in the middle of length uranium plate). Initial parameters of coolant flow are: temperature 50 °C, pressure 132 kPa and speed 1 $m \cdot s^{-1}$

4.1.1 IRT-4M FA Mesh independent study

Table 3. Boundary conditions and calculated CFD data IRT-4M

	Mesh 1	Mesh 2	Mesh 3	Mesh 4	Mesh 5	Mesh 6	Mesh 7	Mesh 8
Number of cells [-]	0.89 M	1.55 M	1.57 M	1.90 M	2.23 M	2.57 M	2.90 M	3.24 M
Minimal cell size surface mesh [mm]	1.560	0.400	0.200	0.150	0.100	0.050	0.010	0.005
Maximal cell size surface mesh [mm]	40	7	4	4	4	4	4	4
Amount of boundary layers [-]	3	3	3	4	5	6	7	8
y+ [-]	10.56	10.38	10.39	7.18	5.90	4.70	4.25	3.96
Skewness MAX [-]	0.837	0.839	0.826	0.844	0.839	0.827	0.849	0.844
Skewness AVG [-]	0.117	0.100	0.103	0.087	0.102	0.092	0.099	0.104
Velocity in hot channel [m/s]	1.582	1.582	1.583	1.582	1.582	1.582	1.582	1.582
Total pressure inlet [kPa]	5.01	4.99	5.00	5.14	5.20	5.34	5.39	5.42
Total pressure outlet [kPa]	0.13	0.16	0.15	0.14	0.13	0.06	0.05	0.05
Pressure drop [kPa]	4.88	4.84	4.85	5.00	5.06	5.28	5.33	5.37
Number of iterations [-]	2875	1402	922	2634	2244	1820	2595	522
Calculation time of one iteration [s]	7.5	20.5	23.4	10.9	11.2	13.8	11.1	55.2
Computational score (cells·iteration)/time [-]	118k	76k	67k	173k	199k	185k	261k	59k

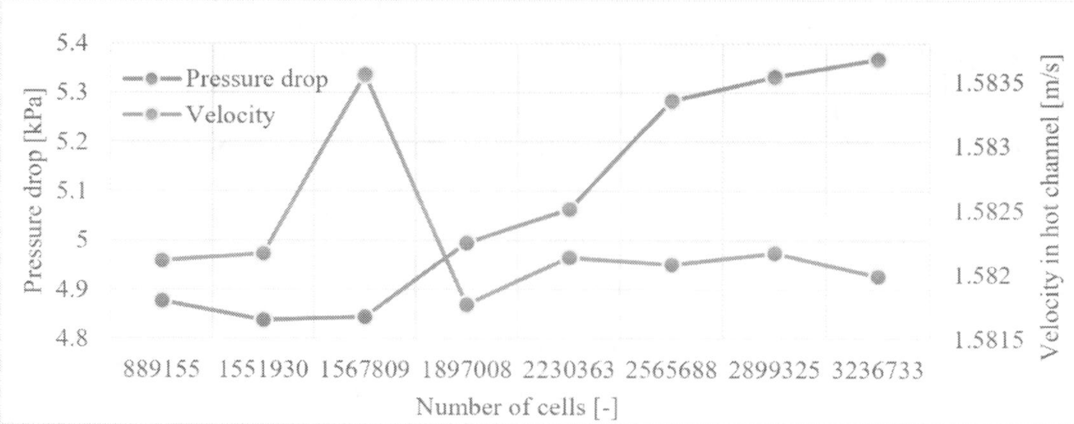

Figure 10. Dependency of the pressure drop and velocity on mesh size IRT-4M

4.1.2 *FPFA Mesh independent study*

Table 4. Boundary conditions and calculated CFD data FPFA

	Mesh 1	Mesh 2	Mesh 3	Mesh 4	Mesh 5	Mesh 6	Mesh 7
Number of cells [-]	1.1M	2.1M	2.4Mk	2.9M	3.5M	4M	4,5M
Minimal cell size surface mesh [mm]	1.56	0.40	0.20	0.15	0.10	0.05	0.01
Maximal size surface mesh [mm]	40	7	4	4	4	4	4
Amount of boundary layers [-]	3	3	3	4	5	6	7
y+ [-]	14.39	13.57	13.53	11.01	7.67	6.45	5.24
Skewness MAX [-]	0.849	0.849	0.846	0.837	0.361	0.844	0.849
Skewness AVG [-]	0.169	0.135	0.134	0.359	0.136	0.139	0.149
Velocity in hot channel [$m \cdot s^{-1}$]	1.795	1.795	1.795	1.795	1.795	1.795	1.795
Total pressure inlet [kPa]	11.06	11.01	11.02	11.4	11.62	11.85	11.97
Total pressure outlet [kPa]	0.40	0.40	0.40	0.40	0.39	0.39	0.39
Pressure drop [kPa]	10.66	10.61	10.62	11.02	11.23	11.46	11.58
Number of iterations [-]	2280	1761	1107	1051	288	202	136
Calculation time of one iteration [s]	12.63	16.35	26.02	27.40	100.00	142.57	211.76
Computational score (cells·iteration)/time [-]	87k	127k	92k	107k	35k	28k	21k

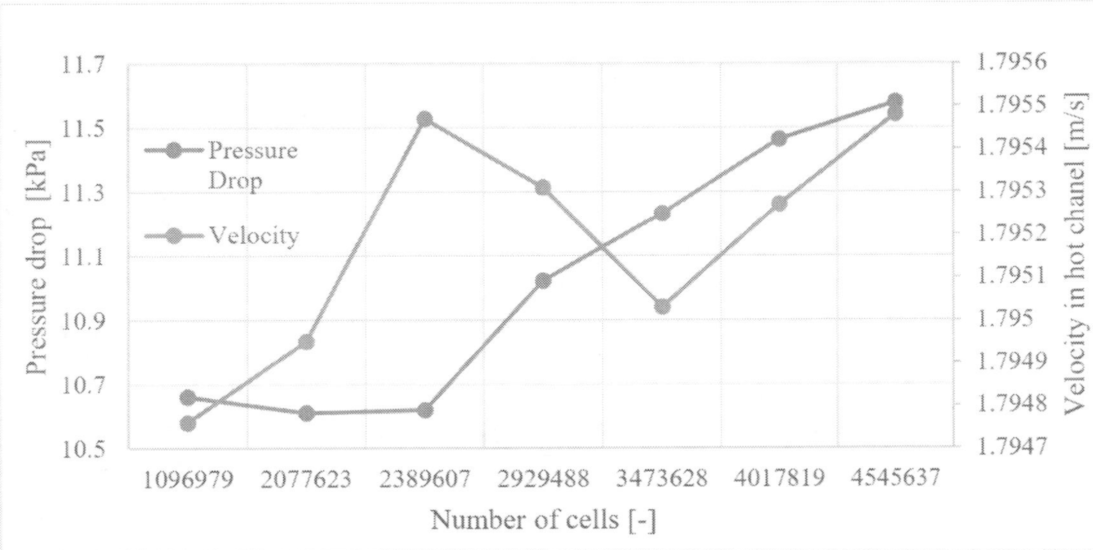

Figure 11. Dependency of the pressure drop and velocity on mesh size FPFA

4.2 Star-CCM+ simulation

For the following CFD simulation was used code STAR-CCM+ in version 2022.1. The mesh was chosen polyhedral in combination with Thin mesher and Prism layer mesher settings in the fuel grid area. The number of cells is $10.2 \cdot 10^6$, the basic cell size is 5 mm. Thin mesher was set to 3 layers, Prism layer mesher also 3 layers. All mesh important parameters are summarized in Table 3 below.

Table 5. STAR-CCM+ FPFA Mesh parameters

Number of cells [-]	Boundary face number [-]	Base cell size [mm]	Minimum surface cell size [mm]	Amount of boundary layers [-]	Amount of thin layers [-]	y+ [-]
10,238k	4,152k	5	0.5	3	3	12

Concerning y+ assuming values under 12, the K-epsilon low Re turbulence model and all y+ Wall treatment model was chosen.
From Table 4 can be seen slight deviations from the experiment. Deviation reduction can be achieved by further refining of numerical mesh and refining the physical or numerical model. The deviations against ANSYS Fluent may be due to different numerical mesh or different physical models.

Table 6. CFD against experiment deviations

Mass Flow [$kg \cdot s^{-1}$]	2	4	6	8	10
ANSYS Fluent [%]	-19.3	-17.4	-12.5	-9.4	3.2
STAR-CCM+ [%]	3.3	-6.0	-8.7	-11.5	-3.0

The future steps of the benchmark are tuning of a physical and numerical model. Then a more detailed calculation with heat transfer will be made and the results will be extended by other

operating parameters (flows, temperatures). Based on verified data of a single element will be made reactor core model.
Deviations from CFD codes are relatively small and their use for further calculations appears to be valid and reasonable. Future work will be focused on fuel element roughness approximation, test channel optimization and benchmark of experimental loop with the orifice plate.

4.3 Fluent fuel assemblies pressure drop comparison

The dependency below is created from Mesh number 2 for FPFA and Mesh number 7 for IRT-4M. The best possible mesh was chosen based on an independent mesh study. The choice is based on the number of cells and the number of iterations performed in 8 hours. Every mesh was calculated with 10 different coolant inlet velocities at the inlet to the model from 0.4. to 2.2 $m \cdot s^{-1}$ with the step 0.2 $m \cdot s^{-1}$. Regard to the inlet velocity, the new FPFA fuel has from 50 % to 90 % higher pressure drop than the old IRT-4M fuel. A higher pressure drop is associated with higher heat transfer.

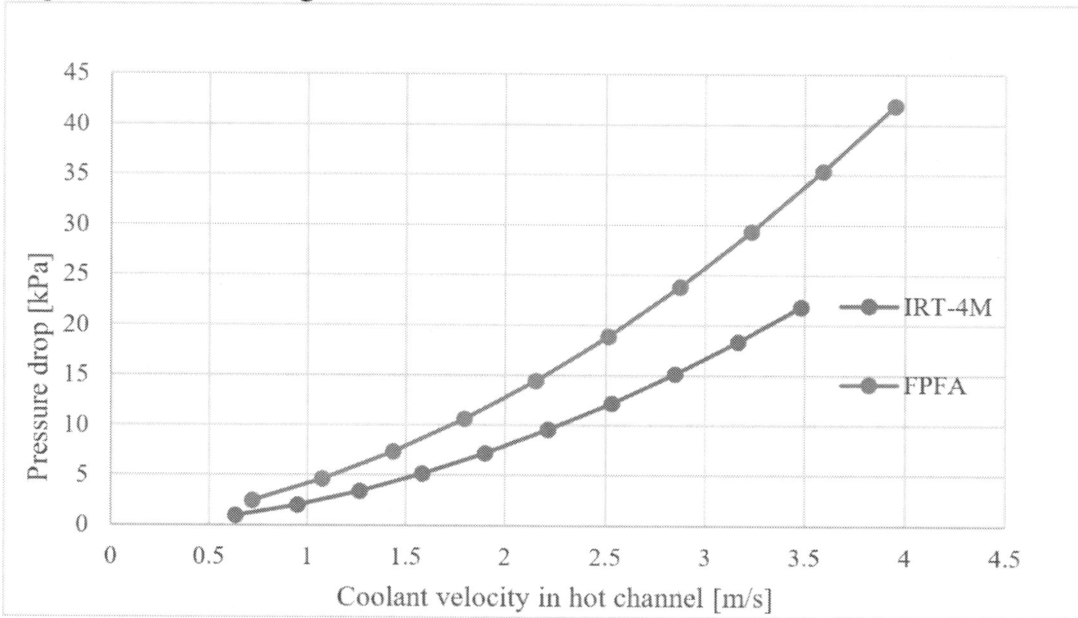

Figure 12. ANSYS Fluent simulation of IRT-4M and FPFA pressure drop dependency on velocity in the hot channel

4.4 Fluent heat transfer simulation

Heat transfer was simulated only in the case of FPFA. The temperature at the inlet is 45 °C and at the outlet 56 °C. Heat is generated in a matrix (U_3Si_2 45 % and Al 1050 55 %) with a thickness of 0.51 mm and passes through a wall (Al 5754) with a thickness of 0.38 mm.

The following simplifying hypotheses are assumed in the CFD simulation:

- Constant thermal conductivity specific heat and density
- Steady state
- Constant heat output of fuel 337 $MW \cdot m^{-3}$: this corresponds to the heat output of 328 kW of one fuel assembly
- In real core configuration the fuel element is surrounded by other elements in experiment and simulation the element is placed in a channel

Table 7. Material physical properties [6]

c_p specific heat at constant pressure [J·kg^{-1}·K^{-1}]	U_3Si_2	0.104·T+199	232 (45 °C)
	Al 1050	0.46·T+892	1038.35 (45 °C)
	Al 5754	960	
α heat conductivity [W·m^{-1}·K^{-1}]	U_3Si_2	0.0118·T+4.996	8.75 (45 °C)
	1050	230	
	5754	130	
ρ density [kg·m^{-3}]	U_3Si_2	12200	
	1050	2700	
	5754	2670	
	U	19100	

CFD simulations are based on conversation mass, momentum, energy and entropy. [5]

Nu … Nusselt number [-]

Nusselt number for forced convection $Nu = f(Re.Pr)$

$$Nu = CRe^{0.8}Pr^{1/3} \;[-] \qquad (4)$$

C = channel geometry coefficient [-]
Re = Reynolds number [-]
Pr = Prandtl number [-]

The Nusselt number was then used for the calculation of the heat transfer coefficient from the wall to the liquid in a turbulent flow.
According to the simulation. at an input speed of 0.6 m·s^{-1} and a mass flow of 3.1 kg·s^{-1}, the coolant heats up from 45.0 °C to 55.5 °C at a fuel assembly heat power 328 kW. The fuel matrix locally reaches a maximum temperature of 77.0 °C (in the middle uranium plate). The maximum surface temperature of the fuel plates is 72.0 °C.

5 Conclusion

This work contained experimental and simulation data of flow characteristics for FPFA fuel and only simulation data for IRT-4M fuel. Pressure drop is significantly higher in the case of FPFA fuel. Higher pressure drop is related to higher heat transfer, FPFA should be safer than the precedent IRT-4M fuel.

The result of the heat transfer simulation says that the coolant flow velocity in hot channel 3,11 m/s is needed to cool the heat output 328 kW of one single fuel element. Coolant temperature increases from 45 to 56°C.

Data obtained by CFD simulation correspond with data obtained by experiment. The deviations between the data are within the measurement uncertainty tolerance range. Simulated data slightly underestimate pressure drop against experimental data, it can be caused by the rougher surface of the mockup model, than was expected. In CFD simulations was assumed default surface roughness.

Based on experimentally obtained data will be built future reactor core model.

Acknowledgments

The authors thank the European Commission for funding the LEU-FOREvER project and for the continued interest and support of research and development of research reactors in EU countries, they thank Cerca for the development of FPFA fuel, which can replace the original IRT-4M fuel, and also CTU in Prague for providing postgraduate education, laboratories and laboratory equipment.

References

1. European Commission [online]. France, Paris: Cordis, 2020 [cit. 2022-03-04]. Available from: https://cordis.europa.eu/project/id/754378
2. J. Koubbi, M. Boyard, F. Huet (Technicatome), V. Romanello, A. Dambrosio, M. Hrehor (Cv Řež). Leu-Forever Project: Preliminary Neutronic Calculations Status. 2019, , 1-11
3. Tashiema L. Wilson, Emily E. Moore, Denise Adorono Lopes, et al. Uranium nitride-silicide advanced nuclear fuel: higher efficiency and greater safety, 2018
4. J.T. White, A.T. Nelson, J.T. Dunwoody, D.D. Byler, D.J. Safarik a K.J. McClellan. Thermophysical properties of U3Si2 to 1773K [online]. Journal of Nuclear Materials, 2015 [cit. 2021-12-29]. ISSN 0022-3115. Available from: https://reader.elsevier.com/reader/sd/pii/S002231151500241X?token=3C451550DF763ADAF1755C0F424FC020129825E93ED0590753B055BF2D9672424D6614D6D06230FB0643C1F8BE0A1C00
5. Luisa Ferroni, a Antonio Natale. Exergy Analysis of a PWR Core Heat Transfer [online]. 2016 [cit. 2021-12-29]. doi:10.18280/ijht.34Sp0239. Available from https://ec.europa.eu/research/participants/documents/downloadPublic?documentIds=080166e5cba98bfa&appId=PPGMS
6. Huet, Francois. LVR-15 Fuel element qualification report. Technic Atome [online]. 2019 [cit. 2021-12-30]. Available from: https://ec.europa.eu/research/participants/documents/downloadPublic?documentIds=080166e5cba98bfa&appId=PPGMS
7. GE Flow Meter AT600 [online]. 2019 [cit. 2022-4-30]. Available from: https://www.instrumart.com/assets/aquatrans-at600-manual.pdf
8. Rosemount pressure transmitter 1154 [online]. 2019 [cit. 2022-4-30]. Available from: https://www.emerson.com/documents/automation/product-data-sheet-rosemount-1154-alphaline-nuclear-pressure-transmitter-en-89502.pdf

CFD Simulation and Flow Study of a Velocity Compounded Radial Re-Entry Turbine - Elektra

Fabian Janischowsky[1,*], *Dominik* Stümpfl[2,**], *Philipp* Streit[2,***], *Andreas P.* Weiß[2,****], and *Andreas* Lesser[1,†]

[1]OTH Regensburg, Turbomachinery Laboratory, Galgenbergstr. 30, 93053 Regensburg, Germany, +49 941 943 9965

[2]OTH Amberg-Weiden, Center of Excellence for Cogeneration Technologies, Kaiser-Wilhelm-Ring 23, 92224 Amberg, Germany, +49 9621 482 3531

Abstract. In order to meet the targets set out in the Paris climate agreement of 2015, the energy sector needs to be restructured. In Germany for example the goal is to move out of both nuclear and coal-fired power. In order to fulfil more demanding energy requests alternative sources of energy must be found and made available. One component is the usage of Waste Heat Recovery.
This paper deals with the evaluation simulation of an Elektra turbine, which concept is a velocity compounded radial re-entry single wheel turbine. Based on the first turbine design, multiple simulations were carried out in order to maximize the efficiency. With these simulations, optimization opportunities are described and identified. The paper concludes with a comparison of the results achieved by numerical simulation and on the experimental test bench. The main focus for this paper is to analyze the actual flow phenomena to identify possibilities for further improvements. All simulations were calculated with the commercial software FINE™/ TURBO 15.1.

*e-mail: fabian.janischowsky@oth-regensburg.de
**e-mail: d.stuempfl@oth-aw.de
***e-mail: ph.streit@oth-aw.de
****e-mail: a.weiss@oth-aw.de
†e-mail: andreas.lesser@oth-regensburg.de

1 Introduction and Motivation

With the purpose of slowing down the climate change, countries like Germany get out of nuclear and coal-fired power [1]. To generate electricity via a Rankine cycle, alternative heat sources like geothermal heat or waste heat need to be tapped. In comparison, these heat sources typically provide heat in a much smaller scale than conventional heat sources. In need of turbines dealing with these low temperatures, spoken in a power generating context, new concepts have to be developed to reach reasonable efficiency levels. In previous papers the usage of the Elektra turbine was evaluated [2, 3]. This turbine is velocity compounded and works according to the concept of a Curtis turbine, with the exception of only consisting of a single wheel with a radial re-entry. Therefore both velocity stages of the Curtis turbine are realised in the same wheel, which makes the Elektra much smaller and hence cheaper to manifacture, compared to the Curtis turbine.
The first Elektra invented by Otto Kolb, was built in 1905. This turbine implemented four wheel passes and is presented in Figure 1[4, p. 234].

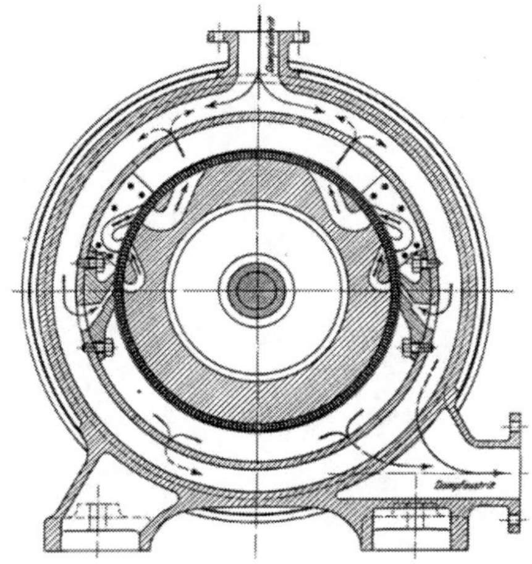

Figure 1: First built Elektra [4, p. 234]

2 Elektra Design

For the available test bench at the OTH Amberg-Weiden an air-driven Elektra with two wheel passes, according to one deflection channel, was built. The geometry of the first built version (V1) is presented in Figure 2. The turbine main design data is determined according to the capacity of the eddy current brake and the installed air compressors at the laboratory. These data is given in Table 1. The last four parameters of this table were calculated with a 1D-Turbine-Design-Tool (1DTDT) at the OTH Amberg-Weiden. [2]
In this paper the Elektra will be parted into four sections: Laval nozzle (Measuring points 0 and 1), deflection channel (2 and 3), outlet (4) and rotor. In the Laval nozzle the fluid is accelerated to supersonic flow. Its highest speed is expected to be close to measuring point 1. After the first wheel pass, the flow gets deflected and enters the wheel a second time, before exiting the turbine through the outlet.
In comparison to previous tests the geometry (V1 in [2]) was adjusted, as can be seen in Figure 3. The final geometry presented in this paper is counted as Version 7 (V7), because 5 other geometries have been tested since the initial geometry V1.
The most important changes are as follows:

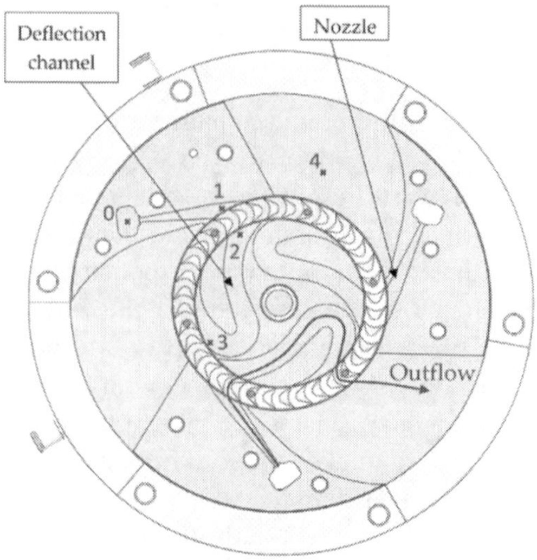

Figure 2: First Elektra design (V1) with the position of the pressure tappings (red cross) and the flow path (red line) [3]

Table 1: Turbine main design data [cf. 2]

Parameter	**Unit**	**Elektra air turbine**
Working fluid	-	pressurized air
Mass flow rate	g/s	65
Total inlet pressure	bar	10
Total inlet temperature	K	≈293
Static exit pressure	bar	≈1
Wheel diameter	mm	80
Rotational speed	rpm	29,000
Degree of admission	%	≈61
Pressure ratio	-	10
Nozzle exit Mach number 1DTDT	-	2.04
Rotor inlet relative Mach number 1DTDT	-	1.58
Expected efficiency η 1DTDT	%	64
Expected shaft power 1DTDT	kW	≈5

1. Laval nozzle: Several shock systems at the end of the Laval nozzle were detected, which indicates an overexpanded nozzle. Therefore the nozzle outlet width has been decreased. This leads to a decrease of the angle which the nozzle sweeps over the rotor from 19.2° to 15.5°, resulting in a lower degree of admission for the turbine.

2. Deflection channel: The deflecting direction of the channel has been changed. Therefore the fluid is first deflected and then flows through a straight channel. The fluid leaves the first rotor passage under highly unsteady and inhomogeneous conditions. The deflection in the channel also results in a inhomogeneous flow field since detaching of the flow is possible. Therefore it is easier to use the more turbulent flow to prevent detaching in the deflection channel by first deflecting it and after that homoge-

nize the fluid by using a longer and less curved channel. With this more homogeneous field it is easier to optimize the second rotor passage [5].

3. Outlet: In V1 a large mass flow rate leaving the nozzle directly to the outlet was detected, which does not contribute to the work of the turbine. To prevent this leakage the beginning of the outlet was shortened.

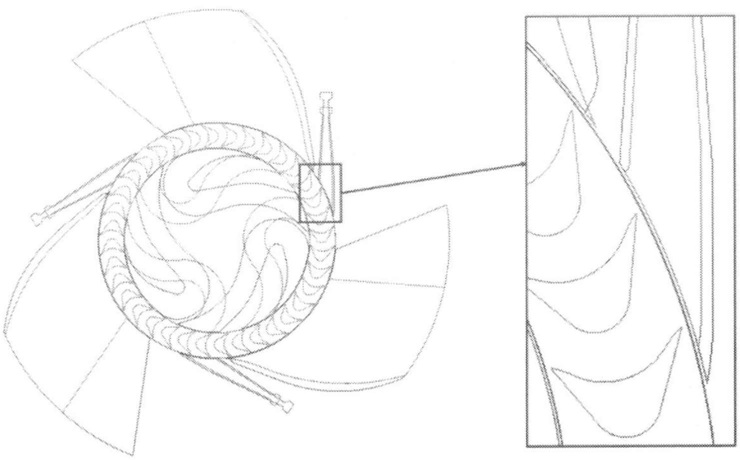

Figure 3: Comparison of initial (black) and final (red) geometry

3 Simulation Approach

The simulation model used for the different geometries is shown in Figure 4. Since the design in the Elektra is rotationally symmetrical, by 120°, it is suitable to only compute one third of the turbine with a periodicity of three. Additionally the inlet geometry was shortened. This is implied due to the assumption that the flow is homogeneous at the narrowest cross section of the Laval nozzle (Ma=1). Therefore only a small part of the geometrical inlet is needed to simulate the subsonic inflow.

Furthermore, compared to the geometry simplifications have been applied to the simulation model to achieve shorter simulation run times and a more stable behaviour. Therefore the tip clearances between rotor and stator, which will lead to a mass flow rate above and below the rotor, were neglected. The chosen Rotor-Stator-Interfaces (RS-Interface), see Table 2, require the same periodicity between rotor and stator. Therefore a 120° stator block (blue in Figure 4) on each side of the Rotor has to be implemented. The thickness of these RS-Interface Blocks can be chosen according to the radial gap between rotor and stator.

The chosen parameters for the steady and unsteady simulations are given in Table 2. The Spalart-Allmaras turbulence model is used in order to further reduce the computing time. The extended wall functions need to be used, since the first cell height is at 1×10^{-5} m, which results in a y_+ of about 20 to 30 at the end of the Laval nozzle. The time step is chosen according to 20 Positions for each blade passage, to evaluate the differences in the flow through the turbine depending on the rotor position. In a rotor with 12 meshed blades (respectively 36 blades in the geometry) this results in 20 positions · 12 blades = 240 time steps per rotation over 120°. The unsteady simulation is carried out with as many periods as

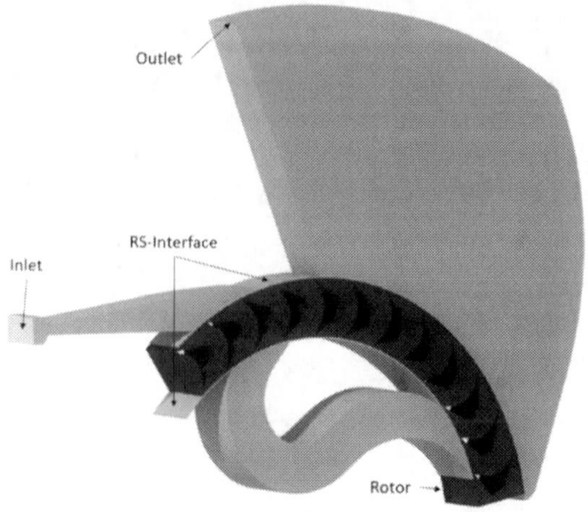

Figure 4: Simulation model

needed until the simulation could be accounted as converged. This convergence is achieved as soon as the difference between the actual and the previous period is no more visualizable. It is typically reached after 7 to 10 periods.

Table 2: Simulation Parameters

Category	Parameter	Value
Flow model	Fluid	Air (perfect gas)
	Turbulence model	Spalart-Allmaras with extended wall function
	RS-Interface	Full non matching frozen rotor (steady), domain scaling (unsteady)
Boundary conditions	Inlet	Subsonic, total pressure (p_t = 1,000 kPa), Total temperature (T_t = 293.15 K)
	Outlet	Subsonic
		Averaged static pressure (p_0 = 100 kPa)
	Walls	Solid, adiabatic
	Rotational speed	29,000 rpm
Numerical control	Time stepping	0.5° per time step
	Number of periods	7

Since the turbine is only partially admitted, highly unsteady behaviour and phenomena are to be expected. Therefore the results of a steady simulation are only used to initialize its unsteady simulation.

4 Flow Study

In the following pictures instantaneous results are shown. The cutting plane is placed at 50 % of the nozzle height. The absolute M ach number for the final geometry version is plotted in Figure 5. The main part of the fluid (red arrows) flows as expected through the first rotor passage, to the deflection channel, through the second rotor passage and into the outlet. The

exact mass percentage, which follows this route depends on the leakage mass flow rates, which are highly influenced by the geometry of the deflection channel. In every simulated geometry this leakage accounts to up to 25 %. Due to the occurrence of such high leakages its origins shall be identified to analyze where further improvement is possible.
In the final geometry (V7) four leakage mass flow rates (pink arrows) were detected. The listing relates to the numbers in Figure 5.

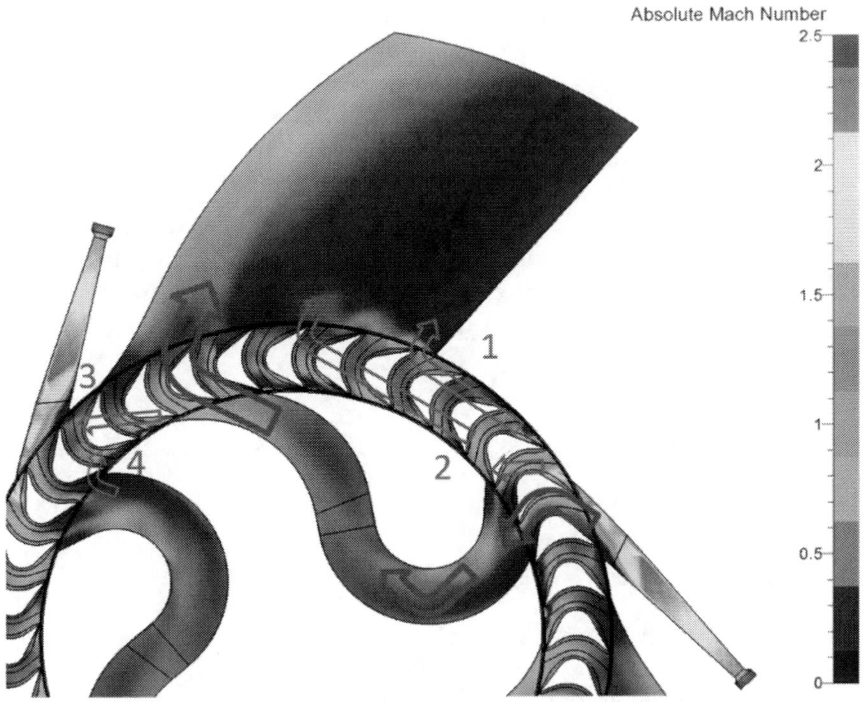

Figure 5: Absolute Mach number distribution of the final geometry V7

1. The first leakage has its origin in the gap between rotor and stator. Here the flow leaves the nozzle and is not able to enter the rotor, but is flowing over the rotor and directly to the outlet. This mass flow rate therefore has no significant contact with any blade and hence does no work.
2. The second leakage mass flow rate follows the main stream through the first rotor passage, but is not able to reach the inlet of the deflection channel. During the time, while the fluid is flowing through the first blade passage, the rotor rotates a couple degrees and transports the fluid in circumferential direction. As the fluids velocity is too slow or the rotational speed is too high, a fraction of the turbines mass flow rate hits the inner wall of the rotor. The fluid of this mass flow rate stays in the rotor and is pressed outwards during the rotation, because of the existence of centrifugal forces. The fluid leaves the rotor approximately in the middle of the outlet right between the first leakage mass flow rate and the main flow.
3. The third leakage mass flow rate is similar to the second one but for the second rotor passage. The fluid enters the second rotor passage after flowing through the deflection channel, but is not able to reach the outlet in time. It then has to remain in the rotor and is pressed against the Laval nozzle, where it is accelerated into the rotor and flows back to the inside of the deflection channel.

4. The fourth leakage mass flow rate occurs in the deflection channel. It leaves the deflection channel through its inlet and flows against the rotational direction through the rotor to the outlet of the deflection channel. There it meets the third leakage mass flow rate and again hits the Laval nozzle. Hence the third and fourth mass flow rate never leave the turbine but keep fluctuating between deflection channel, rotor and nozzle and are therefore accounted as recirculation areas.

Both recirculations (3 and 4) are critical for the nozzle flow, since they now tend to stream inside the nozzle. This results in an early detaching point on the streamwise left side of the nozzle (Point 5 in Figure 6). Most of the mass flow rate of the Laval nozzle is shifted to the right side, since there is no resistance from instreaming mass flow rates. As a result the flow condition of the nozzle is shifted, because it is no longer homogeneous streamed through, but tipped to the right. This tipping is shown in Figure 6.
At point 6 another effect is visualized. Here the Mach number increases locally to around 3 to 4. In this area exists the fastest flow of the Elektra turbine. It depends on the rotor position and only occurs when a blade is reaching the end of the Laval nozzle and cuts off the mass flow rate coming from the nozzle. Hence at this time a small second Laval nozzle is created, which accelerates the flow and results in the existence of another shock system at around 30 % chord length.

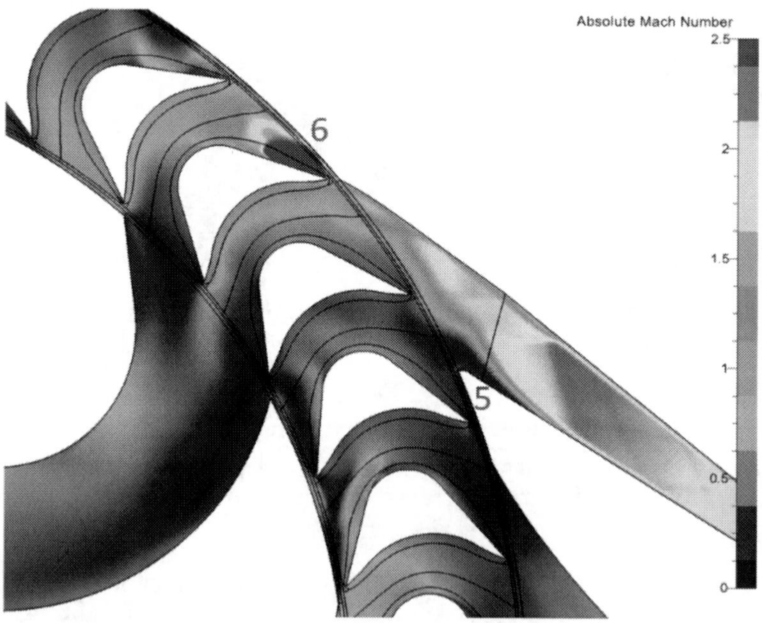

Figure 6: Absolute Mach number distribution of the final geometry V7 at the first rotor passage

In Figure 7 the static pressure is shown. The static pressure for the nozzle outlet (Point 7) is higher as intended at the design conditions (100,000 Pa). Therefore the nozzle cannot expand to the outlet pressure but has to find a condition, where at its outlet it meets a static pressure of approximately 200,000 Pa. This occurs due to pressure losses in both rotor passages and a high pressure state in the deflection channel. The losses lead to the difference in the static pressure between nozzle outlet and turbine outlet, where the boundary condition predefines a static pressure of 100,000 Pa.

Both, the mass flow rate in the nozzle and the high pressure at its outlet, result in bad conditions for the nozzle itself, which leads to the occurrence of a large shock system at around two-third of the length of the nozzle. According to the 1DTDT the nozzle outlet Mach number should be at 2.0 (see Table 1), the simulated Mach number at the nozzle exit is, with around 1.4, significantly lower.

The high static pressure in the deflection leads to another effect presented at point 8 in Figure 7. In this area fluid from the deflection channel is sucked into the rotor, streams against the rotational direction and collides with the second leakage mass flow rate. The suction occurs since the static pressure in the deflection channel is significantly higher than the static pressure in the rotor in that area. The static pressure of the rotor is decreasing while rotating through the outlet area, because fluid is transported from the rotors inner radius to the outlet due to circumferential forces. The collision of both mass flow rates leads to a dissipation indicated by an increase in the entropy as presented in Figure 8. This fluid with high entropy is then mixed in the rotor with more fluid streaming from the deflection channel.

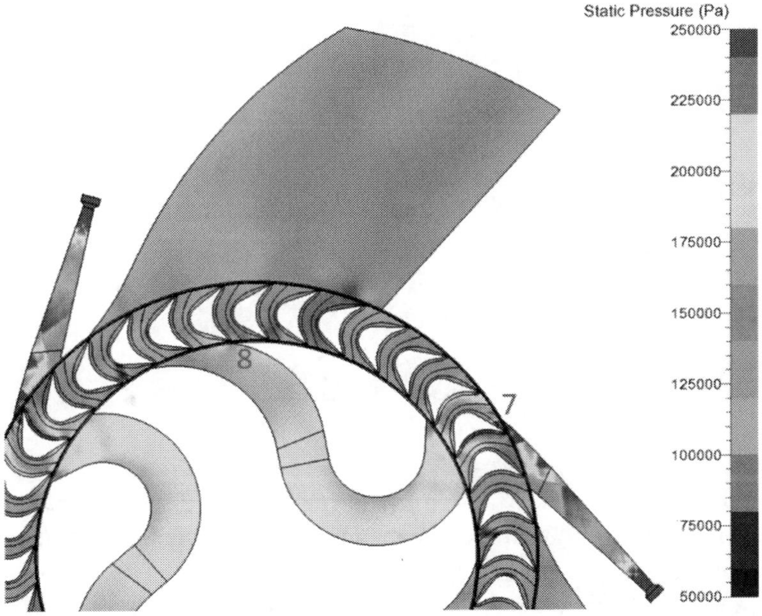

Figure 7: Static pressure distribution of the final geometry V7

Furthermore two more areas with entropy production in the turbine exist, marked with numbers 9 and 10. At Point 9 the leakage mass flow rates 3 and 4 collide and increase the entropy at that position. The fluid with high entropy enters the Laval nozzle, is accelerated through the rotor passage and mixes again with the main flow in the deflection channel.

At point 10 the second Laval nozzle leads to a shock system and hence increases the entropy in the area where the shock occurs. The fluid with high entropy flows in rotational direction with the rotor and leaves the system through the outlet. The entropy in the outlet appears to be lower than at the shock position due to mixing processes with the mass flow rate from the deflection channel.

The appearance of this amount of unsteady phenomena assures the need to only evaluate unsteady simulations.

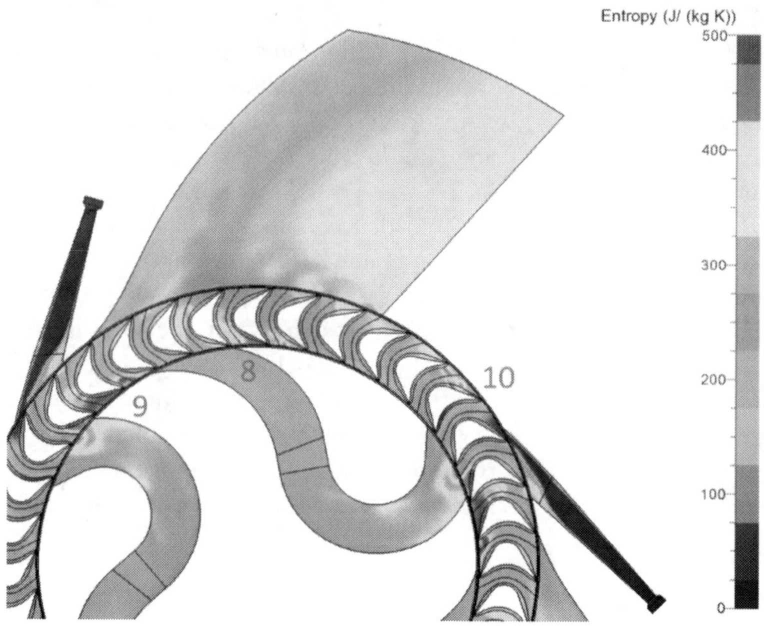

Figure 8: Entropy distribution of the final geometry V7

5 Comparison

Both geometries, V1 and V7, were manufactured and tested on the compressed air test bench of the OTH Amberg-Weiden. The specifications of the test bench are described in [3]. The efficiency is calculated according to Equation 1:

$$\eta = \frac{2 \cdot \pi \cdot n \cdot M}{\dot{m} \cdot c_p \cdot T_{in} \cdot \left[\left(\frac{p_{s,out}}{p_{t,in}} \right)^{\frac{\kappa-1}{\kappa}} - 1 \right]} \tag{1}$$

Where n is the rotational speed, M the torque, $\dot{m}$ the mass flow rate, c_p the specific heat capacity, T_{in} the inlet temperature, $p_{s,out}$ the static pressure at the turbine outlet, $p_{t,in}$ the total pressure at the turbine inlet and κ the heat capacity ratio.

In Figure 9 the measured efficiency depending on the pressure ratio, between inlet and outlet, and the rotational speed for initial and final geometry is given. The green curve describes the calculated efficiency in the simulation for different simulated rotational speeds. First to be noted is that the pressure ratio is not identical but in the range between $PR = 10 - 10.4$. Hence the results for the same PR might differ, but are still considered as comparable. This is justified by the influence on the change of efficiency being significantly higher with the geometry changes than with the pressure ratio differences. Therefore most fraction of the difference between the curves is related to the geometry while the pressure ratio difference is only responsible for a minor portion of efficiency change. Nevertheless the efficiency for the final geometry compared to the initial geometry increased, but its amount is depending on the rotational speed.

The maximum efficiency for the initial geometry was measured at 35,000/min at 33 %.The efficiency for the final geometry reaches a plateau between 22,500/min and 32,500/min at 36.0–36.5 %. As in both cases the design point of the turbine is at 29,000/min the highest efficiency should have been at that rotational speed.

The comparison of both curves leads to the conclusion that the final geometry results in a

6 %p higher efficiency than the initial geometry until its plateau at 22,500/min is reached. After that the efficiencies merge together resulting in a higher efficiency for the final geometry at 37,500/min of 2 %p. The shift to a maximum efficiency at lower rotational speed was observed to highly depend on the geometry of the deflection channel.

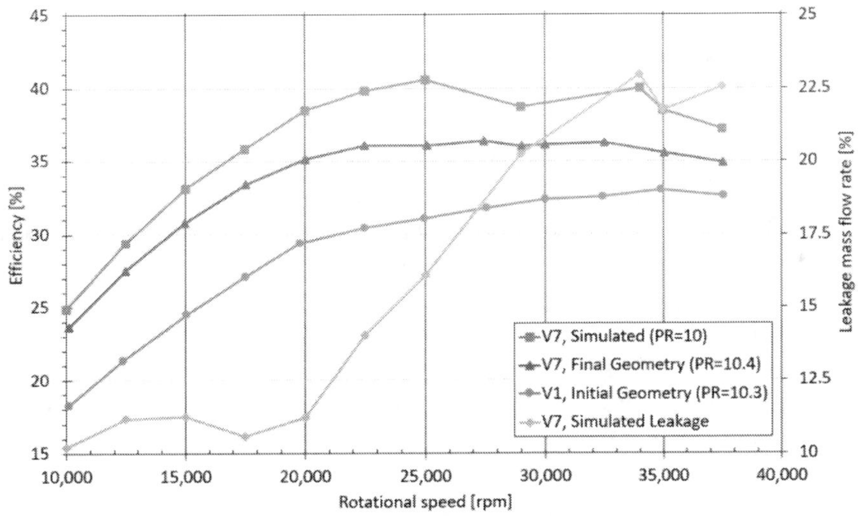

Figure 9: Comparison of efficiency initial geometry vs. final geometry

In the simulation the leakage mass flow rate between V1 and V7 increased from 15 % to 21 % at 29,000/min. The increase in the leakage mass flow rate corresponds to the shift of the point of the highest efficiency to lower rotational speeds. The assumption is, that for lower rotational speeds a higher percentage of fluid is able to enter the deflection channel. Hence the leakage mass flow rates 2 and 3 decrease, since they exist only due to the rotation of the rotor. This assumption is supported by the fact, that the overall massflow leakage increases from 20,000/min at 11 % to around 22 % for the highest rotational speeds, as can be seen in Figure 9. If the rotational speed is increased both losses (2 and 3) increase proportionally and therefore have a more negative impact on the overall efficiency. The comparison of the initial geometry and the final geometry in Figure 3 shows, that the inlet of the deflection channel is larger for the initial geometry and hence reduces leakage mass flow rate 2. The same concept applies for the channels exit with leakage mass flow rate 3. Here the offset between channel outlet and turbine outlet is reduced for the final geometry. For the same velocity the fluid has less time to leave the rotor and enter the turbine outlet. Since the leakage mass flow rate 3 also shifts the Laval nozzles state to earlier shocks and lower Mach numbers, its impact on the turbines efficiency might be relatively huge.
For the comparison of the measured efficiency with the simulated efficiency, the same efficiency definition as in Equation 1 is used. It is clear that the simulation overestimates the measurement by up to 4 %p. This effect is expected since in the simulation some negative effects such as manufacturing inaccuracies, bearing friction or disc friction are ignored. Although the measured efficiency has to be lower than the simulated one, it is still possible that there are additional effects, that influence the turbine on the test bench.

6 Conclusion

In the flow study for the current version of the Elektra turbine 10 different flow phenomena have been identified. Since they all affect the turbine efficiency in a negative way they have to be targeted for further improvements. Comparing the two geometries, the V7 performs significantly better than the V1. For higher rotational speeds its advantage shrinks but still remains by a couple percentage points. The worse performance at higher rotational speeds can be explained by the increased leakage mass flow rates.
Especially the mass flow leakage rates 3 and 4 have a significant impact on the nozzles flow condition and shift the turbine to new, unwanted states. To further verify that assumption it is necessary to find a measurement method that could record the percentage of the leakage mass flow rates during the tests. The simulation model seems to be sufficient to predict the performance of the Elektra turbine. This model could be improved by the usage of hub and shroud gaps in order to simulate the overflowing fluid which will lead to a higher mass flow leakage than the one which is currently simulated. If the assumption for the mass flow leakage is verified, these leakages have to be tackled with in future geometries. Since these losses currently equal a fourth of the turbine's mass flow rate there seem to be many possibilities for improvement.
The geometry of the deflection channel shows high impacts on those leakages and therefore should be improved in a first step. This could be done by for example widening the inlet of the deflection channel in order to reduce mass flow leakage 2. Another option would be to increase the tolerances for the offsets in both rotor passages. By varying the deflection channels inlet and outlet position the mass flow leakages 2 and 3 could be reduced even further. With these improvements the static pressure inside the deflection channel should be decreased leading to reduced outflows in the rotor which should then reduce the mass flow leakage 4 and the entropy increase at phenomenon 8. Concluded the position and angles of the deflection channel leads to most of the described phenomena, therefore a better channel should reduce the mass flow leakages back to around 11 % and increase its efficiency for higher rotational speeds. The mass flow leakage 1 according to the second Laval nozzle (phenomenon 6 and 7) seem to be of a lesser order and should therefore be tackled only after these improvements.

References

[1] Sterchele P., et al., *Paths to a climate-neutral energy system* (Fraunhofer Institute for Solar Energy Systems ISE, Freiburg, 2020)

[2] Weiß A. et al., *Uncommon turbine architectures for distributed power generation - development of a small velocity compounded radial re-entry turbine* (Archives of thermodynamics, 2020)

[3] Weiß A. et al., *Numerical and Experimental Investigation of a Velocity Compounded Radial Re-Entry Turbine for Small-Scale Waste-Heat-Recovery* (MDPI Energies, 14., 2021) (DOI: 10.3390/en15010245)

[4] Kolb O., *Dampf-Turbine* (1900, Patent-Nr. AT000000003865B, Kaiserlich königliches Patentamt)

[5] Sautereau M., *Numerische Optimierung des Umlenkkanals einer Elektra-Turbine* (Master thesis at OTH Regensburg, 2021)

Preliminary Experimental Investigation of the Design-Overloaded Stage in Two-Stage Axial Turbine Test Rig

Tomáš Jelínek[1*], *Martin* Němec[1], *Petr* Milčák[2] and *Richard* Fichtl[2]

[1] VZLU – Czech Aerospace Research Centre, Beranovych 130, Prague, Czech Republic
[2] Doosan Škoda Power, Tylova 1/57, Pilsen, Czech Republic

Abstract. The article deals with the issue of experimental research of the so-called design overloaded stage. A new experimental device enabling measurements on two-stage axial turbine is advantageously used in this research task. The experimental setup is designed so that a preparatory stage is included in front of the examined stage. The stage design introduces some aerodynamic overload already in the flow path design phase. This research program is focused primarily on the effect of leakage flows and their mixing with the mainstream. This paper describes the verification measurements on a newly introduced two-stage arrangement and reveals the first data obtained.

1 Introduction

The global steam turbine market is evolving due to composition of electricity generation sources, the need for greater decentralization of power production, and the requirements for cogeneration of electricity and heat. Demand for large, continuously operated units is moving towards smaller units with greater operating flexibility. High competition in the market pushes to reduce prices, but at the same time it does not slow down the requirements of achieving high efficiency. This encourages the producing companies to research new technical solutions.

The above market requirements direct the design of turbines to smaller, often single-casing, machines (power up to 100 MW), where the pressure on the price affects the design options. One of possible solutions applied by Doosan Skoda Power Company is the deployment of so-called design-overloaded stages. However, it is not an aerodynamic overload stage in the sense of operation in non-design mode. On the contrary, this overload regime is already considered when designing the stage, respectively the whole flow path. This aerodynamic overload is caused, for example, by the design of the flow section with a reduced number of stages compared to the so-called optimal loading. The motivation of such a design can be, for example, a demand to reduce the bearing span and the price reduction mentioned before. The designed stages thus work in the nominal (design) mode, but with certain aerodynamic overload.

* Corresponding author: jelinek@vzlu.cz

The optimal load of the stage is usually understood as a load at which the stage reaches the highest efficiency. When an isolated stages is considering, this mode is close to the regime when the outlet stage velocity is minimal, i.e., in the axial direction. Aerodynamic overload of the stage leads, among other things, to a deviation of the output speed from the axial direction - the output angles reach values of 60° or less. In a multi-stage arrangement of stages, the design of the downstream stage must be subject to such a "design regime". Due to the gradual expansion and construction of the flow path from similar stages (with a similar level of reaction), this stage is also aerodynamically overloaded. The situation for the downstream stage is thus repeated, i.e. the design-overloaded stages usually have a non-axial input and output.

Doosan Škoda Power Company has implemented the extensive development program over the past decade, in which it has developed and tested a wide portfolio of blade geometries for the design of stages ranging from low reaction to full reaction stages. Experimental research of these stages was carried out in cooperation with VZLÚ - Czech Aerospace Research Centre. At the same time, loss models and procedures for selecting a suitable blade type or blade combination for a given application were prepared [1]. The choice of the stage loading resp. the processing rate of the thermal gradient was primarily solved through the selection of the appropriate level of stage reaction in combination with the reference design patterns. The aerodynamic stage overloading was chosen to be rather moderate; if greater thermal gradient processing was required, a reduction in the level of reaction was preferred.

From the above requirements for higher aerodynamic stage loading, questions logically arise regarding the appropriate level of aerodynamic overload compared to the reduction of the reaction of the stage, resp. a combination of both approaches.

2 Theoretical background

The dependency of the stage efficiency on the stage loading represented by a velocity ratio u/c is the basic axial turbine characteristic. The optimal loading usually corresponds to the regime with axial outlet velocity, i.e. smallest leaving loss.

When the stage loading changes, the stage works in an overloaded or underloaded regime; correspondingly the flow with off-design direction hits the profiles. This change is associated with an increase in profile (incidence) losses and thus losses of the entire stage. In the off-design regime, the stage outlet velocity is non-axial and thus the outlet velocity (leaving loss) is increased. These basic ideas can be observed broadly in literature as a chart with the efficiency depending on the u/c ratio, where the efficiency in the total-to-static definition changes more significantly due to the change in outlet velocity. On the contrary, such a dependence of the efficiency in the total-to-total definition is relatively flat, as the change in row efficiency due to incidence is mainly reflected.

Moreover, the change in the stage loading will also induce a change in the stage reaction (distribution of the thermal gradient between the stator and the rotor) and thus other related changes. When the complex stage concept is considered, changes in the seal leakage rates are evident, as also their downstream mixing with the main flow from the blades.

Consider now a multistage flow path consisting of similar stages. The shape of stator blades should correspond to the outlet flow of the upstream stage. This means that a higher outlet velocity, caused by the off-design regime, might be fully utilized as higher inlet energy in the further stage and therefore we can deal only with the total-to-total efficiency. Also, the shape of rotor blades should be appropriately adjusted during the design to the considered stage loading regime. When speaking about design-overloaded stages, it means, these stages are purposely designed for higher loading (compared to the nominal loading with axial outlet flow). Such stages do not suffer from incidences in this design-overloaded regime and a higher outlet velocity means no additional loss. On the other hand, the profiles for design-overloaded stages have higher flow turning which might result in higher both profile and secondary losses. Finally, the behaviour of design-overloaded stages in off-design conditions is also a topic of interest. The assessment of increased stage power from overloading versus increased stress resistance of more bended profiles is also an interesting topic mainly from the point of aspect ratio (blade length to chord) and its effect on losses.

Experimental verification of the behaviour of design-overloaded stages in various operational conditions is the subject of this paper.

3 Experimental setup

3.1 Experimental equipment description

To experimentally verify the design of the design-overloaded stage, it is necessary to achieve a non-axial inlet into the tested stage. For this purpose, it is possible to advantageously use a two-stage arrangement in the test rig, where the design-overloaded stage is preceded by another "preparatory" stage. This stage can also be used to change the inlet conditions for tested stage; for example by changing the clearance in the shaft and shroud seal (see previous research at the single?-stage and 1.5-stage test rings presented in [2] and [3]).

A proper test rig has been created at VZLÚ, which enables the measurement of two axial turbine stages. The device was described in detail in [4] and its cross-section is shown in Fig. 1. Both stages are mounted on the overhanging end of the shaft, which is connected on the opposite side via a clutch to a torque meter. It is more accurate than the water brake (dynamometer) that is used only to set and keep specific testing RPM. In this way, the overall performance of both stages can be measured. To determine the performance of the individual stages, it is also necessary to measure the differential torque between the two stages. For this purpose, a special hub has been designed, on which the wheels of both stages are mounted, and which measures the torque of the second stage using a measuring element equipped with strain gauges. The signal from this measuring element is lead out via a hollow shaft to the side of the dynamometer and is scanned employing a telemetry device.

A detailed investigation of the flow field parameters can also be performed by this experimental device. The flow parameters are measured by the pressure probes in the planes behind the first and second stage, or the planes between the stator and the rotor.

Fig. 1. The cross-section of the two-stage axial test rig.

3.2 Experimental models

The newly designed stage working in aerodynamic design-overloaded mode was installed as the rear stage. The existing stage from the previous research was chosen as the preparatory (front) stage, which works during the tests in the off-design (overloaded) regime so that the output angle is about 70°. The second stage has been designed in two variants of rotor blades to build a full-reaction (HR) overloaded stage and a mid-reaction overloaded stage (MR). Both variants use the same stator blades. This stator wheel (Fig 2) has a special prismatic-blade-based design that corresponds to practical use in small industrial turbines of DSPW. The prismatic shape of blades and their special assembly enabled us to employ 3D printed plastic blades with instrumentation by pressure taps in the bladed wheel.

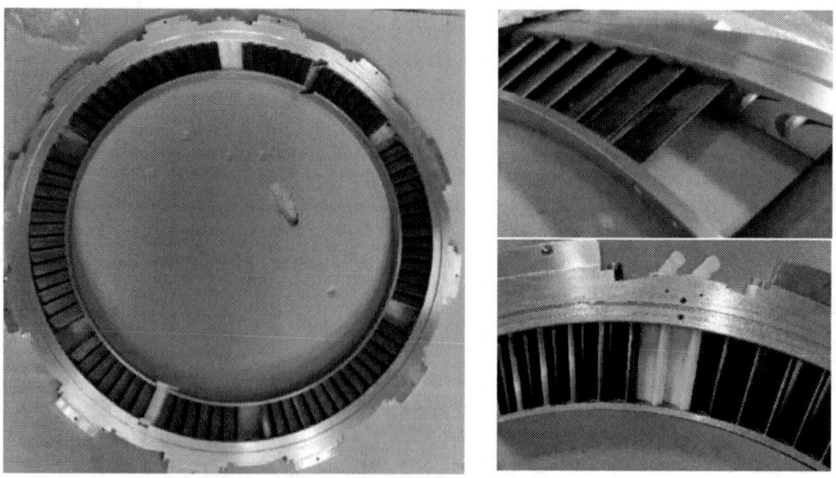

Fig. 2. New nozzle wheel for the design-overloaded stage.

Due to the extensive testing campaign, only the first HR variant tests results are presented. The overview of the stages geometry for the tested HR variant is in Table 1.

Table 1. Summary of stages geometry.

	The existing geometry of 1st stage		The new geometry of 2nd stage	
root diameter [mm]	506.00			
blade length [mm]	45.50	47.00	48.50	50.00
blade count [1]	98	78	96	78
midspan chord [mm]	23.00	28.65	24.00	29.49
stage reaction TS [%]	~ 50		~ 50	
design velocity ratio [1]	~ 0.63			
flow outlet angle [°]	70.00		70.00	

3.3 Experimental work procedure

In the case of measuring a single-stage arrangement, the setting of the operating point (pressure drop) is performed simply based on the measured parameters in front of and behind the stage. Furthermore, to determine the operating characteristics, the stage is gradually loaded and unloaded using a brake (dynamometer) and thus the rotational speed is changed. In the case of a two-stage arrangement of design-overloaded stages, it was essential to find such a pressure drop (mode), when the outlet angle from the 1st stage was identical to the outlet angle from the 2nd stage when the stage load changed. This regime then simulates the design-overloaded stages operation in the flow path. If there was different loading of stages, it will simulate the classic behaviour in off-design states, where the change in load is propagated from the rear stages and only in a few stages. The primary goal is to simulate the design operation of a design-overloaded stage.

For this reason, the experimental research procedure was set up by determining the design pressure ratio first. Here, the aim was to confirm or find the pressure ratio, when the aerodynamic load of both stages is the same, respectively the output angles are the same. Then the design point was determined. For the design pressure ratio, the design point was determined by changing the speed at which the outlet angle from the first stage corresponded to the design angle (identically also the outlet angle from the second stage). This phase also verified the behaviour of both stages in different regimes - the change of output angles should be roughly the same for both stages.

4 Results

4.1 Operating Vibrations

As this measurement was also the first operation of the two-stage configuration of the test rig, it was necessary to measure the operating parameters of the assembly, in particular, the vibrations over the entire operating speed range at first. Both wheels and hub assembly were properly balanced before the operation started.

The initial measurements were aimed primarily at the behaviour of the entire device with the installation of an assembly of two impellers (the weight of the rotor assembly was about 50 kg). In the entire investigated speed range (up to 6 000 rpm), the operation was very calm in terms of vibration. Fig. 3 shows the vibrations as a function of rotational speed for the case of a two-stage turbine in comparison with a single-stage turbine operated during the start-up phase of the device [4]. It demonstrates sufficient dynamic stiffness of the measurement device and indicates the smooth operation of future operation in experiments with a two-stage turbine assembly.

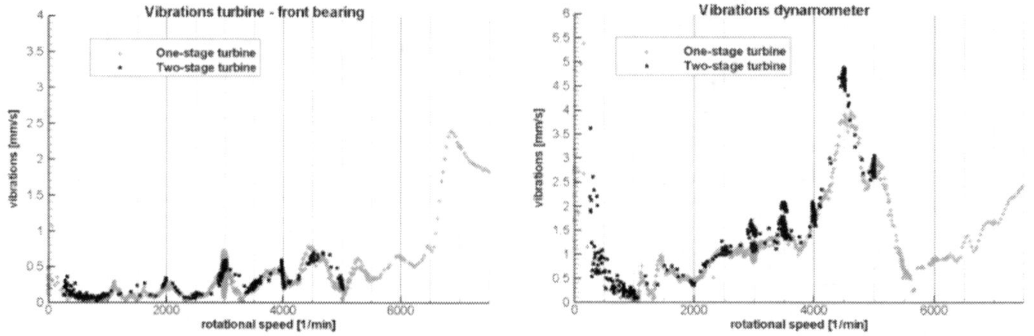

Fig. 3. The measured vibrations as a function of rpm. Comparison of one-stage and two-stage assembly.

4.2 Comparison of first (preparatory) stage parameters

Initial analyses of the results were focused on comparing the flow fields gained from the original measurement on the single-stage turbine with current ones on the two-stage turbine to verify the assumption that results from experiments performed in the past can be used.

Fig. 4 shows the radial distribution of the parameters at the output of the first stage. For this summary, the relative angle at the rotor outlet and the relative mass flow distributions were selected. The usefulness of these parameters is that they blur slight differences in the turbine mode settings. A more significant difference is evident at the blade tip. It is caused by different traversing planes - the current measurement is performed just behind the rotor and therefore there was no mixing of the seal purge flow with the main flow, while in the original measurement the plane is more distant, and the flows were mixed. Despite some differences, the comparisons show that the results of the original measurement can be used to solve the overall efficiency distribution between the two stages.

Fig. 5 shows the comparison of the efficiency obtained from the original measurement on a single-stage turbine and the current results from the two-stage turbine. In the case of the two-stage turbine, the torque of the first stage is computed from the total torque and the second stage torque measurement.

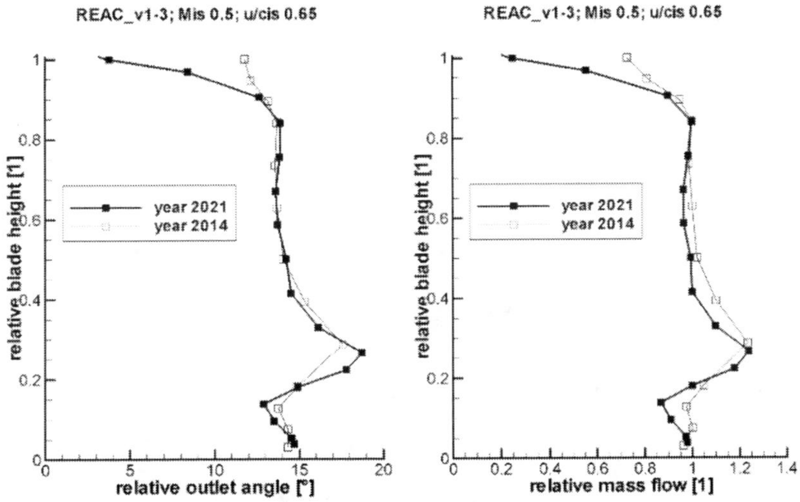

Fig. 4. Radial distribution of relative outlet angle (left) and relative mass flow (right) at the first stage outlet in comparison with the original (2014) results on a single-stage turbine.

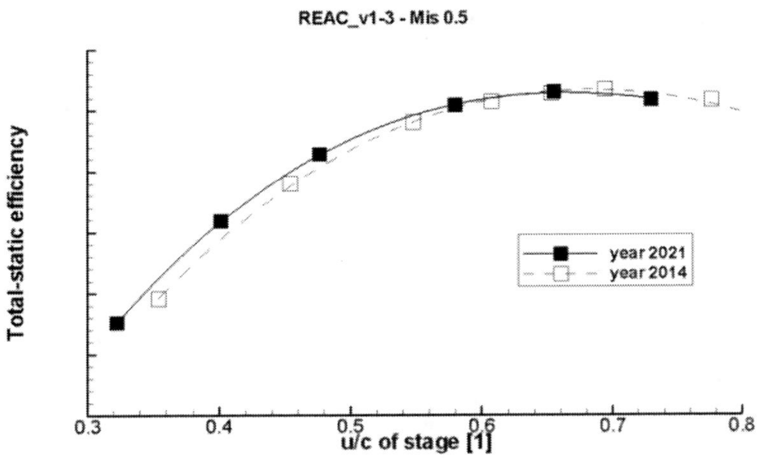

Fig. 5. Comparison of efficiency gained during the original measurement and current results using the differential torque meter

4.3 Design pressure drop determination

It was mentioned in previous paragraphs that the essential initial phase was to determine the design pressure drop where the outlet angles from both stages were the same. The result of this phase is a function of both stages outlet angles shown in Fig. 6. It is seen that at low pressure drops the aerodynamic load of the first stage is higher than the second stage (assessed from the value of the outlet angle) and vice versa for the second stage. The balanced load occurs at a pressure ratio of approx. 1.40–1.43, which corresponds very well to the design value. From the curves of the values of the output angles as a function of rotational speed, it is also possible to see that both stages are overloaded, resp. underloaded simultaneously. At the design pressure ratio, the design RPM (i.e. operating point) is based on values of approximately 3700-3800 RPM, which is only slightly higher than the design considers.

The efficiencies as a function of rotational speed for both stages are shown in Fig. 7. Two approaches to the distribution of torque between stages are compared here. One approach is the distribution based on the calculation of the characteristics of the first stage (the calculation is based on the previous measurement results on a single stage in 2014), the second approach is the determination of the second stage torque using the differential torque meter.

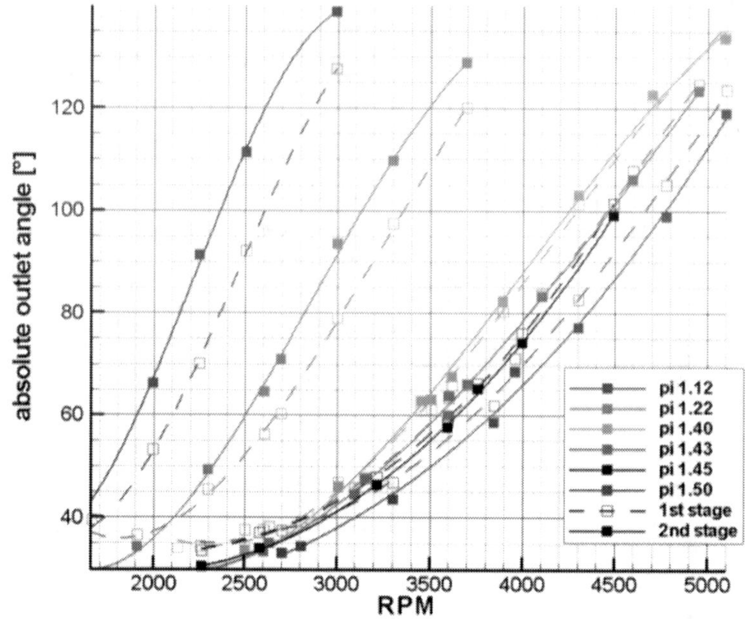

Fig. 6. Output angles from both stages as a function of rotational speed and inlet-to-outlet static pressure ratio

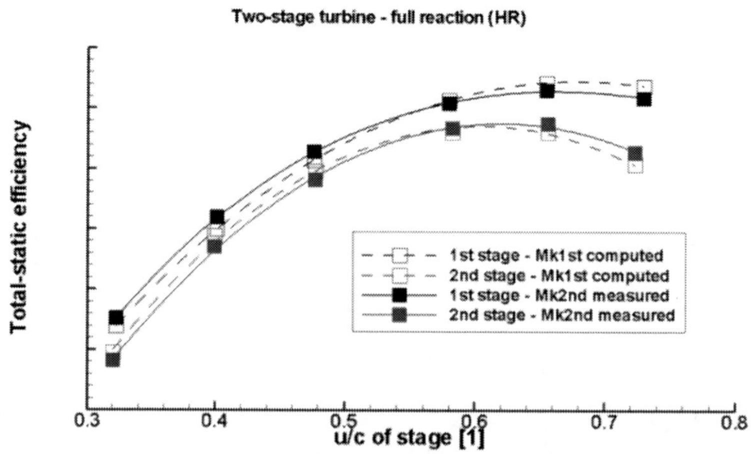

Fig. 7. Total to static efficiency: comparison of determination by dividing the overall torque measurement using the differential torque meter and using the data from the one-stage configuration.

5 Conclusion

This paper reveals part of the present research works that are conducted in the cooperation of Doosan Skoda Power and VZLU research institute. It is focused on design-overloaded stages that represent one of the answers to the changing trends in the steam turbine market.

This new experimental program is possible thanks to a new test rig available at VZLU laboratories – the two-stage axial turbine. New blading has been designed, manufactured, and assembled with some parts from the single-stage test rig to form the design-overloaded stage. First experiments have been performed to verify trouble-free operation of the test rig and to confirm the design point of this new two-stage turbine. An overview of acquired experimental data is presented.

Further experiments are going to focus on the investigation of tip leakage flow mixing with the main stage flow. Various configurations of the tip leakage flows in the first stage will be tested to evaluate the impact on the performance of the second design-overloaded stage. The tests will be performed for several operation regimes and for two stages reaction levels: full-reaction and mid-reaction. The importance of the shroud leakage flow mixing has been recognized and broadly investigated for example by Rosic et al. [5] or Shibata et al. [6].

This extensive experimental program is realized with the financial support of the Technology Agency of the Czech Republic within the THETA 3 program, project No TK03020061.

References

1. J.H. Cheon, P. Milcak, A. Pacak, C.R. Kang, M. Stastny, *Profile Loss Prediction for High Pressure Steam Turbines*, in Proceedings of ASME Turbo Expo 2016, GT2016-56114. (2016)
2. J. Uher, P. Milcak, R. Skach, D. Fenderl, P. Zitek, M. Klimko, *Experimental and Numerical Evaluation of Losses from Turbine Hub Clearance Flow*, in Proceedings of ASME Turbo Expo 2019, GT2019-90726. (2019)
3. M. Nemec, T. Jelinek, J. Uher, P. Milcak, *Effect of Stage Reaction and Shaft Labyrinth Seal in a Stage of an Axial Steam Turbine,* in Proceedings of the ASME Turbo Expo 2020, GT2020-14741 (2020)
4. M. Němec, T. Jelínek, *Adaptable Test Rig for Two-Stage Axial Turbine*, MATEC Web of Conf., 345, 00022. (2021),
5. B. Rosic, J.D. Denton, G. Pullan, *The Importance of Shroud Leakage Modeling in Multistage Turbine Flow Calculations,* J. of Turbomach., 128, 699–707. https://doi.org/10.1115/1.2181999 (2006)
6. T. Shibata, H. Fukushima, K. Segawa, *Improvement of Steam Turbine Stage Efficiency by Controlling Rotor Shroud Leakage Flows: Part I — Design Concept and Typical Performance of a Swirl Breaker*, GT2018-75269 (2018)

Air Drying in an Industrial Compressor

Matěj Jeřábek[1*], *Michal Volf*[1], *and Lukáš Richter*[1]

[1]University of West Bohemia, Faculty of Mechanical Engineering, Department of Power System Engineering, Univerzitní 22, Pilsen, Czech Republic

Abstract. In many industrial applications where compressed air is used it is necessary to increase the quality of the air by decreasing its humidity. This is done by either a condensation or adsorption dryer in the compressor circuit. In this article, the focus is on an industrial adsorption air dryer. The dryer is briefly described together with its function and a selected range of adsorbents suitable for air drying. The paper further deals with the development of a mathematical model of air humidity adsorption which is based on the Pseudo Gas-side Controlled (PGC) model. This paper contains a detailed description of the model and describes its implementation in MATLAB. Finally, the results of the simulations are discussed.

1 Introduction

Compressed air is used in many industries and, with the expansion of automatization and the use of pneumatic equipment, it could be said that it has become an essential sources of power. In many crucial applications, it is not only the precise pressure, but also the quality of the gas that needs to be maintained. The quality of compressed air is determined by its humidity, i.e. the lower the humidity, the higher the quality. It goes without saying that water condensation and its subsequent leakage from the compressor via the distribution piping or from the equipment itself is not only undesirable in clean and sterile applications such as food processing plants (for example in meat or milk processing where water condensation occurs frequently due to the low temperatures), but it is also unwanted in all industrial plants.

Therefore, condensing or adsorption dryers are used for drying compressed air. Condensation dryers are cheaper, but the air does not reach a sufficiently low relative humidity (given by a low dew point value) and thus cannot prevent condensation at temperatures approaching the freezing point of water. It follows that to achieve high quality compressed air, i.e. to achieve the lowest possible humidity, adsorption drying is required.

According to [1], the adsorption process is characterized by the adhesion of molecules or ions of a given substance, usually called an adsorbate, to a solid or liquid surface, usually called an adsorbent. The adsorbate is usually present in the gaseous or liquid phase. It is a process that affects only the surface of the adsorbent. If the molecules do not adhere to the surface but permeate the volume, it is absorption. The opposite process to adsorption is called desorption. In terms of the long-term reliability of the dryer operation, it is essential to ensure that desorption is efficient enough to avoid the accumulation of water (or adsorbed substances in general) over time, which would ultimately lead to dryer failure and could also cause failure of the following technology. Adsorption occurs commonly in nature and has many industrial applications (for example air drying) [2].

This article provides a brief overview of industrially used adsorbents for air drying, together with a description of the dryer and its role in the compressor circuit. A* 1D

*Corresponding author: volfm@kke.zcu.cz

mathematical model of the cycle is briefly described with a detailed description of the 1D adsorption model. Both models are implemented and numerically solved using MATLAB software. The derived model should be able to predict the outlet moisture content, pressure and temperature of the compressed air after passing through the dryer column which is in the compressor circuit. Thus, this model is able to predict the compressed air properties not only for adsorption but also for desorption that is necessary in order to maintain the long-term reliable operation of the entire circuit.

Nomenclature

A_b	bed cross section area [m^2]
a	volumetric surface are [m^2/m^3]
c	specific heat [J/kg K]
H_A	heat of adsorption [J/kg]
h	convective heat transfer coefficient [W/m^2 K]
h_m	convective heat transfer coefficient [kg/m^2 s]
k	thermal conductivity [W/m K]
$\dot{m}$	mass flow rate [kg/s]
m	mass [kg]
q	average water content [$kg_w/kg_{dry\ adsorbent}$]
T	temperature [°C]
t	time [s]
v	superficial velocity of air [m/s]
w_a	air humidity [kg_w/kg_a]
w_s	equilibrium humidity [kg_w/kg_a]
z	axial position [m]

Greek

ρ	density [kg/m^3]
ε	bed porosity

Subscripts

0	initial condition
a	air side
b	bed
s	adsorbent side
v	water vapor
w	water
z	axial direction

2 Dryer in compressor cycle

As mentioned above, the adsorption dryer is part of the compression circuit, as shown in Fig. 1. A fabric air filter is located at the inlet of the entire circuit to filter particulate matter from the intake air. This is followed by a shut-off damper which prevents the exhaust of pressurized air from the circuit and allows the air supply to the circuit to be regulated. Next comes the compressor, in this case a two-stage screw compressor with intercooling. The compressed air then travels to the oil separator where it is stripped of cooling oil by centrifugal force and then passes through a fine fabric filter. Next comes the cooler. Since the air is hot after being pushed, it must be cooled in the heat exchanger. This exchanger is a plate exchanger, composed of two parts. In one part the hot oil flows and in the other the compressed air from the compressor flows. The heat is dissipated by the ambient air which is driven through the exchanger by a fan.

The cooling of the compressed air causes some of the humidity to condense and therefore the heat exchanger can also be considered as a condensation dryer. This means that the performance of the adsorption dryer is largely influenced by the performance of the heat exchanger, which is dependent on the temperature of the ambient air. The more water condenses in the exchanger, the more water will be removed in the next part of the cycle, the separator, and the less moisture will need to be removed in the adsorption dryer itself.

Finally, the compressed gas goes into the adsorption dryer where it is dried to the desired humidity. After leaving the dryer, when the air has the desired properties, it is distributed through pipes for final use.

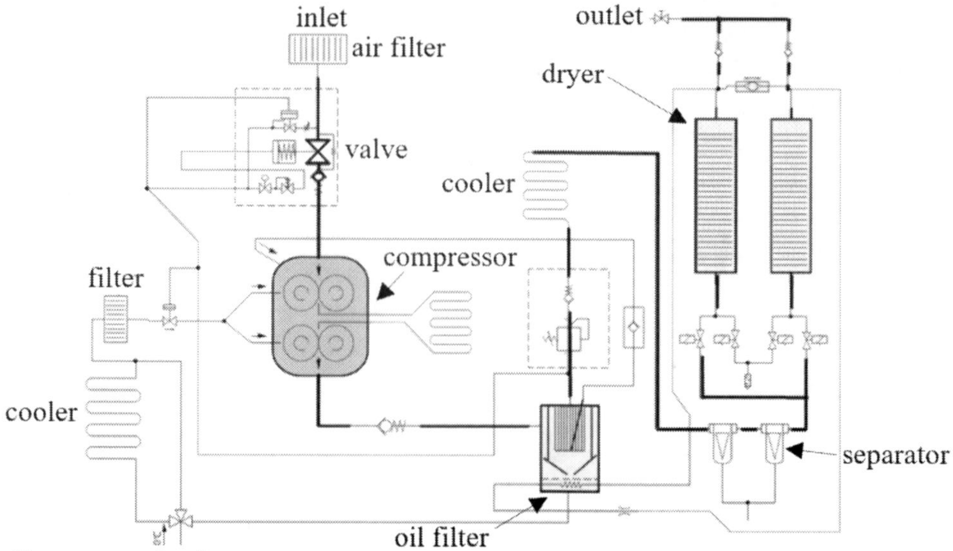

Fig. 1. Compressor circuit

3 Air dryers

According to [3], there are three basic types of air dryers: chemical, refrigeration and adsorption dryers. However, for all types of dryers, the air must be properly cooled and free of condensate before entering, meaning that the already condensed humidity must be removed in a moisture separator.

3.1 Chemical air dryers

Chemical air dryers contain materials that bind to the air or absorb moisture from the air when in close contact. There are two types: one that uses a spray in the form of pellets or granules, and the other type uses liquid ethylene glycol that absorbs water vapour. The glycol is regenerated (i.e. dried) in the plant using fuel gas or steam as the heating medium. The released moisture is discharged into the atmosphere. The regenerated glycol is recirculated by a pump through a water-cooled cooler which lowers the temperature of the glycol before it returns to the drying vessel [3].

3.2 Refrigeration air dryers

Refrigerated air dryers are used in the air system before and after compression. In the pre-compression system, the air must be cooled to a lower temperature for a given final dew point pressure in the duct. This requires more cooling capacity for the same end result. Two systems are used for cooling the air after compression. The flow in smaller and medium sized systems is predominantly air flow through directly cooled coils. Larger systems cool water that circulates through coils and cools the air. Such a system can reach a lower dew point [3].

3.3 Adsorption air dryers

According to [3], adsorption is the ability of some extremely porous materials to hold vapour in their pores, until the desiccant is heated or exposed to a drier gas. The material is always in a solid state and undergoes alternating cycles of drying and regeneration (there is no change in its composition). The most frequently used materials for air drying are activated alumina and silica gel, and molecular sieves are also sometimes used. Adsorption is suitable for achieving the lowest feasible humidity values.

The reverse process is called regeneration and is crucial in terms of long-term reliable operation of the dryer. Regeneration is used to remove the adsorbed humidity and is usually carried out by diverting some of the already dried air through a pressure reducing valve or orifice, thereby reducing the pressure to atmospheric levels and subsequently removing part of the adsorbed moisture. This air along with the recovered moisture from the saturated drying bed is discharged into the atmosphere. To increase the efficiency of the regeneration process, the drying air is often heated before it passes through the drying bed, or the bed itself is heated: the air then removes much more moisture from the system.

For our simulations, both silica gel and activated aluminium F-200 are used as adsorbents. The former is used so that we can validate the mathematical model with experimental data from literature, the latter is modelled because it is used in the compressor circuit's adsorption unit. We assume that these adsorbents are poured into the dryer in the form of beads.

4 Mathematical model of adsorption

The mathematical model of air drying is based on the law of conservation of mass and the law of conservation of energy. These equations are applied to both the adsorbent and the adsorbate. The conservation law of mass for the adsorbate yields the following equation:

$$\frac{\partial w_a}{\partial t} = -\frac{m_a}{\rho_a \varepsilon A_b}\frac{\partial w_a}{\partial z} - \frac{h_m a}{\rho_a \varepsilon}(w_a - w_s) \tag{1}$$

The law of conservation of mass for the adsorbent can be written as follows:

$$\frac{\partial q}{\partial t} = \frac{m_a}{m_s}(w_{a,z} - w_{a,z+dz}) \tag{2}$$

Furthermore, the energy equation for the adsorbate can be expressed in the form:

$$\frac{\partial T_a}{\partial t} = -\frac{m_a}{\rho_b \varepsilon A_b}\frac{\partial T_a}{\partial z} + \left[\frac{ha}{\rho_a c_a \varepsilon} + \frac{c_v h_m a}{\rho_a \varepsilon}(w_a - w_s)\right](T_s - T_a) \tag{3}$$

Finally, we can write the energy equation for the adsorbent as:

$$k_b \frac{\partial^2 T_s}{\partial z^2} + H_A h_m a(w_a - w_s) + ha\,(T_a - T_s) = c_s \rho_s (1-\varepsilon)\frac{\partial T_s}{\partial t} \tag{4}$$

These equations [4], are suitable for modelling adsorption using the first-order Pseudo Gas-side Controlled (PGC) model, which is suitable for emphasizing the physical characteristics of the adsorption mechanism [5]. Moreover, for the mathematical modelling of adsorption, it is necessary to know the adsorption isotherm model in addition to the equations mentioned above. In principle and according to [6], the models can be divided into:

- a model based on the potential theory of Polanyi

- a single-layer chemical model of adsorption
- a multilayer physical model of adsorption
- an ion exchange model

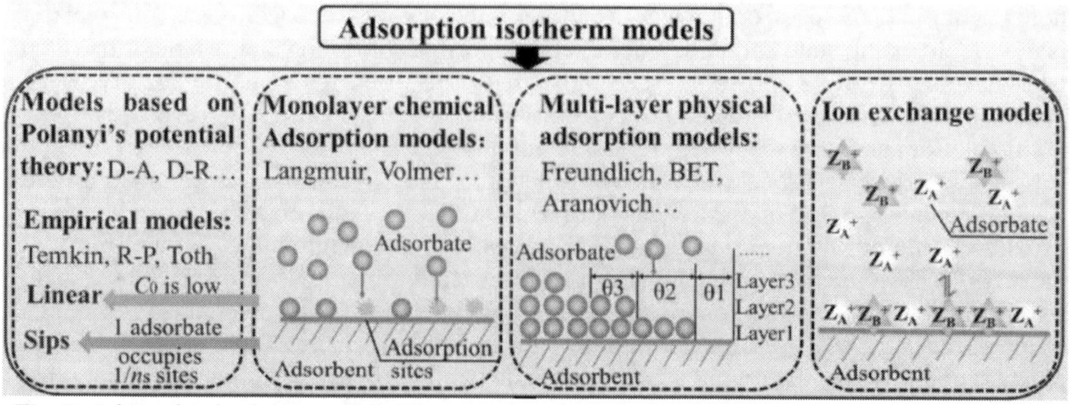

Fig. 2. Adsorption isotherm models [6]

These models are compared in Fig. 2. Another important equation in the mathematical model is the adsorption isotherm. According to [6], the isotherm represents the relationship between the equilibrium concentration of the adsorbate and the equilibrium amount of adsorption at a given temperature. Using isotherms, information about adsorption such as the adsorption mechanism, maximum adsorbent capacity as well as adsorbent properties can be investigated. In principle, we can have three mechanisms of adsorption [6]:

- monolayer chemical
- multilayer physical
- ion exchange

For our mathematical model, the adsorption isotherm from [7] was used for the silica gel. This adsorption isotherm is an empirically generated polynomial in which the relative humidity is calculated from the adsorbent saturation. For the activated alumina, the same polynomial was derived using a data sheet from the adsorbent manufacturer (BASF). Thus, in both cases empirical models of the adsorption isotherm are used.

4.1 Simplification and assumptions

For the mathematical modelling we use a cylindrical dryer with a height of 0.8m and a diameter of 0.13m. A non-isothermal, adiabatic adsorption model is chosen for the simulation. The following assumptions were considered in the derivation of the model:

- Pressure drop in the column is neglected
- Moist air contains water vapour as the only adsorbable component
- The particles of the adsorbent are round and have a uniform diameter
- Uniform porosity is assumed at all points in the column
- Heat transfer is considered only in the column height direction, i.e. the walls are considered to be adiabatic
- No chemical processes take place in the column

4.2 Implementation and numerical solution

The mathematical model was implemented and numerically solved in MATLAB. The air dryer column was divided into a defined number of layers (see Fig. 3) and for each layer the system of differential equations mentioned above was applied using the central difference scheme for numerical solution of the first and second derivative, which in general form yields:

$$f'(z) = \frac{f(z+h) - f(z-h)}{2h} \tag{5}$$

$$f''(z) = \frac{f(z+h) - 2f(z) + f(z-h)}{h^2} \tag{6}$$

where *h* is the height of the layer and *z* is the axial coordinate of the column. This yields a large system of ODEs that was solved using the commonly used *ode45*, i.e. the explicit Runge-Kutta method with variable time-step.

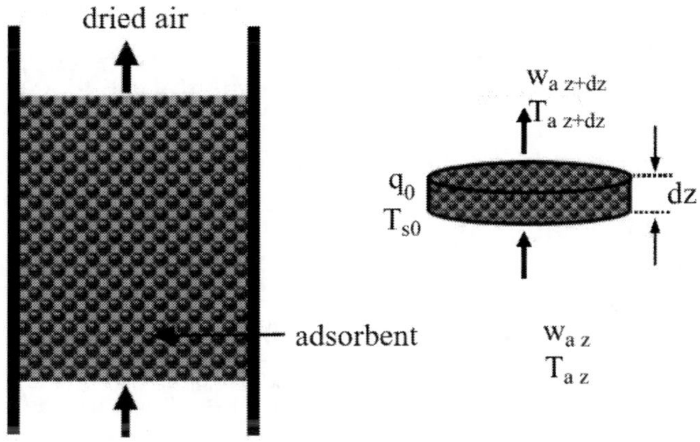

Fig. 3. Discretization of adsorber column [7]

The base equations are accompanied by auxiliary equations (for example for saturation water pressure, etc.) and also by boundary and initial conditions. These conditions are summarized below:

$$T_a\,(z, t=0) = T_{a0}, \quad T_s\,(z, t=0) = T_{s0}, \quad w_a\,(z, t=0) = w_{a0}, \quad q\,(z, t=0) = q_0 \tag{7}$$

$$T_a\,(z=0, t) = T_{a,\,in}, \quad w_a\,(z=0, t) = w_{a,\,in}, \quad \left.\frac{\partial T_s}{\partial z}\right|_{z=0} = \left.\frac{\partial T_s}{\partial z}\right|_{z=L} = 0 \tag{8}$$

5 Adsorbents

In connection with the mathematical modelling of adsorption, we conducted a review of adsorbents commonly used in industrial applications. Only adsorbents suitable for air drying are listed here. According to [5], adsorbents can be used in several forms, as powder, small pellets, or granules. The most important property of an adsorbent is porosity which is associated with high surface area. Another important property of adsorbents is mechanical and wear resistance. Last but not least, the adsorbent should allow easy regeneration. Many adsorbents are amorphous and contain networks of interconnected pores. The types of pores are shown schematically in Fig. 4. In adsorption, water molecules must first penetrate the film of liquid occurring on the surface of the adsorbent particle. Then they pass through the macropores to the micropores where the molecules are adsorbed.

According to [8], silica gel and activated alumina are generally suitable for adsorption of moisture from humid air.

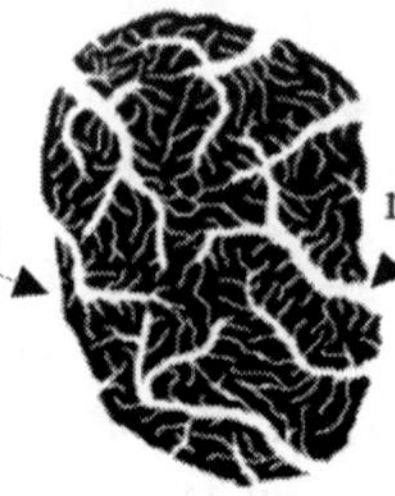

Fig. 4. Types of pore in adsorbent [5]

5.1 Silica gel

Silica gel is the most widely used adsorbent in industrial applications. Its main advantage is its high adsorption capacity and easy regeneration, meaning low regeneration temperature [8]. It is prepared by coagulation of colloidal silicic acid, which leads to the formation of porous and nanocrystalline granules of various sizes. The surface area of the granules ranges from 250-900 m^2/g [5].

5.2 Activated alumina

Activated alumina is a synthetic porous crystalline gel with a surface area ranging from 200 to 300 m^2/g [5]. Activated alumina has a higher capacity for water than silica gel at elevated temperatures, on the other hand has lower adsorption for elevated relative humidity (as seen in Fig. 5). Activated alumina has a low adsorption heat for water and other polar substances [8]. However, the disadvantage of this adsorbent is its higher regeneration temperature.

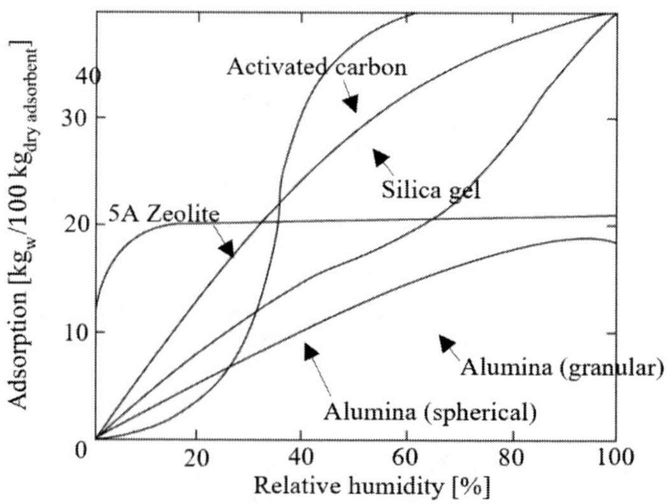

Fig. 5. Adsorption isotherms of water vapour from atmospheric air [8]

6 Results

In this chapter, the results obtained from the mathematical model are presented. First, we compare the results with the experimental data from literature and then we discuss the results for the compressor circuit and the adsorption dryer.

6.1 Comparison with experimental data

The mathematical model was validated using experimental data from [7] which gives measurements of adsorption using silica gel. The comparison shows that the mathematical model corresponds very well to both adsorption (Fig. 6) and desorption (Fig. 7). The agreement is not just for the output humidity of the air, but also for its output temperature. In the graphs below, the orange dots indicate the measurements from [7], in blue you can see the resulting variables over time from the mathematical model.

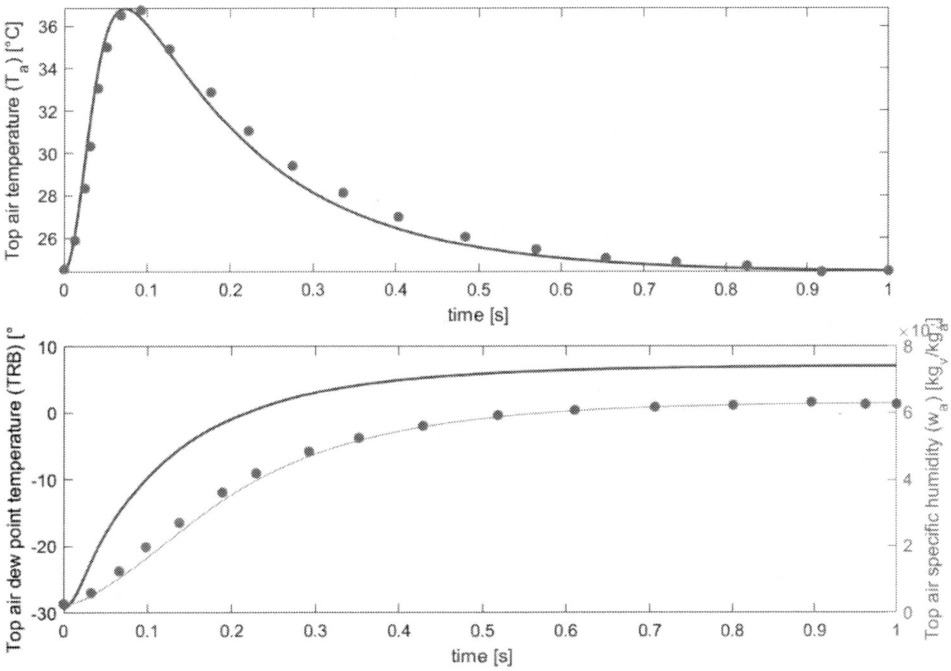

Fig. 6. Comparison of the adsorption process using silica gel

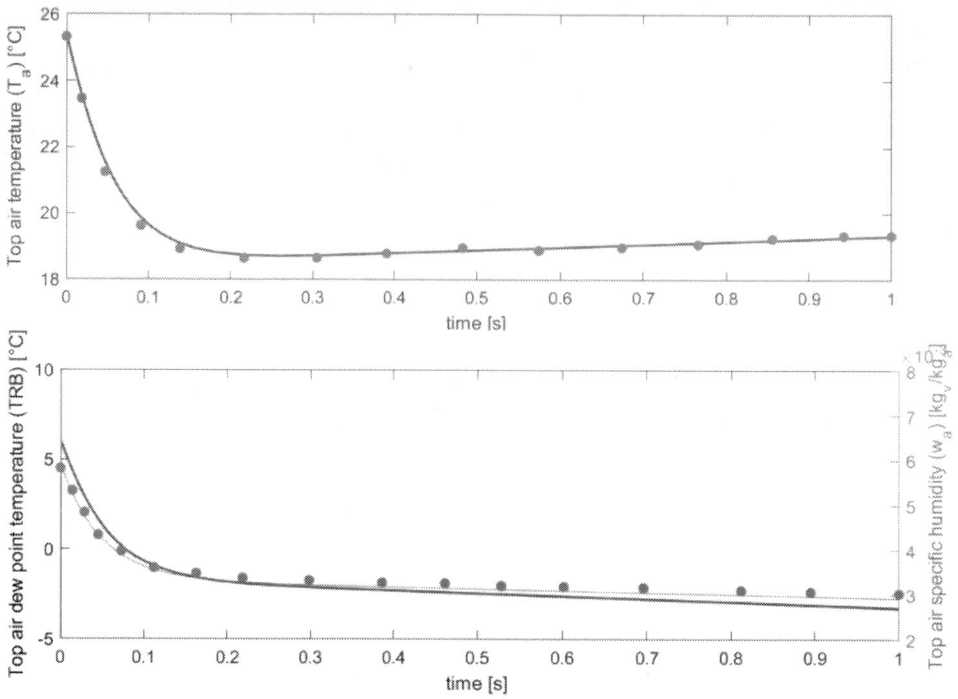

Fig. 7. Comparison of the desorption process

6.2 Results for the adsorption dryer in the compressor circuit

In this subchapter, the results concerning the adsorption dryer in the compressor circuit are presented. In the dryer, the adsorbent is activated alumina, so the adsorbent properties are different from those used in the previous simulation. The resulting outlet humidity and temperature curve is shown in Fig. 8. The orange line in the graph of the output humidity is there because in the mathematical model the output humidity is also plotted using the dew point temperature (blue plot). Since the dew point temperature is pressure dependent, these variables have a different path.

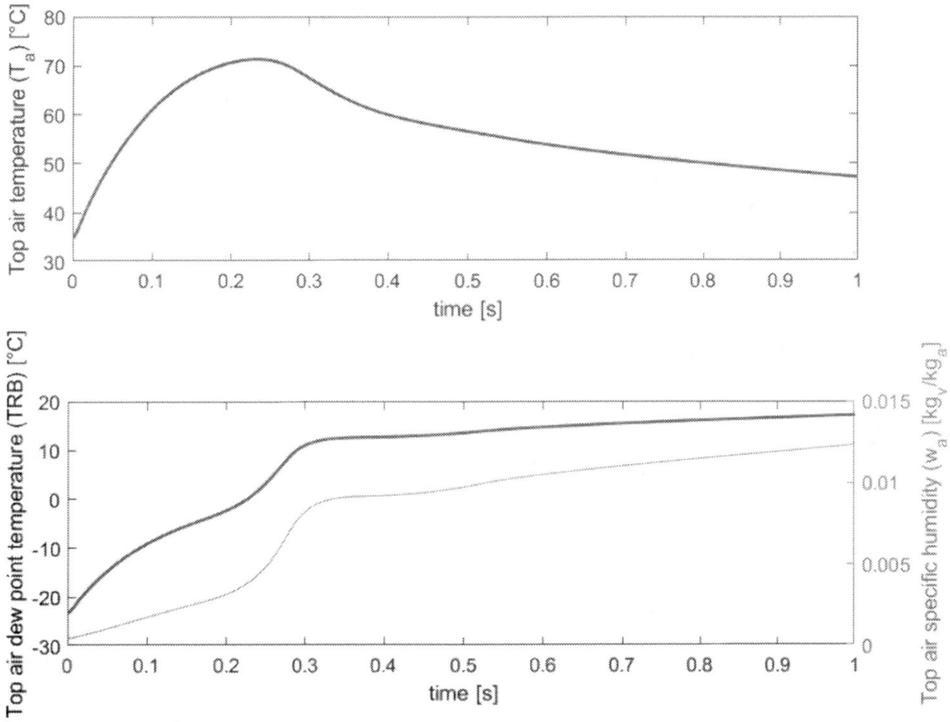

Fig. 8. Plot of the adsorption process

The results were obtained using the mathematical model validated by silica gel; however, this does not necessarily mean that these results are validated too. For that, we would need to perform measurements using activated alumina so that we could validate the equations that we use to model properties of this adsorbent. Unfortunately, no suitable data were found in the literature and thus we cannot proceed with that as of now. Nevertheless, as part of the ongoing project, measurements will be carried out at a later stage. So far it is fair to say that the temperature and humidity curves qualitatively provide the expected results.

7 Conclusion

In conclusion, the proposed mathematical model gives very accurate results when using silica gel as the adsorbent, so it can be assumed that the simplifications and assumptions are valid. However, the model cannot be verified when using activated aluminium as no suitable experimental data in the literature could be found. It is therefore necessary to obtain experimental data from measurements on a real dryer using an adsorbent identical to the one defined in the mathematical model. For this purpose, in the ongoing project aimed at the development of an industrial compressor, we are also building an experimental stand on which the required measurements will be carried out.

This project is co-financed from the state budget by the Technology Agency of the Czech Republic under the TREND Programme, project number FW03010376 - Development of an industrial compressor and its digital twin. The presented work was also financially supported by student project SGS-2022-023 (Research and development of power machines and equipment).

References

1. EL-EMAM, Salad and MOSTAFA M., Awad. ResearchGate. *Theoretical and Experimental Investigation on the Performance of Hybrid Desiccant Air Conditioning System Using Activated Alumina in a Radial Flow Packed Bed.* [Online] 2010. [Cited: 4. 19., 2022.] Available from: http://dx.doi.org/10.1109/THETA.2010.5766425.

2. Y, Artioli. ScienceDirect. *Adsorption.* [Online] 2008. [Cited: 4. 18., 2022.] Available from: https://doi.org/10.1016/B978-008045405-4.00252-4.

3. Mobley, R. Keith. ResearchGate. *Air Dryers.* [Online] 2000. [Cited: 4. 25., 2022.] Available from: http://dx.doi.org/10.1016/B978-075067174-3/50066-4.

4. MURATHATUNYALUK, Siripan and STICHNVIVHIT, Koranut.. ScienceDirect. *Modeling of Non-Isothermal Adsorption Process in a Silica Gel Desiccant Packed Bed.* [Online] 2018. [Cited: 4. 19., 2022.] Available from: https://doi.org/10.1016/B978-0-444-64235-6.50011-5.

5. EL-NAAS, Muftah. ResearchGate. *Modelling of Adsorption Processes.* [Online] 2013. [Cited: 4. 19., 2022.] Available from: https://www.researchgate.net/publication/288773735_Modelling_of_adsorption_processes.

6. Jianlong Wang, Xuan Guo. ScienceDirect. *Adsorption isotherm models: Classification, physical meaning, application and solving method.* [Online] 2020. [Cited: 5. 4., 2022.] Available from: https://doi.org/10.1016/j.chemosphere.2020.127279.

7. Ramzy A. K., T.P. Ashok Babu. ResearchGate. *Semi-analytical method for heat and moisture transfer in packed bed of silica gel.* [Online] 2010. [Cited: 5. 3., 2022.] Available from: http://dx.doi.org/10.1016/j.ijheatmasstransfer.2010.09.007.

8. SHUGUANG, Deng. ResearchGate. *Sorbent Technology.* [Online] 2006. [Cited: 4. 16., 2022.] Available from: https://www.researchgate.net/publication/267778978_Sorbent_Technology.

Investigation of heat treatment process using Particle Image Velocimetry

Dániel Koós[1], *Dávid* Faragó[1*], and *Péter* Bencs[1]

[1]University of Miskolc, Department of Fluid and Heat Engineering, H-3515 Miskolc-Egyetemváros, Miskolc, Hungary

Abstract. Martensitic hardening is a technology widely used in mechanical engineering practice to increase the hardness and abrasion resistance of ferrous alloys. To achieve martensitic hardening, the workpiece must be cooled the workpiece very quickly, which can be achieved through using various refrigerants. The quality and the properties of these coolants have a great influence on the success of the process, so their investigation is essential. Throughout the investigation method – according to ISO 9950 standard –various aqueous suspensions used as refrigerants are studied by placing a heated cylindrical specimen in a refrigerated container (which will be described later) and by recording the temperature change over time. The flows generated in the tank during the heat treatment influence the nature of the heat transfer, thus the hardening process itself. The task is to map these flow conditions to give a comprehensive description of the qualification process in question. With the advancement of measurement technologies, there are more and more possibilities to determine the velocity of flowing media, even at several points simultaneously. Using the available PIV (Particle Image Velocity) system, the two-dimensional vector field of the flow can be determined on a complete plane surface. In the present research, the applicability of seeding particles required for the PIV measurement technique is also investigated as part of the mentioned project, establishing further research possibilities. The classification of eight different seeding particles is presented. As a final result, seeding particles suitable in the current arrangement for proper flow display are listed.

1 Introduction

In industrial applications, it is important to know the effects of certain factors that influence heat transfer during the application of a given refrigerant (heat treatment). In case of water-based polymer solutions used as refrigerants, the heat removal characteristics vary within certain limits due to the simultaneous influence of temperature, concentration, and flow rate. Knowing the trend and extent of the change, heat removal can – to a certain extent – be optimized for the given process [1-2]. Their quality and properties can greatly influence the success of the whole process, so their investigation is essential. According to ISO 9950 standard, the use of different aqueous suspensions employed as refrigerants is studied by

* Corresponding author: david.farago@uni-miskolc.hu

placing a heated cylindrical specimen in a refrigerated container designed for this purpose, then recording the temperature change over time. The flows generated in the tank during the heat treatment influence the nature of the heat transfer, thus the hardening process itself. To study the flow processes, a planar or a spatial measurement can be performed. A conventional PIV measurement is a non-intrusive optical method that can provide an instantaneous flow velocity field in a single plane. Although non-intrusive, the technique requires seeding particles to operate. Particles should be neutrally buoyant, and usually micron sized. The exact size, type and density of particles mixed with the working fluid can vary for every measurement. Seeding particles must meet two important requirements. First, they must follow the flow well. Second, they must reflect the light emitted by the light source (usually a LASER or high-power LED) effectively. Since the recording cameras need a clear view of particles, the medium must also be transparent. If seeding particles can't properly reflect light in the given conditions, then the obtained images are not reliably processable and will provide spurious results. It is possible to increase the intensity of illumination and/or the sensitivity of the camera to improve image quality [3]. For the same flow medium, the flow properties of smaller particles are more favorable, but larger particles reflect light to a greater extent. Considering these two effects, the flow properties, and the properties of the light source, the optimal diameter of seeding particles can be obtained [4].

2 Measurement system

Particle Image Velocity (PIV) is an optical measurement method, with which planar images of the velocity field can be obtained. Seeding particles are fed into the main flow. Using a precisely timed light source, the area of interest is illuminated twice with a short time gap between the two pulses. Seeding particles reflect some of the light, which then can be captured with a recording unit (one or more high-speed cameras, matrix detector). Particle images are recorded simultaneously with the light pulses. The displacement of a given particle can be determined from subsequent images, from which its velocity can be derived. The time gap between two pulses can vary depending on the peak velocity and seeding properties of the measured flow. An appropriate displacement of particles is essential for a good correlation. In the present research, illumination is provided by a double-pulsed laser, and the recording is done by a CCD camera. After processing and post-processing, the results of the completed images, valuable information about the given flow can be obtained [5].

The whole PIV system consists of several subunits. These include lasers and other optical devices for illumination, seeding particles and their feeding apparatus, a camera for capturing images, and post-processing hardware and software. Figure 1 shows the elements of a typical PIV measurement set-up [5].

When dealing with low-velocity liquid flows, it is very important that the settling rate of the particles be as low as possible, thus tracers with a density similar to that of the flow medium should be used. An example is polystyrene seeding particles, which have a density of 1050-1090 [kg/m^3], but their reflectivity is rather poor. Silver-coated polystyrene particles could be used to overcome said disadvantage, but their price is a notable drawback. At relatively low cost, hollow glass spheres can be used with the highest efficiency within the size range of 1-100 μm. With air filling the inside of the glass spheres, their density can be controlled during production. The air / glass transitions also have a high refractive index; hence they provide good visibility. When investigating gaseous flows, some form of water aerosol or oil can be used as a seeding medium. The density of these particles are several times higher than that of the gas, the settling rate can be reduced by setting a small diameter for the particles. Seeding particles are usually of the order of 1 μm are used in this case [3].

In case of liquid flows, a trivial solution is most often present, namely the solid particles can be mixed in the liquid to form a suspension. In case of gaseous media in particular, the challenge is to introduce the seeding particles into the area of investigation in a sufficient quantity and in a sufficiently homogeneous distribution without disturbing the flow itself. Possible solutions to feed the seeding particles into the main flow if the particles are liquid can be atomization or condensation, while if the particles are solid, the atomization of their dispersion can be a solution.

Acquired images are divided into smaller parts, called the interrogation zones. For each interrogation zone, one vector field is calculated. Particles that leave the interrogation zone are lost, reducing the number of particles that contribute to the correlation. Particles inside an interrogation zone must move homogeneously in the same direction, else the statistically calculated vector fields will show spurious results. Usually, there should be about 10-15 particles in each interrogation zone, which may increase at higher flow rates [4].

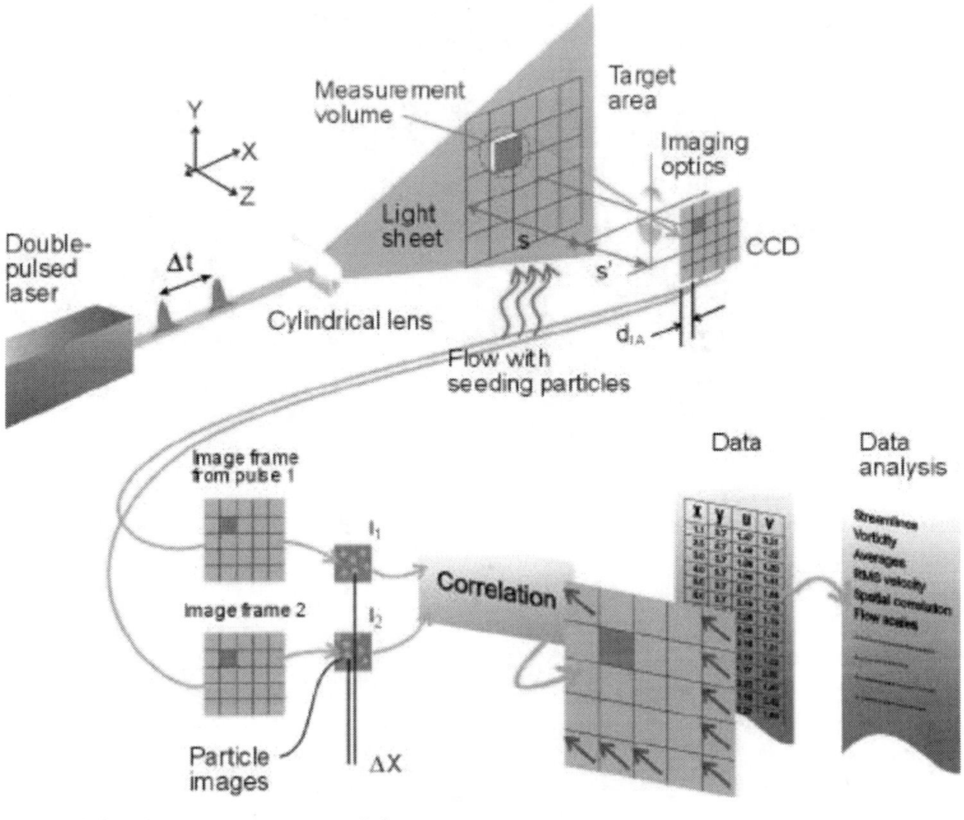

Fig. 1. Principle of PIV measurements [5].

3 Measurement setup

The measurements were performed in a container designed for the qualification of heat treatment media according to the ISO 9950 standard. The wall of the container was made of plexiglass to make it possible to use PIV measurement technique. During the heat treatment process, a steel rod heated to 850 °C was placed in the opening fashioned for this purpose, and the temperature change of the rod over time is recorded. The flow within the container is enhanced by a propeller stirrer with adjustable speed [2]. The container and its associated accessories are shown in Figure 2. The aim of our investigation was to select the optimal seeding particles to use, then use them to examine the nature, velocity, and homogeneity of the flow in relevant volumes.

During our investigation, the specimen, its fixation, and the temperature measurement devices were removed. The flow regime created by the mixer was divided into two regions by a plexiglass sheet, as a result of which the flow transpired continuously in the formed channel in a well-defined manner.

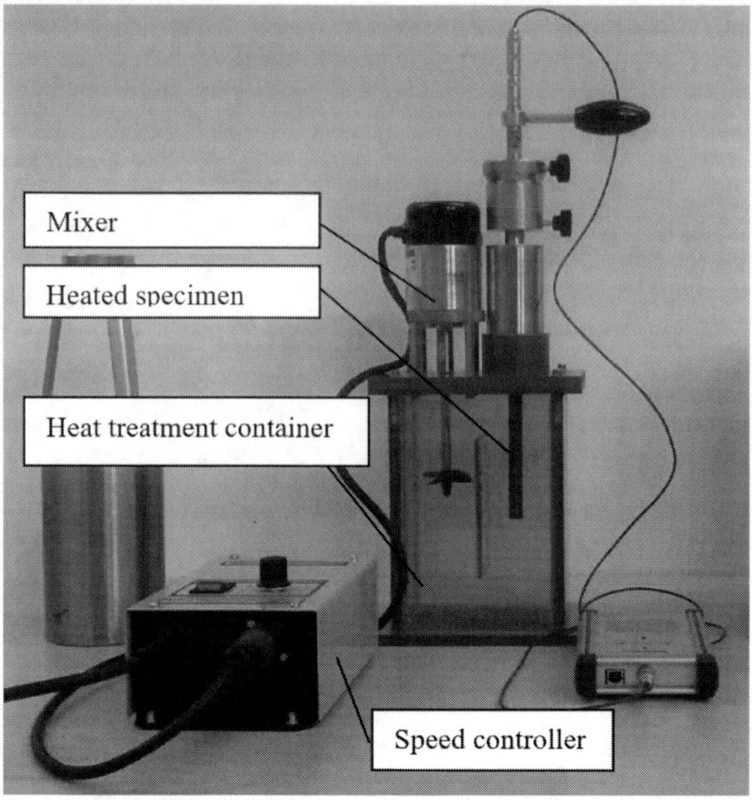

Fig. 2. The container and its associated accessories.

One of the main elements of the PIV measurement technique is the laser light emitted by the high-energy laser, which illuminates the investigated area in a plane section after proper optical imaging. The light of the Nd: YAG double-pulsing laser was refracted twice at angles of 90° on plane mirrors, then by directing it through the light sheet optics the necessary 2D planar light sheet was generated. The test area was located on the symmetry plane "x" of the container, and its size covered the entire cross-section of the flow. Figure 3 shows a schematic of the applied measurement arrangement with the laser, light paths, heat treatment container, and the camera.

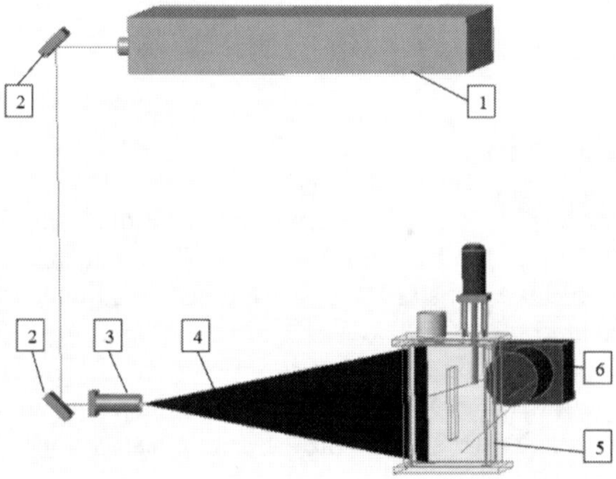

Fig. 3. Schematic of the applied measurement arrangement: 1: double-pulse laser; 2: plane mirrors; 3: light sheet optics; 4: 2D planar light sheet; 5: heat treatment container; 6: camera.

Three operating parameters of the laser were controlled. These were: Δt time difference, laser frequency and intensity. The value of Δt shows the time elapsed between two laser pulses. Based on preliminary measurements, Δt was chosen as 50 ms. Evaluation accuracy is increased during post-processing by changing the size of the computational area. The recording frequency of the camera was 7.5 Hz. Hereinafter, five subsequent image-pairs form a measurement series. The strengths of the two laser pulses can be controlled independent of one another within the range of 0V-10V. These parameters were optimized during experiments pro re nata.

The requirements for the facilitated camera are: they must be able to take images at a sufficiently high resolution and with a sufficiently short shutter speed. The TSI Power View 630059 CCD camera we use perfectly meets these requirements; it captures images at 4 MP resolution with shutter speeds of up to 200 ns [7]. Figure 4 shows an image recorded by the camera of the liquid with seeding particles.

Fig. 4. Image of the mixture of liquid and seeding particles (glass powder) taken by the camera.

Ideally, the contrast between the seeding particles and the black background is high, and the particles are in focus, so the image's overall clarity is high. Most of the defects in our case are probably caused by the refraction of plexiglass, which could be mitigated by using clear glass.

For the CCD camera we used, it was possible to control the focal length and aperture diameter of the lens. Prior to recording, these two parameters always had to be set for the respective seeding particle, as they had different light reflection and geometric properties (diameter). When adjusting the focal distance, the goal was to change the sharpness of the images, while changing the aperture diameter was to change the depth of field of the images. Experience has shown that by increasing the aperture diameter, the intensity of the laser light also had to be increased to obtain high quality images.

In our experiments, eight types of seeding particles were compared using the described equipment. Selection was based on availability, adequate light reflectance, and presumed flow properties (size, density – particles need to be able to move with the flow). The available particles and their characteristics are listed in Table 1.

Table 1. Investigated seeding particles.

ID	Composition, characteristics, features
T1	Al_2O_3 - 100-200 μm size range
T2	Al_2O_3 - 40-60 μm size range
T3	Glass powder - 0-50 μm size range; grinding material for nanotechnological grinding
T4	Talcum powder - commercially available
T5	„Light powder" – commercially available glitter powder with aluminium content
T6	Styrofoam powder – made by coffee grinder, from standard Styrofoam board, with a characteristic particle size of ~500 μm
T7	„Mica powder" –commercially available glitter powder, ground mica shale and titanium oxide
T8	Glitter powder – commercially available, small particle size glitter powder

4 Results

The applicability of the previously listed seeding particles in an aqueous medium was investigated using the described equipment (container with stirrer) and measurement technique (PIV). Changing the speed of the propeller of the stirrer was made possible by a frequency controller. Experiments were performed at the lowest speed, and by placing an equal volume of seeding particles in the water every time. The examination of the region shown in Figure 5 was the primary goal, and the vector field of this area was determined and evaluated during subsequent post-processing.

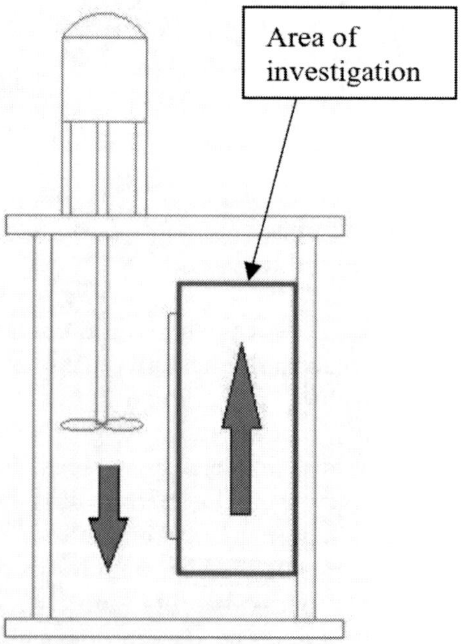

Fig. 5. The area of investigation.

To obtain an accurate relation of the location of particles compared to the measurement domain, a calibration procedure is needed. During the process, a target with a pattern of equidistant dots is placed in a way that its plane is parallel with the laser light sheet. The parameters and the exact location of the calibration target are well known; hence the recording and post processing software (*Insight 3G*) can be calibrated.

Images resulting from the eight measurements series were processed, and evaluated for three interrogation zone sizes, and for each of the three evaluations, the average velocities, and the velocities at a fixed point of each seeding particle were determined. Control velocities were measured to validate the obtained results using ultrasound ultrasonic velocity measurements and by adding 6, relatively big pearls to the system, and calculating their displacement using MATLAB.

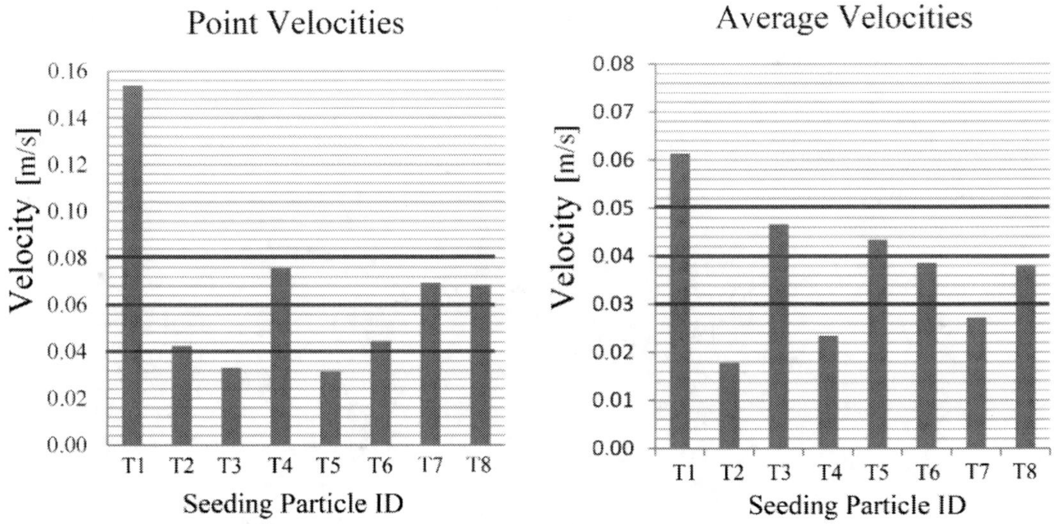

Fig. 6. Fixed point and average velocities of different seeding particles in case of a 64x64 pixel sized computational area, with the control velocity at a fixed point (red horizontal line) and average control velocity region (between blue horizontal lines).

The diagrams show that both the average and fixed-point velocity values of particle T1 (Al_2O_3, 100-200 μm) are outside of the control velocity range, so this seeding particle was declared unsuitable, and no further studies were performed with it. Both the fixed-point and average velocities of T6 (Styrofoam powder) and T8 (glitter powder) fell within the control range, hence under the given conditions, based on our investigations, the flow was followed best by these two seeding particles. In addition, the fixed-point velocity values of T2, T4 and T7, and the average velocity values of the T3, proved to be adequate, so further studies were performed for these seeding particles.

A statistical analysis of the suitable seeding particles followed. Five subsequent image-pairs form a measurement series, and a vector field was created based on each image pair. The variance of the vector components u and v are known. The variance values of the five vector fields were averaged and plotted for the seven different tracer particles (Fig. 7).

We are trying the find the seeding particles that produce the lowest variance values, as measurements performed with them should provide the most reliable results. The trends in Figure 7 show that the variance in velocities for component v are always greater than that of component u, which is expected since the flow is mostly vertical (v direction). This implies a greater degree of variation in values. Seeding particles with the smallest variance values were T2, T3, T7 and T8, for these the cross-correlation procedure using *Insight 3G* software proved to be successful.

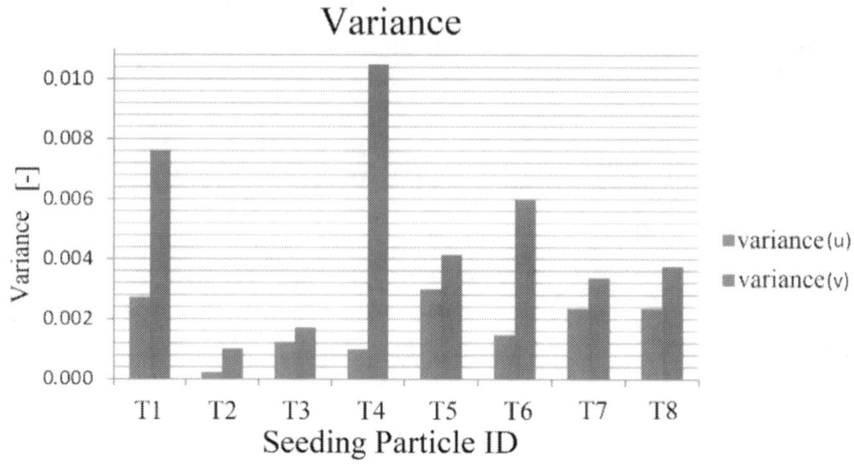

Fig. 7. Variance values for different seeding particles.

T8 (glitter powder) passed well in both control phases (velocity and statistical), and particles T2, T3 and T7 also showed good results in the statistical test, so flow visualization was performed for these four seeding particles.

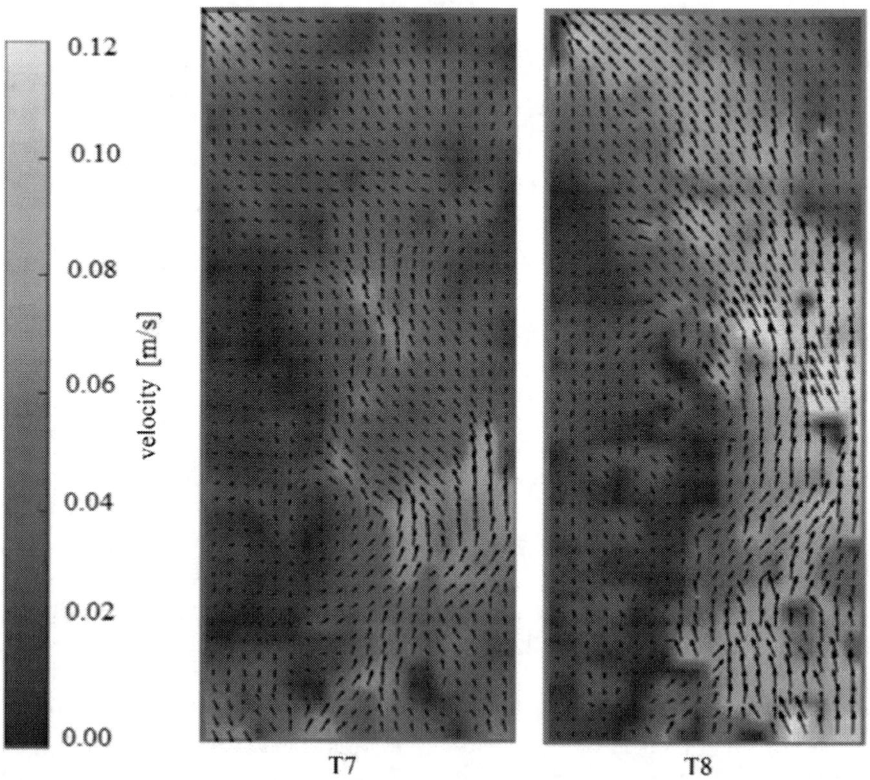

Fig. 8. Velocity field with vectors obtained by using mica powder (T7), glitter powder (T8).

5 Conclusion

The applicability of different seeding particles in a container designed for the qualification of heat treatment media according to the ISO 9950 standard has been investigated. Eight – mostly commercially available – materials were investigated as seeding particles for PIV measurements. Seeding particles must meet to prerequisites: they should follow the flow well, so their size should be adequately small, and their average density should not differ greatly from the density of the medium. They must also reflect the light emitted by the light source effectively to provide clear, relatively high contrast images for the PIV processing software to avoid spurious results. The continuous flow of the mixture of medium and seeding particles were granted by a propeller stirrer placed inside the container.

Post-processing of the images taken during the measurements was performed using the *Insight 3D* software. The size of the computational area was optimized by checking the velocity values after processing. At this time, a 64x64 pixel computational area proved to be the most efficient, so we performed further data processing according to this.

During the evaluation of the resulting vector fields, we compared the fixed point and average velocities of the validation measurements (executed prior to PIV investigations using MATLAB and ultrasonic velocity measurement) with the velocities obtained by the use of different seeding particles. Some seeding particles with false results were filtered. Under the given conditions, T6 (Styrofoam powder) and T8 (glitter powder) performed the best at following the flow. It is likely that their effectiveness is due to their density being similar to that of water. The density of styrene (raw material of Styrofoam) is ρ=1040 kg/m^3. If a sufficiently fine powder is produced from it, most of the air bubbles are removed, so it can be used effectively as a tracer particle in aqueous media. The base material of the glitter powder we obtained was polyethylene terephthalate (PET), density of which is ρ=1380 kg/m^3, which is slightly higher than the density of water.

The appropriate reflectance of the investigated particles was inferred from the variance of the velocity vector files. The lowest variance values were produced by T2 (Al_2O_3) and T3 (glass powder). T7 (mica powder) and T8 (glitter powder) particles also showed adequate variance values. Based on these results, it can be stated that by using these four seeding particles, the cross-correlation procedure during data processing created a correct vector field with few errors.

Overall, four of the eight seeding particles were found to be suitable for the current scenario during data processing. These are Al_2O_3 (T2), glass powder (T3), mica powder (T7) and glitter powder (T8). Using these four seeding particles, the velocity distribution of the area of investigation was obtained as shown in Figure 8. Based on obtained the images, it can be said that although the Al_2O_3 (T2) powder gave the smallest values of standard deviation and variance during the statistical evaluation, it created an insufficient vector field in the lower part of the investigated area. The images also show that the generated flow nonhomogeneous, a high velocity region occurs. This flow pattern will, of course, be altered if the specimen is inserted, but it is likely that the visible trend will remain, resulting in a higher velocity flow along one side of the rod. This causes the specimen to cool asymmetrically, resulting in thermal stresses and other structural defects. In the future, velocity fields obtained during the cooling of the rod can be examined in a similar manner, using the experience of the present research to our advantage. Although mica powder and glitter powder have proven to be the most suitable under the current circumstances, introducing a specimen of 850 °C into the water during the heat treatment will massively alter the whole filter process. Seeding particles can melt, air bubbles can form around the specimen, and a whole set of new factors should be considered. Glass powder with a particle fraction of 0-50 μm can provide a viable alternative that has shown good properties in our studies and is also resistant to high temperatures.

Acknowledgements

The authors are grateful to the Hungarian-German Intergovernmental S&T co-operation programs TKA-DAAD (160552), for the financial support of this research.

References

1. B. Liščić, H. M. Tensi, and W Luty, Theory and technology of quenching: a handbook. Berlin; London: Springer-Verlag, (1992)
2. G. Kerekes, M. Kocsisné Baán, and I. Felde, "A hőmérséklet, az áramlási sebesség és a koncentráció együttes hatása vízbázisú polimeradalékos közeg hűtőképességére," in OGÉT 2014: XXII. Nemzetközi Gépészeti Találkozó, pp. 178–181., (2014)
3. A. K. Prasad, "Stereoscopic particle image velocimetry," Experiments in Fluids, vol. 29, no. 2, pp. 103–116, Aug. (2000), doi: 10.1007/s003480000143.
4. A. Melling, "Tracer particles and seeding for particle image velocimetry," Measurement Science and Technology, vol. 8, no. 12, pp. 1406–1416, Dec. (1997), doi: 10.1088/0957-0233/8/12/005.
5. "Measurement Principles of PIV - Dantec Dynamics | Precision Measurement Systems & Sensors," Dantec Dynamics | Precision Measurement Systems & Sensors, (2019). https://www.dantecdynamics.com/solutions-applications/solutions/fluid-mechanics/particle-image-velocimetry-piv/measurement-principles-of-piv/.
6. "PowerView Plus 4MP Camera 630059 | TSI," tsi.com. https://tsi.com/product-components/powerview-plus-4mp-camera-630059/ (accessed Mar. 15, 2022).

ORC and sCO_2 cycle for high temperature WHR applications

Tereza Kubíková[1*]

[1]Department of Power System Engineering, Faculty of Mechanical Engineering, University of West Bohemia in Pilsen, Univerzitni 22, 306 14, Pilsen, Czech Republic.

Abstract. This article suggests a suitable closed thermodynamic cycle for waste heat from a cement plant by using alternative working fluids. A pair of closed working cycles is compared: supercritical CO_2 cycle (sCO_2) and the Organic Rankine Cycle (ORC). In the case of ORC, it was necessary to choose a suitable working fluid. The goal is to minimize the ODP (Ozone Depletion Potential) and GWP (Global Warming Potential), to maximize the thermal efficiency and to optimize the minimum working temperature difference (pinch points, approach points), the temperature profiles of the heat exchangers and their working pressure and temperature, which affects the cost. Both cycles (ORC and sCO_2) including a detailed component analysis are calculated using Python in Spyder IDE, which includes all the libraries for this task. According to the results of the calculations, the ORC cycle was chosen for further calculations of the components (condenser and turbine). The most suitable working fluids with high efficiency are hexamethyldisiloxane (MM), ethylbenzene and toluene, from which toluene was selected.

1 Introduction

The advantage of a closed cycle in comparison to an open cycle (e.g. the Ericsson-Brayton cycle of a gas turbine [1] [2]) is that it can work with a different fluid than the combustion product and ambient air. On the other hand, a closed cycle needs a heat exchanger, which transforms heat from the combustion products to the cycle working fluid. The most common example of a closed cycle is the Rankine-Clausius cycle using steam as a working fluid; it is used mainly in regular power plants. However, the thermodynamic properties of steam are not ideal for all applications. The waste heat recovery (WHR) typically has to deal with relatively low temperature, which implies lower thermal efficiency even in an ideal Carnot cycle (which uses the hard to achieve isothermal heat exchange). Some examples of thermal cycles and their working temperature gradients and heat powers are plotted in Fig. 1 (from Liu [3])

* Corresponding author: tereza.kubikova@doosan.com

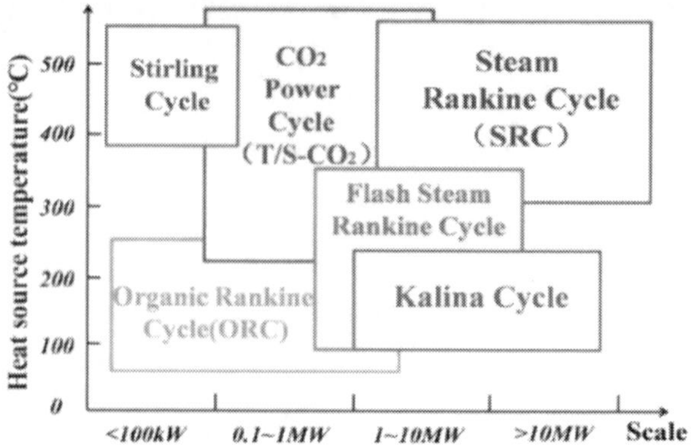

Fig. 1 Diagram of closed thermal cycles - comparison of their powers and temperatures.

1.1 WHR usage

Waste heat is lost into the atmosphere or a condenser if not recovered. It can be recovered by adding a new cycle, e.g. ORC, sCO2, CCGT (Closed Cycle Gas Turbine), Kalina cycle etc. and the heat can be used for electric power generation or for heating. The heat can be recuperated locally, i.e. the hot combustion products at the turbine outlets are used to pre-heat the working fluid before the heater inlet. Thus the amount of energy needed is decreased. In the case of an alternative cycle (e.g. due to better thermodynamic properties of an alternative working fluid for the low-potential heat), the combustion product (source) transfers the thermal energy to the working fluid of the external cycle in a heat exchanger [3], which increases the total power output and reduces the specific emissions.

Fig. 2 shows the classification of WHR sources according to Liu [3]. Alternative classification (Table 1) is suggested by Johnson [4] in his study according to the heat quality, i.e. the usable temperature gradient, but Jouraha [5] suggests slightly lower temperature values:

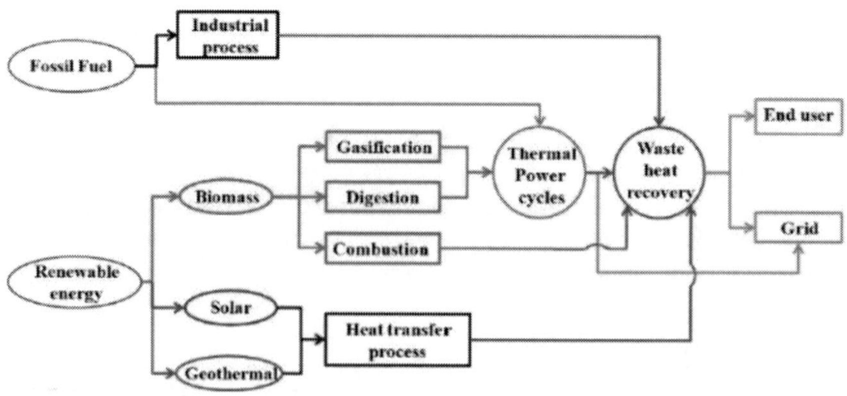

Fig. 2 WHR source classification. Figure from Liu [3]

Table 1. Temperature Classification of WHR sources.

High (>650 °C) or by [4] (>400°C)	Nickel refining furnace Steel electric arc furnace Glass melting furnace Coke oven Iron cupola Copper refining furnace
Medium (230-650°C) or by [4] (100-400°C)	Steam boiler exhaust Gas turbine exhaust Cement kiln Heat treating furnace Reciprocating engine exhaust
Low (<230°C) or by [4] (<100°C)	Cooling water from: Furnace doors Air compressors Internal combustion engine

Low temperature applications need different materials and types of heat exchangers. The reason is that corrosion can form at lower temperatures. The methods of WHR include the transfer of the waste heat into the cycle as a new external heat source. Again, a higher quality of heat produces a higher thermal efficiency of the entire cycle.
The heat recuperation consists mainly of several types of heat exchangers [5]:

- *Regenerators for air* (rotational, recuperative) – they use the waste heat of combustion products of low and middle potential sources (turbine outputs, recovery of furnaces or compression boilers)
- *Recuperation or economization* – heat exchangers with tubes or fins, which recover the middle and low potential heat for preheating of incoming fluid (e.g. water)
- *Waste heat boilers* – recovery of high potential heat to produce steam
- *Heat tubes* – heat pump which transfers the heat along a tube by using evaporation and condensation of working fluid (water, acetone, methanol, ammonia)
- *HRSG* – Heat Recovery Steam Generator – system of the previously mentioned exchangers for heating up, boiling and overheating a different working fluid (steam). Such a device is used in gas-steam turbines, where it uses the heat from the output of a BC turbine to the Rankin steam cycle
- *Device for direct transformation of heat to electricity* – thermocouples or thermophotovoltaic cells (not yet in mass use due to their low efficiency)

1.2 Organic Rankine Cycle (ORC)

The Organic Rankine cycle (ORC) is a closed cycle similar to the Rankine cycle with steam, but it uses an organic compound as a working fluid. ORC is suitable for middle and low potential waste heat sources, which are not suitable for the steam cycle, and for renewable energy sources such as solar, geothermal or biomass. [7]
Fig. 3 shows the T-s diagram and scheme of ORC cycle according to Colonna [7]. The following processes can be seen in the T-s diagram:

- **1-2** Pump – isentropic pressurizing of liquid fluid to higher temperature and pressure
- **2-3** Cold side of Recuperator – heat intake from expanded working fluid in process 5-6, it preheats the working fluid before the main heat input
- **3-4** Heat exchanger – isobaric heat input from primary WHR source, heating up, boiling and eventually overheating of the working fluid
- **4-5** Isentropic (according to Fig. 3 polytropic) expansion in turbine (or expander) to get mechanical energy. At the stage 5, the working fluid is typically hot enough to use the recuperation effectively
- **5-6** Hot side of recovery heat exchanger – the expanded fluid gives the heat to compressed fluid 2-3. Without this part, the heat would be lost in the condenser
- **6-1** Condenser (water, air) – isobaric-isothermal heat sink

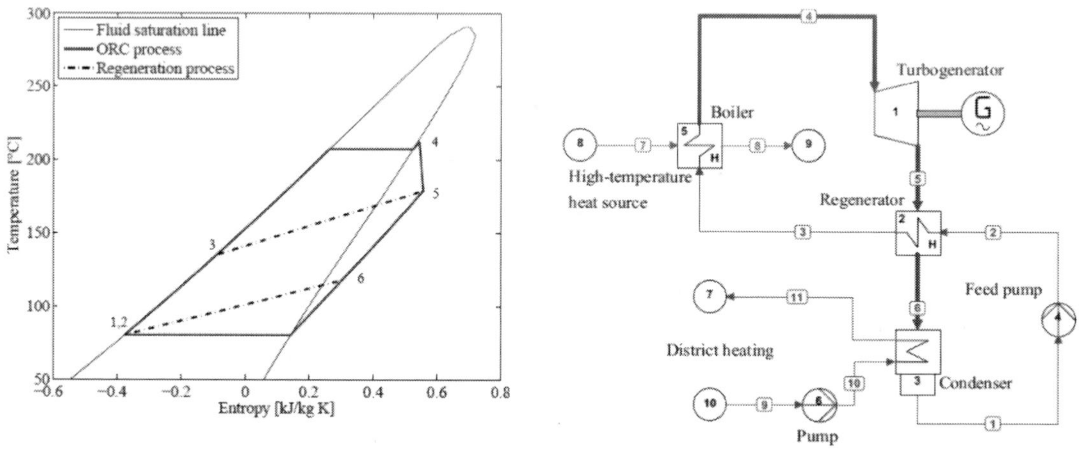

Fig. 3. (left) T-s diagram of ORC cycle; (right) scheme of ORC cycle. Both figures from the work of Colonna [7]

The organic fluid may lose stability or ignite due to high temperatures. To avoid this, a secondary heat transfer loop is added between the primary source and the boiler. This loop transfers the heat into the ORC cycle. A thermal oil is used as the transfer fluid; according to CoolProp tables [8], this oil is stable up to 400°C [7].

1.3 Supercritical CO_2 cycle (sCO_2)

The supercritical CO_2 cycle is a closed cycle similar to the Brayton or Rankine cycle, the working fluid is carbon dioxide (CO_2) and both heat input and output take place at supercritical conditions (in contrast to ORC, where the pressures, relative to the working fluid, are similar to conventional cycles). The critical temperature of CO_2 is 30.98 °C and the critical pressure is 73.8 bar. Thus this working fluid is suitable for low-potential heat sources. Fig. 4 shows the T-s diagram two variants of the CO_2 cycle: the supercritical Brayton cycle with all points above the critical one (red line in Fig. 4) and the classical Rankine cycle using phase transitions (blue line in Fig. 4). The maximum usable pressure is limited by the mechanical stability and it is usually 200 – 300 bar. Therefore the pressure ratios are expected to be low, $\pi \approx 2 - 3$, maximum $\pi \leq 4$. There is no phase transition during the cycle. The main advantage is the saving of compression work due to advantageous density dependence on the pressure near the critical point [9] [10] [11]. Fig. 5 shows the scheme and the cycle diagram according to Machionni [10].

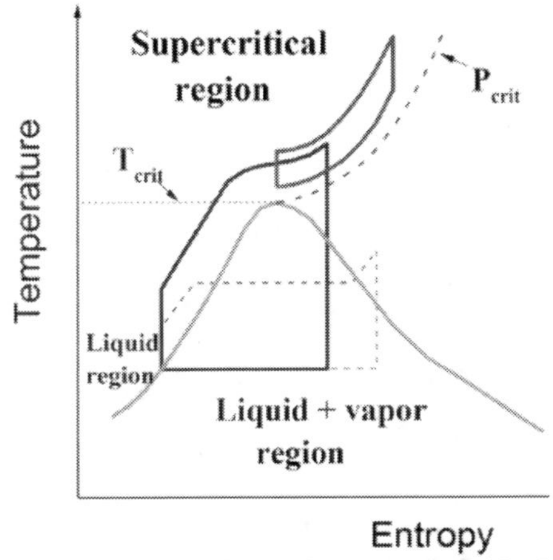

Fig. 4. Figure 4 T-s diagram of CO_2 supercritical Brayton (red) cycle and of classical CO_2 Rankine (blue) cycle. (from Wu [11])

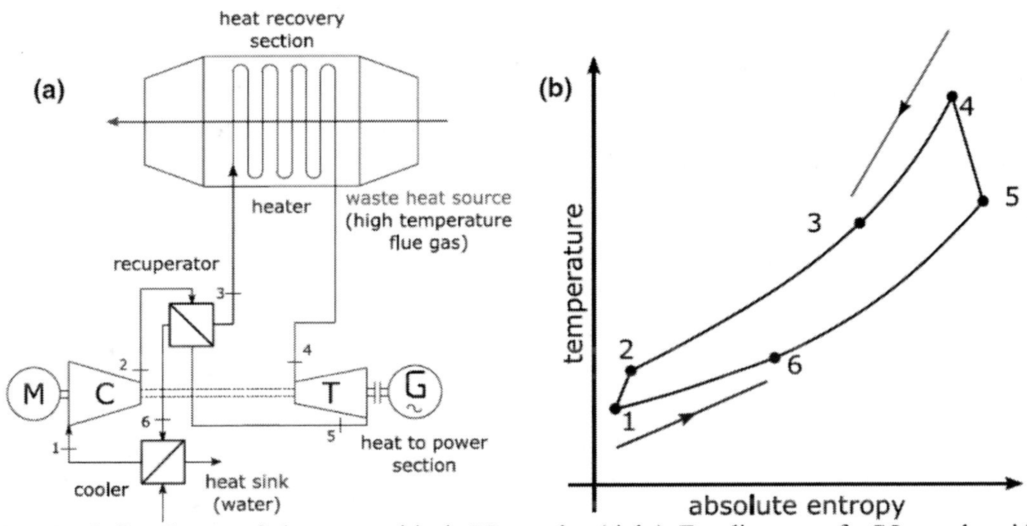

Fig. 5. (left) scheme of the supercritical CO_2 cycle, (right) T-s diagram of sCO_2 cycle with recuperation. (from Marchioni [10])

- **1–2** Compression
- **2–3** Cold side of Recuperator
- **3–4** Heat intake from primary source (WHR)
- **4–5** Expansion in a turbine or expander
- **5–6** Hot side of Recuperator – heat transfer to the colder compressed fluid (**2–3**)
- **6–1** Cooler – Heat sink from the cycle. The goal is to approach the critical point in order to save the compression work (the main advantage against the classical Brayton cycle).

2 Comparison and evaluation of both cycles

The main aim of this article is to compare the applicability of the sCO_2 and ORC cycles for the recovery of waste heat from a cement plant with the parameters listed in Table 2.

Table 2. Cement plant parameters considered for this study

Waste heat power (combustion products)	15 MW_t
Inlet temperature at primary exchanger	380 °C
Outlet temperature at primary exchanger	180 °C
Cooling air (atmosphere)	Temperature ~ 20 °C, humidity ~ 15 %
Power net frequency	50 Hz

2.1 Supercritical sCO_2 cycle

Fig. 6 shows the T-s diagram of a supercritical CO_2 cycle with recuperation by using the specific parameters for this study. Temperatures and pressures of the main points are listed in Table 3.

The red dashed line in Fig. 6 represents the process in the primary heat exchanger, which is heated by using the combustion products from the cement plant with a thermal power of 15 MW. The combustion products cool from 380 °C down to 180 °C. The approach point (the temperature difference between the cold and hot sides of the primary heat exchanger at the turbine inlet) is 20 °C; the pinch point (ΔT at exchanger inlet) is 8 °C. The blue dashed line represents the primary heat sink which is an air heat exchanger with an intake air temperature of 20 °C and outlet temperature difference of 10 °C (thus the outlet air temperature is 30 °C).

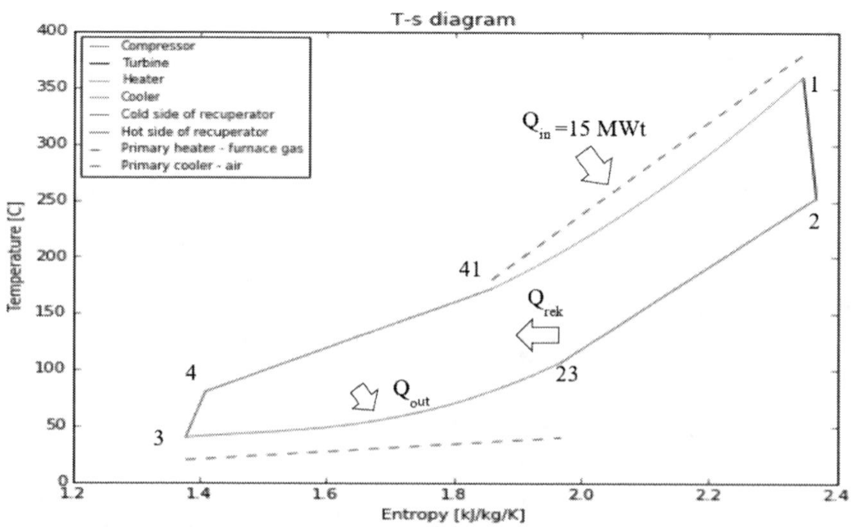

Fig. 6. T-s diagram of sCO_2 cycle with recuperation (green lines).

Fig. 7, Fig. 8 and Fig. 9 show the processes inside the exchangers: primary exchanger between combustion products and working fluid before turbine inlet (Fig. 7), recuperation exchanger between expanded and compressed working fluid (Fig. 8) and the heat sink to the ambient air (Fig. 9). The primary heat exchanger (Fig. 7) cools the combustion products from 380 °C down to 180 °C with a total heat power of 15 MWt to study the parameters (Table 2).

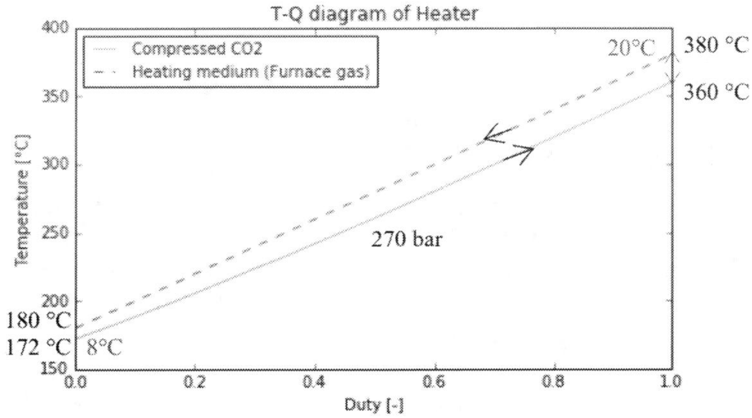

Fig. 7. T-Q diagram of primary and secondary heat input loop of the sCO_2 cycle.

Fig. 8 shows the T-Q diagram of the cold and hot side of the recuperator. The pressure on the hot side is 96.4 bar (i.e. supercritical). It is cooled from 252.6 °C down to 106.7 °C. The working fluid on the cold side is heated from 80.3 °C up to 172 °C at a pressure of 270 bar. The total heat power of this recuperator is 10.3 MW_t. The minimum temperature difference is 26.4 °C and it occurs at the cold end of the exchanger. Therefore, it is not necessary to solve the issues connected with a possible pinch point appearing near the centre of the exchanger due to the strong heat capacity variations near the critical point [9] [19]

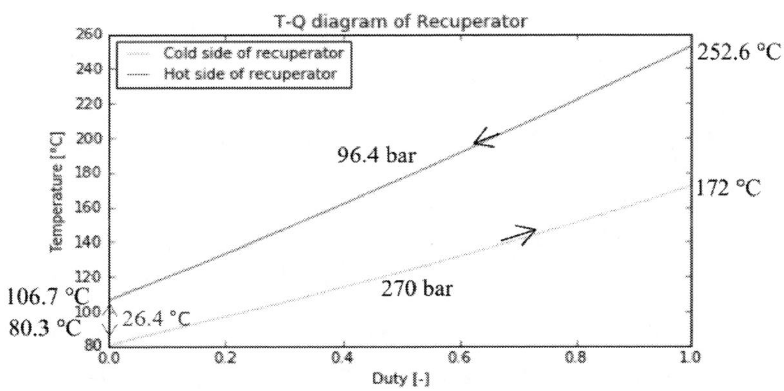

Fig. 8. T-Q diagram of sCO_2 cycle – cold/hot side of the Recuperator

Fig. 9 displays the temperature profile at the primary air heat exchanger. The supercritical CO_2 is cooled from 105.8 °C down to 40 °C (temperature close to critical). Note the temperature profile is strongly non-linear due to steep changes of heat capacity near the critical point. The pinch point appears at the cold end of the exchanger reaching a value of 20 °C. The exchanged heat power is 11.5 MWt.

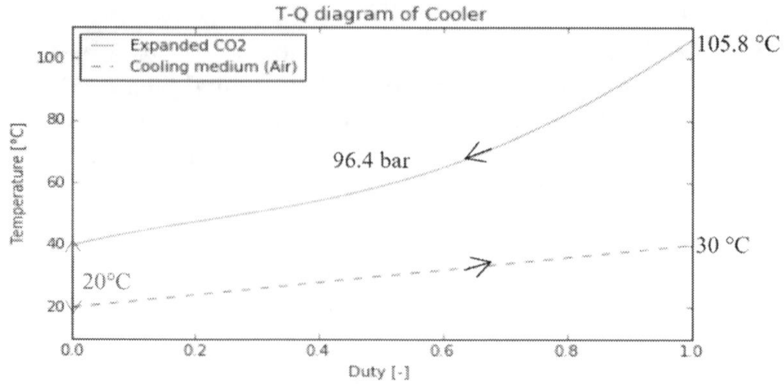

Fig. 9. T-Q diagram of sCO2 cycle – primary heat sink into the ambient air

Properties of sCO_2 during the cycle

By using CoolProp tables we are able to calculate the working fluid properties (e.g. enthalpy, entropy, density, temperature, pressure…) at any point of the cycle. Below, we publish the properties in the main points of the cycle (1 – 41).

Table 3. Selected working fluid properties at main points of the sCO_2 cycle

Point	Volumetric flow	Temperature	Pressure	Specific heat	Thermal conductivity	Density	Dynamic viscosity	Kinematic viscosity
	[m³/s]	[°C]	[bar]	[J/kgK]	[W/m/K]	[kg/m³]	[mPas]	[mm²/s]
1	0.258	360	270	1267	0.052	225.2	0.034	0.15
2	0.571	252.6	96.4	1150	0.039	101.9	0.026	0.26
23	0.338	106.7	96.4	1428	0.031	172.3	0.022	0.13
3	0.097	40	96.4	6989	0.074	596.7	0.045	0.07
4	0.082	80.3	270	2165	0.078	710.9	0.059	0.08
41	0.146	172	270	1625	0.051	398.8	0.035	0.09

Table 3 shows these properties along the cycle. The maximum deviations can be observed near the critical point (i.e. between the heat sink and compressor) [9] [10].Table 4 shows the integral characteristics, i.e. power and efficiencies of the entire cycle. The total generated power is 3.5 MWe and the thermal efficiency calculated by both alternative formulae (1) and (2) is 23.5 % without the losses inside the cycle. Without the recuperation, the efficiency would be below 20 %.

Note, that it is not discussed here the details of flow within the system. The flow near the corners in heat exchangers produces secondary flow [13], [6], which increases the losses on the one hand and on the other, it increases the heat transfer by producing larger-scale structures travelling through the flow [14]. Such a complicated analysis would also depend on the exact design of the components, because even small geometry modifications can have an effect on the turbulent flow [15], which can switch the flow into different turbulent states [16].

The reason for the relatively low thermal efficiency of the studied cycle is the low temperature of the input thermal power combined with the relatively high temperature of the heat sink (180 °C). If it were possible to cool to a lower temperature than the suggested one, the cycle efficiency would be better.

The total efficiency can be computed by using two independent formulae (1), (2), which may produce the same result:

$$\eta(1) = (\dot{Q} - Q_{\text{rem}})/\dot{Q} \qquad [-] \qquad (1)$$

$$\eta(2) = P_{\text{net}}/\dot{Q} \qquad [-] \qquad (2)$$

$\dot{Q} = 15$ MWt is thermal output supplied to the cycle from the flue gas from the cement plant, Q_{rem} is the heat sink calculated as a difference between enthalpies (h_2 enthalpy behind the turbine, h_3 enthalpy in the condenser, h_{23} enthalpy on the hot side of the recuperator) multiplied by the mass flow according to (4). P_{net} is the plant net power output according to (5) without considering losses. In this equation is h_1 enthalpy at the turbine inlet, h_{2id} ideal turbine outlet enthalpy and h_{4id} is enthalpy of ideal compression in the pump. The total mass flow is according to formula *(3)*: $\dot{m} = 58.171\ kg/s$.

$$\dot{m} = \dot{Q}/((h_1 - h_4) - (h_{41} - h_4)) \qquad [kg/s] \qquad (3)$$

$$Q_{\text{rem}} = ((h_2 - h_3) - (h_2 - h_{23})) \cdot \dot{m} \qquad [W] \qquad (4)$$

$$P_{\text{net}} = \Big(\big((h_1 - h_{2id}) \cdot Turbine_{eff}\big) - \big((h_{4id} - h_3)/Compressor_{eff}\big)\Big) \cdot \dot{m} \qquad [W] \qquad (5)$$

The turbine efficiency contains too many non-linear phenomena [12], therefore it has been estimated only roughly: $Turbine\_eff = 0.9$. Similarly, the compressor efficiency: $Compressor\_eff = 0.7$. The heat added by using the recuperator is calculated as (6), where h_41 is enthalpy on the cold side of the recuperator and h_4 is pump enthalpy.

$$Qrekup = (h_{41} - h_4) \cdot \dot{m} \qquad [W] \qquad (6)$$

Volumetric flow rate $\dot{m}v$ is calculated by using the total mass flow (eq. (3) and the density ρ in the investigated points of the cycle as (7):

$$\dot{m}v = \dot{m}/\rho \qquad [m^3/s] \qquad (7)$$

Table 4 Overview of the sCO_2 cycle powers and efficiency

Cycle analysis without auxiliary losses		
Power	[kW]	[MW]
Net power:	3521.91	3.5
Compressor input power:	2130.97	2.1
Turbine output power:	5652.88	5.7
Recuperator power:	10344.45	10.3
Heater power:	15000.00	15.0
Cooler power:	11478.09	11.5
Thermal efficiency	[%]	/
Cycle efficiency η(1):	23.48	/
Cycle efficiency η(2):	23.48	/

2.2 Organic Rankine Cycle (ORC)

The first step in ORC cycle sketch is to select a suitable working fluid. There is a long list of possible candidates containing hydrocarbons (alkanes, arenes), alcohols, ethers, siloxanes, and even more complicated cooling fluids, which can be used in a pure form or as a mixture. The criteria for the working fluid selection are according to Darvish [17] or Quoilin [18], [8]: Environmental sustainability, mainly the Ozone Depletion Potential (ODP) and the Global Warming Potential (GWP).

- Safety issues (carcinogenity, non-flammable, non-toxic, etc.)
- Critical point (temperature and pressure)
- Thermal stability
- High evaporation heat and vapour density
- Low viscosity
- High thermal conductivity
- Low freezing temperature
- Corrosive / anticorrosive properties
- Chemical compatibility with other materials (mainly bearings)
- Good availability and low cost

2.2.1 Working fluid selection

Table 5 gives an overview of some of the high-temperature working fluids studied for the ORC cycle. The choice has been done in order to minimize the ODP and GWP criteria. Another important issue is the back pressure, which would be an issue mainly for siloxanes (e.g. M2DM, MDM, alkanes, etc.). The best compounds with these parameters are the hydrocarbons, which have health and fire-safety issues. Therefore, very good sealing is required for the cycle and simultaneously strict operation rules, which will affect the final cost of the device. Except for the listed working fluids in the Table 5, even more fluids have been investigated, (n-octane, R365mfc, R245fa), whose critical points are not suitable for the current application.
Looking at Table 5, we find the best efficiency for hydrocarbons of **toluene**, **ethylbenzene** and siloxane **MM.** Siloxane MM (hexamethyldisiloxane) has lower efficiency, but it is safer and easier to use. The highest efficiency is achieved with toluene (similarly ethylbenzene or pure benzene). A similar effect is found when using arenes p-xylene and m-xylene respectively. Those arenes are not listed in the selection because they are not yet in commercial use and there is no experience with them. Other options are alkanes: cyclopentane, cyclohexane or n-heptane, which despite their lower critical temperatures offer efficiencies similar to hydrocarbons and are already used commercially. Ethanol is interesting as it does not need recuperation, which would lower investment costs. On the other hand, there is not yet enough experience with using ethanol as a working fluid for the ORC cycle. Siloxane MDM (octamethyltrisiloxane) displays similar efficiency to MM, only the condensation pressure is lower, which makes the construction of pumps and last stage blades more complicated than for compounds with higher condensation pressures. Under different working conditions, these siloxanes and alkanes are commonly used working fluids.[20] [21]

Table 5. Possible ORC working fluids. p1 is the pressure at the turbine inlet, p2 is at the turbine outlet (with losses), p3 is condensation pressure. "Eta without rec" signifies efficiency without recuperation, "Eta rec" is efficiency with recuperation without or with losses

Working fluids	Chem. group	Tcrit [°C]	pcrit [bar]	p1 [bar]	p2 [bar]	p3 [bar]	Eta without rec [%]	Eta rec [%]	Eta rec [%]
								without losses	w/ losses
Toluene (methylbenzene)	Arene	318.2	41.3	35.0	0.120	0.099	27.3	31.2	29.68
Ethylbenzene	Arene	344.0	36.2	14.2	0.045	0.037	-	30.0	28.62
MM	Siloxanes	245.6	19.4	15.0	0.170	0.142	-	25.6	24.25
p-Xylene (para-xylene)	Arene	342.9	35.3	19.0	0.040	0.034	27.2	31.0	29.51
m-Xylene (meta-xylene)	Arene	343.7	35.4	15.5	0.039	0.033	27.3	30.9	29.43
Benzene	Arene	289.0	49.1	44.0	0.360	0.298	-	29.9	28.41
CycloHexane	Cyklo alkanes	280.0	40.8	34.0	0.360	0.300	23.9	29.6	28.17
Ethanol (Bez rekuperace)	Alkohols	241.6	62.7	50.0	0.178	0.148	28.2	-	-
CycloPentane	Cyklo alkanes	239.0	45.7	36.0	1.059	0.879	23.3	27.5	26.13
n-heptane	Alkanes	267.0	27.4	18.0	0.185	0.153	21.1	27.6	26.26
MDM	Siloxanes	291.0	14.2	8.5	0.020	0.017	15.8	24.7	23.44

3 ORC cycle with toluene as a working fluid

The heat input and heat sink are creted by using secondary circuits with suitable working fluids. In the case of heat input, a closed loop with thermal oil (Therminol VP1) is used. This oil is heated by combustion products from a cement plant. The heat sink loop is filled with a mixture of 40 % ethylene glycol (MEG) and 60 % water. This non-freezing mixture is commonly used in power plants. The heat sink was designed with a secondary loop both for the easier calculation of the water condenser and for the reasons of costs and probably a very large area of the air exchanger in case the cooling was only air according to the specifications. The suggested cycle T-s diagram is shown in Fig. 10. Here, the process **1-2** shows expansion in turbine, **2-23** - Hot side of the recuperator, **23-3** - Condenser (23-23sat is the desuperheater – superheated steam cooler, integrated in the condenser), **3-4** - Pump, **4-41** - Cold side of the recuperator, **41-1** represents the heater, which takes heat from the secondary heating loop (red and orange dashed lines in Fig. 10). Main points (temperature and pressure) of the ORC are given in Table 5.

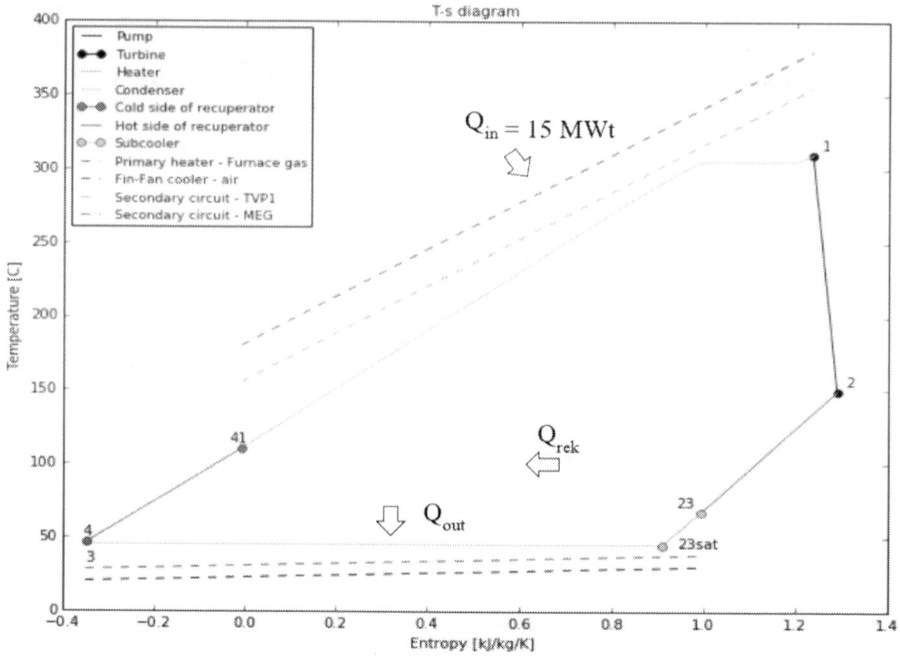

Fig. 10. T-s diagram of toluene ORC cycle with recuperation

The input temperature T_1 is calculated by using the temperature difference between the primary and secondary loop ($\Delta T = 25$ °C) and the approach point between the heated toluene and the heating loop with thermal oil ($\Delta T = 45$ °C). The turbine intake temperature is chosen in order to minimize overheating of the toluene (see Fig. 10), which increases the efficiency of the cycle. Temperature T_3 is calculated by using the temperature of the ambient air (20 °C) and the chosen TTD (Terminal Temperature Difference) of the condenser (7 °C) or the TTD between the primary and secondary cooling loop (8 °C) respectively. The temperature T_{41} is chosen in relation to the approach point and pressure p_1 and it is limited by the output temperature past the primary (and secondary) heat exchanger. This temperature directly affects the pinch points of the recuperator, primary and secondary heat exchanger as well. Pressure p_1 at the turbine inlet is chosen according to the conclusions of works [22] and [23] in order to keep away from the critical point (NRB – Near-critical Region Boundary). Condensation pressure p_3 corresponds to the condensation temperature T_3.

T-Q diagrams of ORC cycle with toluene as a working fluid

We show the T-Q diagrams similarly as in the case of sCO_2 in the previous section. These diagrams allow us to detect pinch/approach points and help to better design the heat exchangers.

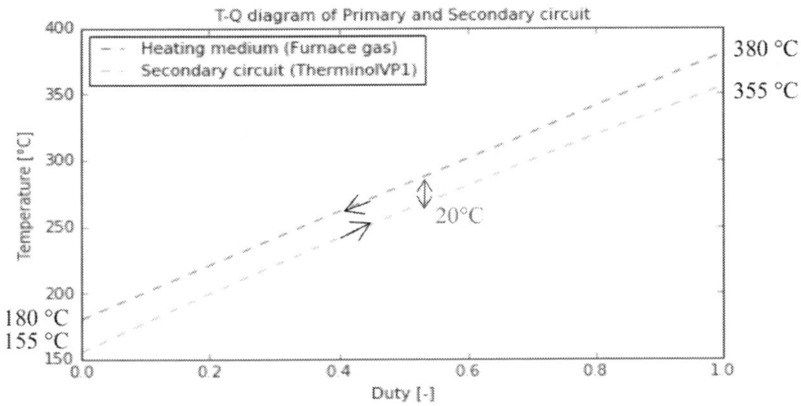

Fig. 11. T-Q diagram of primary and secondary heat input loop of the ORC cycle.

Fig. 11 shows the cooling of the combustion products from the cement furnace and heating of the thermal oil in the primary loop. The temperature difference is chosen to be 25 °C to keep the temperature gradient for heat transfer. The minimum ΔT (the pinch point) appears in the middle of the exchanger, but the heat transfer is effective enough even with this ΔT.

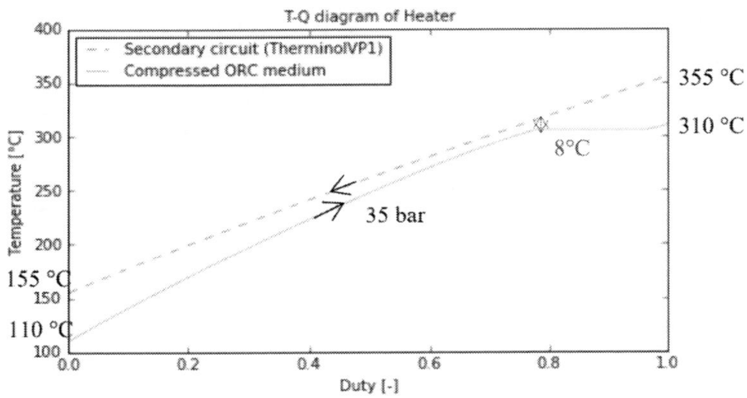

Fig. 12. T-Q diagram of secondary heating loop (Therminol VP1/toluene)

The T-Q diagram in Fig. 12 shows the heating of toluene in the secondary loop by thermal oil (dashed line). The approach point between the working fluids has been optimized to ΔT = 45 °C, thus the admission temperature T_1 is 310 °C and the pinch point has been calculated to be 8 °C. To the right of the pinch point can be observed the isothermal evaporating curve followed by very small overheating of the toluene. Generally, it is better to overheat the organic fluids less (mainly the dry types). First, overheating affects the admission temperature T_1, second, these materials may not be stable enough at higher temperatures and, third, lower overheating increases the efficiency (mainly of the dry materials).

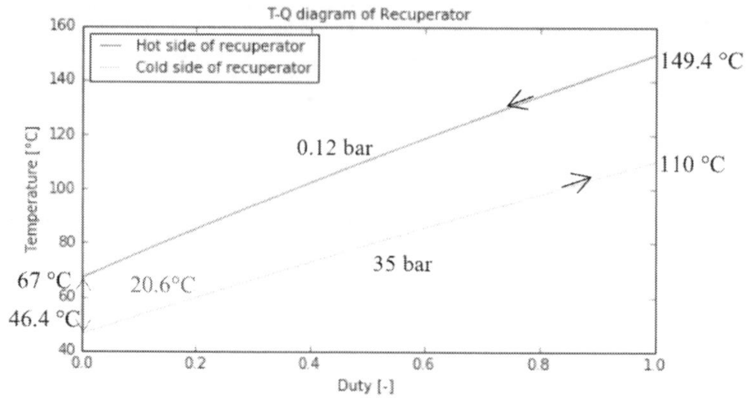

Fig. 13. T-Q diagram of recuperator

Fig. 13 shows the recuperation, which cools the expanded hot toluene and heats the compressed cool toluene before the main heater. The hot toluene cools from 149.4 °C down to 67 °C. Compressed cold toluene heats from 46.4 °C up to 110 °C. The pinch point of $\Delta T = 20.6$ °C appears at the cold end of the exchanger.

The heat sink is formed by a condenser (isothermal light blue line in Fig. 14), the heat transfers into the MEG mixture in a secondary loop, which is cooled by air. The toluene cools from 67 °C down to 45 °C, at which it condenses, heating the MEG from 28 °C to 38 °C with a pinch point of 7 °C. The temperature difference at the primary and secondary exchanger is kept constant at 8 °C and the cooling air heats from 20 °C to 30 °C.

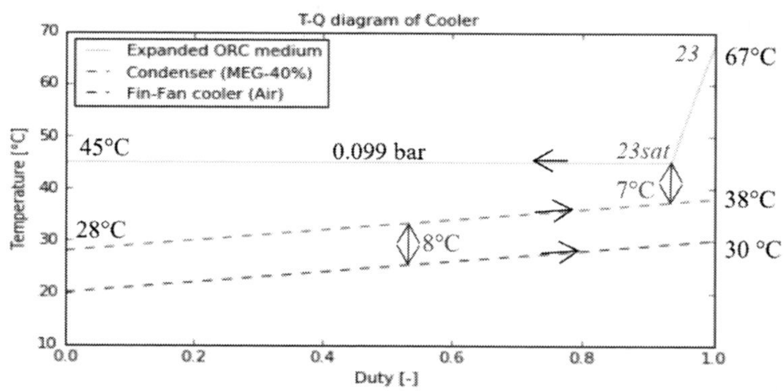

Fig. 14. T-Q diagram of ORC cycle – primary (air) and secondary (MEG) heat sink loop

The minimum temperature differences in the exchangers are summarized in Table 6 for clarity:

Table 6. Minimum temperature differences along the cycle

Minimum temperature difference [°C]	
Minimum dT of Primary heater:	19.89
Minimum dT of Secondary heater:	8.25
Minimum dT of Recuperator:	20.62
Minimum dT of Primary Cooler:	8.00
Minimum dT of Secondary Cooler:	7.66

Properties of ORC with toluene during the cycle

Table 7 summarizes the properties of the working fluid (toluene) at the main points of the cycle. The mass and volumetric flow rates are calculated similarly as for sCO_2 cycle (eq. (7)). The mass flow rate is 24.07 kg/s.

Table 7. Selected working fluid properties at the main points of the cycle

Point	Volumetric flow	Temperature	Pressure	Specific heat	Thermal conductivity	Density	Dynamic viscosity	Kinematic viscosity
	[m^3/s]	[°C]	[bar]	[J/kg/K]	[W/m/K]	[kg/m^3]	[mPas]	[mm^2/s]
1	0.180	310	35	1599	0.0517	130.6	0.0162	0.124
2	76.520	149	0.12	1297	0.022	0.315	0.0097	30.97
23	74.280	67	0.12	1213	0.014	0.324	0.0079	24.36
23sat	0.029	45	0.12	1767	0.012	0.347	0.0074	21.32
3	0.029	45	0.099	1767	0.125	843	0.4408	0.52
4	0.028	46	35	1991	0.126	845	0.4475	0.53
41	0.030	110	35	3590	0.109	784	0.2580	0.33

The properties of toluene at the displayed points is calculated according to CoolProp tables with pressure and temperature or dryness as the control parameters. The heat capacity of toluene near condensation is much lower than that of water. The thermal conductivity and viscosity of toluene reach lower values near the condensation point than water. The density at point 23sat (saturation) is 5× smaller than the density of steam, while its density at point 3 (fully condensed) is close to the density of water. From the comparison of the densities during the ORC and CO_2 cycle, it can be seen that the density of sCO_2 is several times higher, because CO_2 is single-phase and its behaviour is close to the properties of water, while the ORC medium undergoes heating and evaporation (possibly overheating). It follows from the properties of the working fluids, that exchangers and a suitable working medium must be thoroughly designed for a WHR power plant.

Thermal efficiency and power

Table 8 shows the total net power of the ORC cycle and the powers of the individual components. The pump has much smaller input power than the sCO_2 cycle. The cooler power is separated into the Steam superheat cooler power and the Steam condenser power. The thermal efficiency of the entire cycle is calculated according to formulae (1) and (2); both methods return the same value, as expected.
Table 9 shows the values of the estimated efficiencies of the components and their impact on the total power of the cycle and its components. The turbine power is estimated to decrease to 4.6 MW_e (4.8 MW_e without losses), the pump input power increases slightly when considering losses. The net power plant power decreases to 4.5 MW_e (4.7 MW_e without losses).

Table 8. The powers of ORC cycle components without losses.

Cycle anylysis without auxiliary losses		
Power	[kW]	[MW]
Plant net power output:	4680.6	4.7
Heater power:	15000.0	15.0
Turbine output power:	4804.9	4.8
Pump input power:	124.3	0.1
Recuperator power:	2873.9	2.9
Cooler power:	10319.4	10.3
Steam superheat cooler power:	665.4	0.7
Steam condenser power:	9653.9	9.7
Thermal efficiency	[%]	
Cycle efficiency (1):	31.2	/
Cycle efficiency (Basic cycle):	31.2	/

Table 9. Power and efficiency for the ORC cycle with losses

Cycle anylysis including auxiliary losses		
Turbine generator and gearbox efficiency:	0.973 [1]	97.3%
Pump electrical motor efficiency:	0.985 [1]	98.5%
Mechanical losses:	100.0 kW	0.1 MW
Power	[kWe]	[MWe]
Turbine output power at generator terminals:	4577.9	4.6
Pump power electrical consumption:	126.2	0.1
Plant net power output:	4451.7	4.5
Thermal efficiency	[%]	
Cycle efficiency (Cycle with losses):	29.68%	/

4 Conclusion

The primary aim of this work was the analysis and evaluation of two closed cycles for the energy use of waste heat from a cement plant (15 MW_t in flue gas cooled from 380 °C to 180 °C, cooling to ambient air at 20 °C): supercritical CO_2 and ORC. Based on the results of the calculations, the ORC cycle was chosen for further calculations of the components. The difficulty of the ORC cycle was selecting a suitable working medium. Calculations are performed for many working fluids (see Table 5) and are evaluated primarily according to their properties and achieved efficiency. The most suitable media with high efficiency are MM (hexamethyldisiloxane), ethylbenzene and toluene, of which toluene was chosen after critical evaluation, and whose properties across the cycle are summarized in Table 7. For the ORC cycle, calculations are performed first without losses and then including losses, e.g. mechanical and with defined efficiencies of the gearbox, generator and electric motor of the pump, see Table 9. The generated electric power of the ORC unit reaches 4.5 MWe with a cycle thermal efficiency (including included losses) of 29.68%. The individual parts will be designed in detail in the future.
The second aim of this work is to prove that the author is ready to achieve a master degree in mechanical engineering.

Acknowledgements
I thank my supervisor Ing. Jiří Kučera, Ph.D. for significant help with data analysis and discussions. I thank my former supervisor RNDr. Daniel Duda for help translating the article into English.

References

1. T. Adefarati a R. Bansal, "Energizing Renewable Energy Systems and Distribution Generation", v Pathways to a Smarter Power System, 1. editor, Yildiz Technical University, Turkey: Academic Press (2019), https://doi.org/10.1016/C2017-0-03015-X
2. A. Razak, Industrial Gas Turbines, UK: Woodhead Publishing (2007), ISBN 978-1-84569-205-6
3. L. Liu, "Supercritical Carbon Dioxide(s-CO2) Power Cycle for Waste Heat Recovery: A Review from Thermodynamic Perspective", Processes, vol. 8, n. 1461 (2020), https://doi.org/10.3390/pr8111461
4. I. Johnson, "Waste Heat Recovery. Technology and Opportunities in U.S. Industry",U.S. Department of Energy (2008), https://doi.org/10.2172/1218716
5. H. Jouhara, Waste heat recovery technologies and applications, 6 editor, London: Thermal Science and Engineering Progress, pp. 268-289 (2018), https://doi.org/10.1016/j.tsep.2018.04.017
6. P. Boruta, T. Bujok, Ł. Mika a K. Sztekler, "Analysis of Designs of Heat Exchangers Used in Adsorption Chillers", Energies, vol. 14, n. 23, p. 8038 (2021), DOI:10.3390/en14238038
7. P. Colonna, "Organic Rankine Cycle Power Systems: From the Concept to Current Technology, Applications, and an Outlook to the Future", Journal of Engineering for Gas Turbines and Power, vol. 137, n. 10 (2015), https://doi.org/10.1115/1.4029884
8. S. Quoilin, "Techno-economic survey of Organic Rankine Cycle (ORC) systems", Renewable and Sustainable Energy Reviews, vol. 22 (2013), https://doi.org/10.1016/j.rser.2013.01.028
9. K. Brun, Fundamentals and Applications of Supercritical Carbon Dioxide (sCO_2) Based Power Cycles, Sawston,UK: Woodhead Publishing (2017), ISBN 978-0-08-100804-1
10. M. Marchionni, "Review of supercritical carbon dioxide (sCO2) technologies for high-grade waste heat to power conversion", SN Applied Sciences, n. 611 (2020), https://doi.org/10.1007/s42452-020-2116-6
11. P. Wu, "A review of research and development of supercritical carbon dioxide Brayton cycle technology in nuclear engineering applications", Nuclear Engineering and Design, vol. 368, (2020), https://doi.org/10.1016/j.nucengdes.2020.110767
12. D. Duda, T. Jelínek, P. Milčák, M. Němec, V. Uruba, V. Yanovych a P. Žitek, "Experimental Investigation of the Unsteady Stator/Rotor Wake Characteristics Downstream of an Axial Air Turbine", International Journal of Turbomachinery, Propulsion and Power, vol. 6, n. 3, p. 22 (2021), https://doi.org/10.3390/ijtpp6030022
13. D. Duda, J. Bém, V. Yanovych, P. Pavlíček a V. Uruba, "Secondary flow of second kind in a short channel observed by PIV", European Journal of Mechanics, B/Fluids, vol. 79, pp. 444-453 (2020), https://doi.org/10.1016/j.euromechflu.2019.10.005

14. D. Duda, V. Yanovych a V. Uruba, "An experimental study of turbulent mixing in channel flow past a grid", Processes, vol. 8, n. 11, pp. 1-17 (2020), https://doi.org/10.3390/pr8111355
15. V. Yanovych, D. Duda, V. Uruba a P. Antoš, "Anisotropy of turbulent flow behind an asymmetric airfoil", SN Appl. Sci., vol. 3, p. 885 (2021), DOI:10.1007/s42452-021-04872-2
16. D. Duda, V. Yanovych, V. Tsymbalyuk a V. Uruba, "Effect of Manufacturing Inaccuracies on the Wake Past Asymmetric Airfoil by PIV", Energies, vol. 15, n. 3, p. 1227 (2022), https://doi.org/10.3390/en15031227
17. K. Darvish, "Selection of Optimum Working Fluid for Organic Rankine Cycles by Exergy and Exergy-Economic Analyses", Sustainability, vol. 7, n. 11, pp. 15362-15383 (2015), https://doi.org/10.3390/su71115362
18. S. Quoilin, Experimental Study and Modeling of a Low Temperature Rankine Cycle for Small Scale Cogeneration, Lutych: University of Liege - Aerospace and mechanical engineering (2007)
19. White, Martin T. Review of supercritical CO2 technologies and systems for power generation. Applied Thermal Engineering., vol. 185 (2021), https://doi.org/10.1016/j.applthermaleng.2020.116447
20. E. Macchi a M. Astolfi, Organic Rankine Cycle (ORC) Power Systems, 1 editor, Milano: Woodhead Publishing, p. 698 (2016), ISBN 9780081005101
21. T. Tartière, "A World Overview of the Organic Rankine Cycle Market", Energy Procedia - 4th International Seminar on ORC Power Systems, vol. 129 (2017), https://doi.org/10.1016/j.egypro.2017.09.159
22. Y. P. Wang, "Performance Analysis of Near-Critical and Subcritical Organic Rankine Cycle", Applied Mechanics and Materials, vol. 1, site 2448-453 (2013), https://doi.org/10.4028/www.scientific.net/AMM.448-453.3270
23. L. Pan, "Performance analysis in near-critical conditions of Organic Rankine Cycle", 7th Biennial International Workshop "Advances in Energy Studies", vol. 37, n. 1, (2012), https://doi.org/10.1016/j.energy.2011.11.033
24. A. M. Ahmed a A. R. Imre, "The effect of recuperator on the efficiency of ORC and TFC with very dry working fluid", MATEC Web of Conferences, vol. 345, p. 00012 (2021), https://doi.org/10.1051/matecconf/202134500012

MATEC Web of Conferences **367**, 00017 (2022) https://doi.org/10.1051/matecconf/202236700017
Power System Engineering 2022

Vorticity statistics in the near wake of asymmetric prismatic airfoil NACA 64-618 at negative angle of attack –10° at Reynolds numbers 1.6 · 10^4 and 1.6 · 10^5 in distance 0.0 – 0.4 × chord past trailing edge measured by Particle Image Velocimetry

Jan Narovec[1], *Daniel* Duda[1*], *Vít* Horáček[1], *Tetjana* Tomášková[1], *Václav* Uruba[1,2], *Vitalii* Yanovych[1,2]

[1]Faculty of Mechanical Engineering, University of West Bohemia in Pilsen, Univerzitní 22, 306 14, Pilsen, Czech Republic
[2]Institute of Thermomechanics, Czech Academy of Sciences, Dolejškova 5, 180 00, Prague, Czech Republic

Abstract. The airfoil NACA 64-618 is realized by using two levels of quality: first the theoretical profile is printed on a 3D printer in a direct, "naive" way, this geometry has been scanned by using 3D scanner and based on the deviations, a better model has been processed. The flow within the turbulent wake is measured by using Particle Image Velocimetry (PIV) technique at two velocities separated by one order of magnitude.

1 Introduction

Vorticity is a local derivative invariant of velocity field $\vec{u}$ defined as [1]

$$\vec{\omega} = \nabla \times \vec{u}\,, \qquad \text{i.e. } \omega_i = \epsilon_{ijk}\frac{\partial u_j}{\partial x_k} \tag{1}$$

where ∇ is spatial gradient vector operator, × is vector product, therefore vorticity is, correctly speaking, *pseudovector*, thus in two dimensions, it is a scalar (1D) field, while in four dimensions, it would be a second order tensor with 6 independent components. Only in 3D space, number of its independent components matches the dimension. The more exact expression calls the total antisymmetric *Levi-Civita tensor* ϵ_{ijk}. Vorticity is invariant in respect to Galilean transformation and it can be used to eliminated pressure from the Navier-Stokes equations (NSE), however, it is not possible to solve NSE with vorticity only, as the vorticity itself is advected by velocity field [1].

The term "vorticity" is little bit misleading, as presence of vorticity does not proof existence of vortex in a such coordinate. E.g. shear flow displays nonzero vorticity, although there is no vortex and oppositely, there is zero vorticity in a vortex envelope.

* Corresponding author: dudad@kke.zcu.cz

2 Experimental setup

2.1 Airfoil printing and scanning

A pair of airfoil realizations is compared here. The first one is created by "naive" approach: download coordinates from public database [2] and print by using a commercial 3D printer Prusa Mk2.5 from PLA [3]. The GOM Atos 3D scanner [4] [5] [6] is used to measure the shape and to compare with the design. The largest discrepancy is found at the thin trailing edge, which is not printed correctly and is shorter. The second, "processed", variant gains from the errors observed in the first case, thus its trailing edge is slightly remodeled and the surface is smoother by using sandpaper. Additionally, the surface roughness R_q and roughness high R_t is measured by using device Taylor Hobson Sulfronic Duo.

Tab. 1 Basic parameters the studied airfoil realizations compared with the ideal design

	Chord [mm]	Thickness [mm]	Roughness high R_t [μm]	Roughness R_q [μm]
Design	80.00	14.36	0.0	0.0
Naïve	77.01	14.34	70.5	11.35
Processed	80.08	14.47	60.1	9.77

2.2 PIV measurement system

Particle Image Velocimetry (PIV) [7] system at the University of West Bohemia (UWB) is a standard one supplied by the Dantec company. It consists of pair of FlowSense MkII cameras with resolution of 4 MPix, solid-state double pulse laser New Wave Solo with emitting energy of 0.5 J during 5 ns pulse (i.e. power during pulse is around 100 MW). The tracer particles are very small droplets made by fog generator Safex by using liquid of unknown composition. The dimension of the droplets has not been measured, but its usage is proved by wide experience with this method [7]. The system takes a pair of photos separated by time 100 μs (for Re = $1.6 \cdot 10^4$) or 10 μs (for Re = $1.6 \cdot 10^5$) respectively. The repeating frequency of these pairs is 7.4 Hz. Data acquisition is performed by Dantec Dynamic Studio. 2D velocity vector field is calculated by using the "Adaptive PIV" function built in the Dantec Dynamic Studio. The post-processing is made by custom-made software.

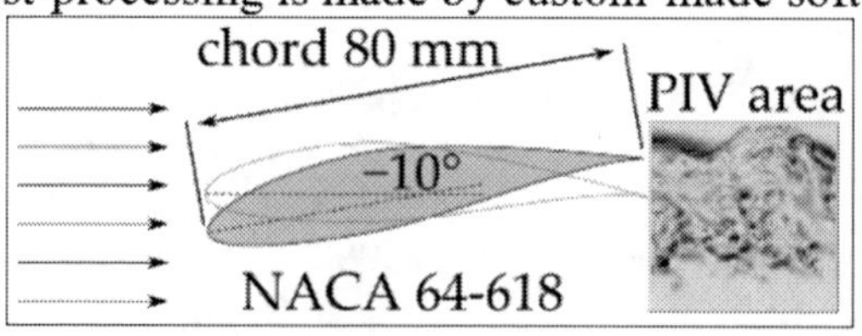

Fig. 1. Sketch of the airfoil profile with measured PIV area.

3 Results

3.1 Statistical moments of vorticity

Figure 2 shows the example of few instantaneous data and the mean vorticities in the four discussed cases. The airfoil realizations are denoted "naive", or case A in our work [8], and "processed", or case C in [8]. We see that both realization display similar topology at same Reynolds numbers, while the Reynolds number (controlled via velocity) has much stronger effect on the mean wake shape. At lower Re, the flow is separated at the leading edge forming a massive wake structure covering the entire Field of View (FoV). At higher Re, the flow is attached, the wake is much narrower, but still a strong asymmetry between pressure (bottom in figures) and suction sides can be observed.

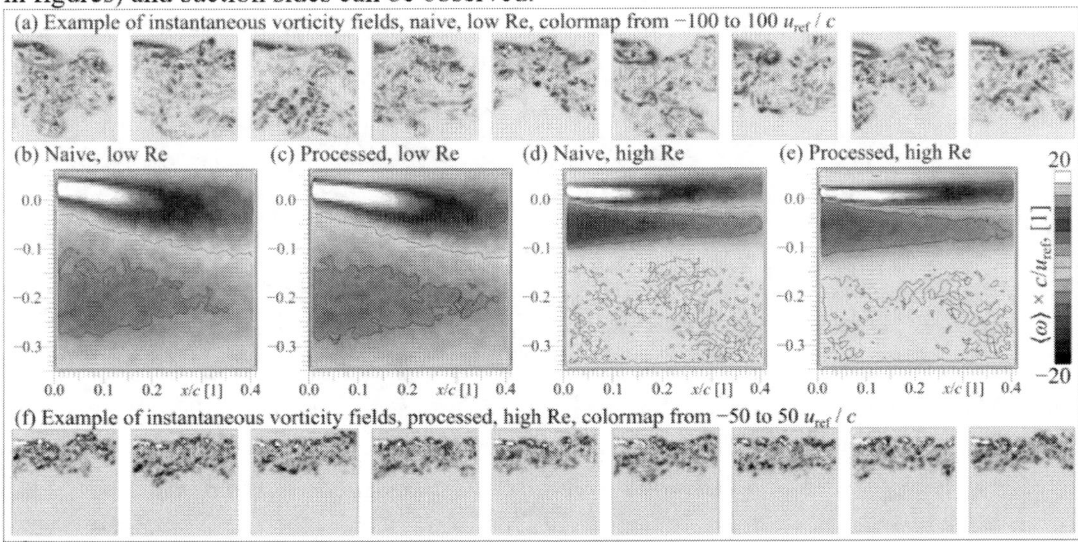

Fig. 2. Vorticity maps. Top line (a) shows examples of instantaneous in-plane vorticity maps measured past the "naive" realization of airfoil, the better realization is too similar to observe differences in instantaneous fields. Middle line (b – e) shows the mean vorticity maps in all discussed cases, low Re = · 10^4, high Re = 1.6 · 10^5. Interval between isolines is 5. The bottom line (f) shows examples of instantaneous vorticities past the better processed airfoil at higher Re.

The standard deviation field is calculated as

$$\sigma_\omega(\vec{x}) = \sqrt{\langle \omega(\vec{x})^2 \rangle - \langle \omega(\vec{x}) \rangle^2} \tag{2}$$

where $\langle \omega \rangle$ plays for ensemble average over the measured snapshots (whose number is 650). The second power of standard deviation is sometimes referred as *Enstrophy*, but the correct enstrophy needs all three component of vorticity variance, while the 2D PIV measurement can offer only the component perpendicular to the measured plane. Figure 3 shows a pair of high-activity strips in the airfoil wake – top one behind the suction side and the bottom one behind the pressure side, where the bottom one is weaker and more diffused. At higher Re, weaker fluctuations can be observed behind the processed airfoil (Fig. 3c vs. Fig. 3d).

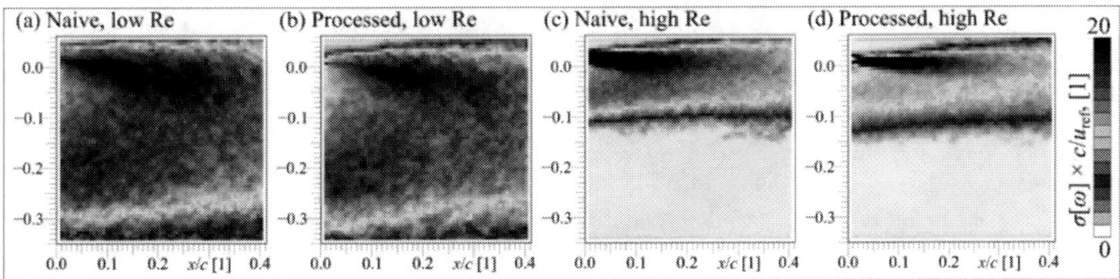

Fig. 3. Maps of standard deviation of vorticity past both discussed airfoil realizations and at both discussed Reynolds numbers.

The *skewness* (or *asymmetry coefficient* or *3rd statistical moment*, česky *koeficient šikmosti*) is calculated as

$$S_\omega(\vec{x}) = \frac{\langle \omega'^3(\vec{x}) \rangle}{\sigma_\omega^3(\vec{x})} \tag{3}$$

where the prime denotes fluctuation components, i.e. $\omega'(\vec{x}) = \omega(\vec{x}) - \langle\omega(\vec{x})\rangle$. In the case of symmetric distribution, the skewness is zero. The positive skewness means, that there is lower number of stronger positive events, while the negative events are weaker and of higher amount. The negative skewness does oppositely. In the case of wake, the vortices can be ejected into the corresponding boundary more easily than to the other side. Therefore, the wake is bounded by the sharp cliff of skewness in the most distant area, in which a vortex can be ejected. This property can be used as a wake boundary [9].

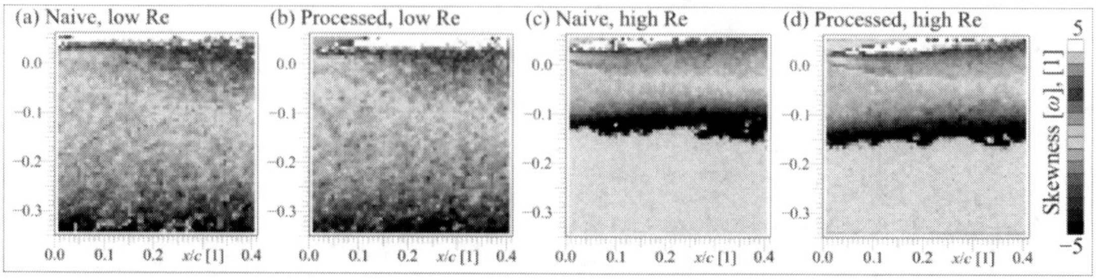

Fig. 4. Maps of skewness (i.e. third statistical moment, sometimes asymmetry coefficient) of vorticity past both discussed airfoil realizations and at both discussed Reynolds numbers.

The *flatness* (or *4th statistical moment* or *kurtosis*, česky *koeficient špičatosti*) is

$$F_\omega(\vec{x}) = \frac{\langle\omega'^4(\vec{x})\rangle}{\sigma_\omega^4(\vec{x})} \tag{4}$$

Its value is always positive and it can be proven, that the flatness of Gauss distribution is equal to 3 (pure red color in Fig. 5), while the flatness of any distribution with polynomial tails diverges [10]. Physical interpretation is, that higher flatness signifies occurrence of strong rare events. Typically, at the wake boundary, there is generally quiet flow, which is time-to-time visited by a vortex from the shear layer region. Such events increase the standard deviation only little, while its effect on flatness is strong. In the central part of the wake, the vorticity flatness is slightly higher than 3. Its expected value for homogenous turbulence is not known, but it has been observed, that in grid-generated turbulence, the vorticity flatness lies between 3 and 4.5 with increase with increasing Re [11], but current results suggest higher flatness at lower Re; however it is questionable, if it can be compared, when the flow state is different.

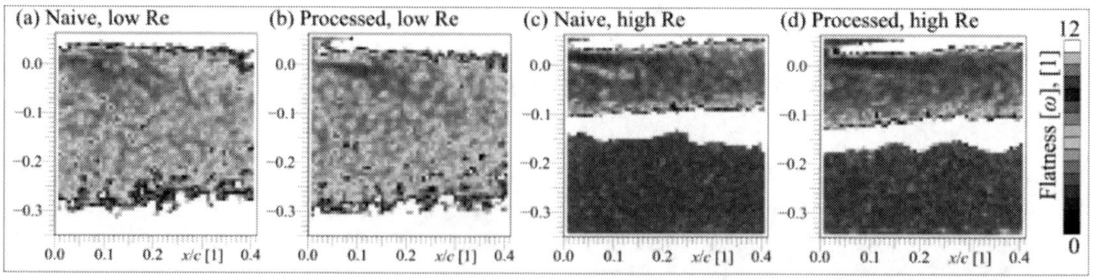

Fig. 5. Maps of flatness of vorticity field convoluted by Gaussian function of width $\sigma = 0.007c$ past both discussed airfoil realizations and at both discussed Reynolds numbers. The pure red corresponds to the value of 3.

3.2 Spectral dependence

Alexakis & Biferale [12] discuss the spectral flux of energy and enstrophy. Unfortunately, the flux is invisible by using our equipment with low time resolution. The change of sign in the energy and enstrophy flux correspond to the transition between direct and indirect cascade at the integral lengthscale. The current studied flow is 3D, but the geometry is prismatic, hence there can be expected 2D behaviour (of course, it has been proven by Wilkinson [13] and others [14] in nineteens, that cylinder perpendicular to the flow produces fully 3D flow since very small Reynolds numbers). However, it has been observed by Solís-Gallego [15] and later confirmed by us [16], that the flow behind prismatic airfoil exhibits two slope energy spectrum similarly as some 2D flows do (e.g. the atmosphere of Earth and Venus [17]). The best we can do with our current data in this direction is to explore the spectral dependence of vorticity statistics, mainly its standard deviation, which relates to enstrophy.

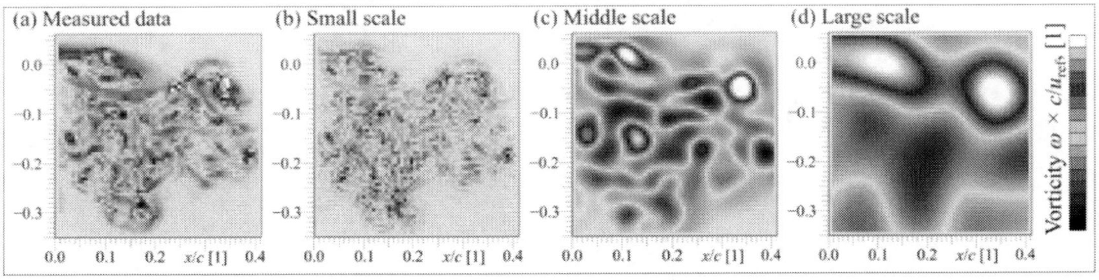

Fig. 6. Example of single instantaneous snapshot (a) decomposed into three length-scale intervals (b-d). The "small scale" corresponds to lengths 0.006 – 0.010 c, "middle scale" means 0.020 – 0.027 c and the "large scale" denotes fluctuations of sizes 0.05 – 0.08 c. the limits of colormap are different for each panel adapting the actual values.

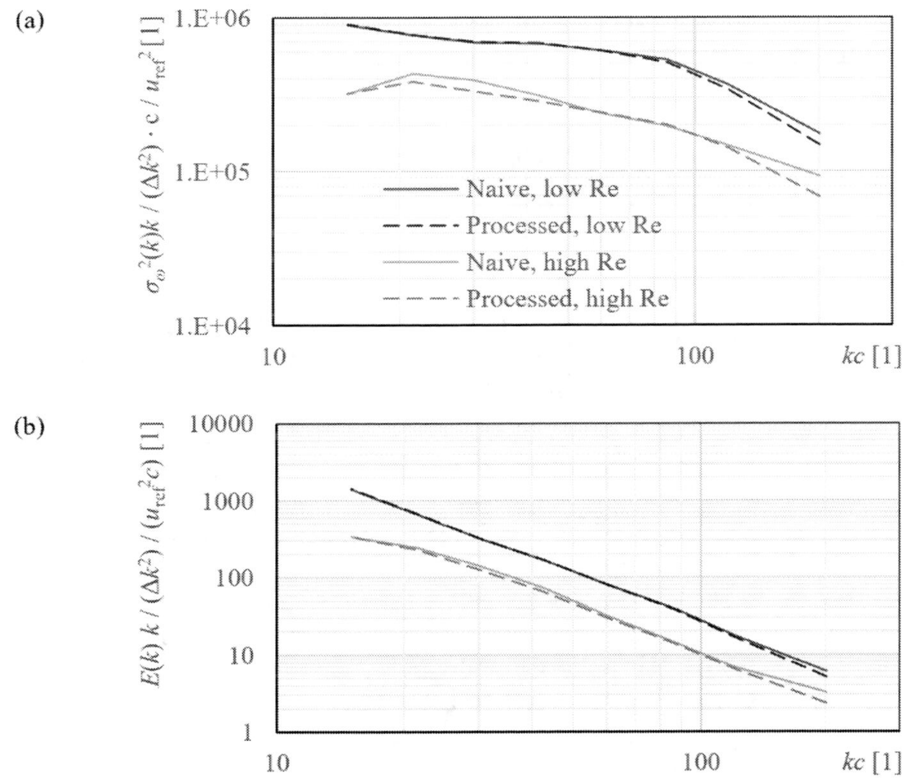

(c)

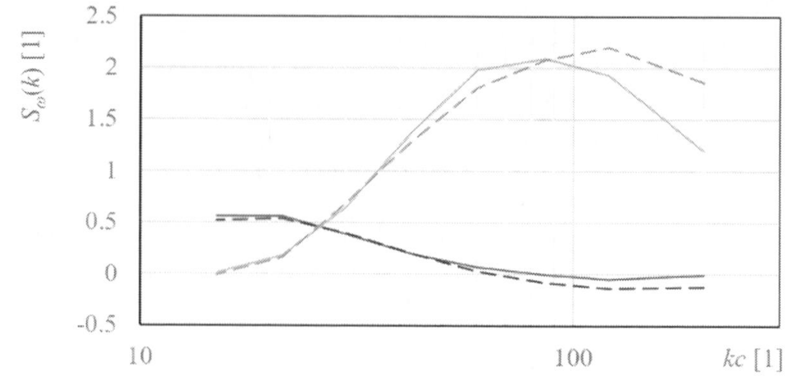

(d)

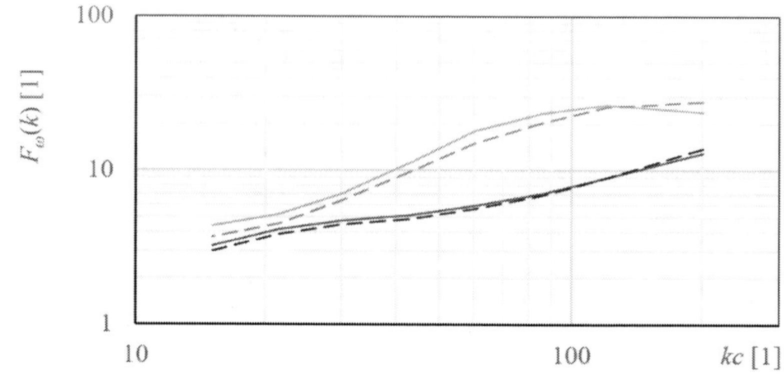

Fig. 7. Spectral dependence of selected quantities. (a): vorticity variance (σ^2) as a function of wavenumber k. (b) power spectral density. (c) vorticity skewness S_ω as a function of wavenumber. (d): vorticity flatness F_ω. Solid red line: naïve at lower Re, dashed blue: processed at lower Re, solid yellow: naïve at higher Re, dashed azure: processed at higher Re.

The decomposition to different lenghtscales is done by using consecutive convolutions with band-pass filters

$$P[\sigma_h, \sigma_l](\xi) = \frac{\exp\left(-\frac{\xi^2}{2\sigma_l}\right)}{2\pi\sigma_l} - \frac{\exp\left(-\frac{\xi^2}{2\sigma_h}\right)}{2\pi\sigma_h} \quad (5)$$

with σ_l and σ_h spanning the range of PIV data (i.e. from the single grid point 0.5 mm to the field of view 34 mm). More details about this method can be found in our work [18]. Figure 6 shows this decomposition to three lengthscales.

The spectral dependence of vorticity fluctuations is plotted in Fig. 7 top left. It is normalized in a similar manner as the standard power spectral densities (Fig. 7 top right), i.e. the variance (second power of σ) at probed wavenumber k is multiplied by k and divided by the square of the band width Δk, more details in [18]. There is an increasing tendency with lengthscale, which does not seem to follow some simple scaling law. At larger Reynolds number, there is decrease at largest scales (smallest k) as the wake width is smaller in that case. The lower occupied area is also responsible for generally lower values at higher Re (normalization by velocity). The spatial distribution in Fig. 8 shows very similar topology as at all scales (Fig. 3) at small and middle scales, while the large scales fluctuation can emerge only since certain distance past trailing edge (lower Re, Fig. 8 ch, i), or it cannot live in wake smaller than its scale (Fig. 8 j, k).

Surprisingly, the skewness shows very different behaviour at different Reynolds numbers: the decreasing tendency with k switches to increasing tendency at higher velocity. The processed airfoil exhibits larger difference between the begin and end of the probed lengthscale interval. The vorticity flatness originates at small scales and this effect is stronger at higher velocity. Again, this effect is stronger at processed airfoil, although it displays lower vorticity fluctuations than the unprocessed naïve airfoil realization.

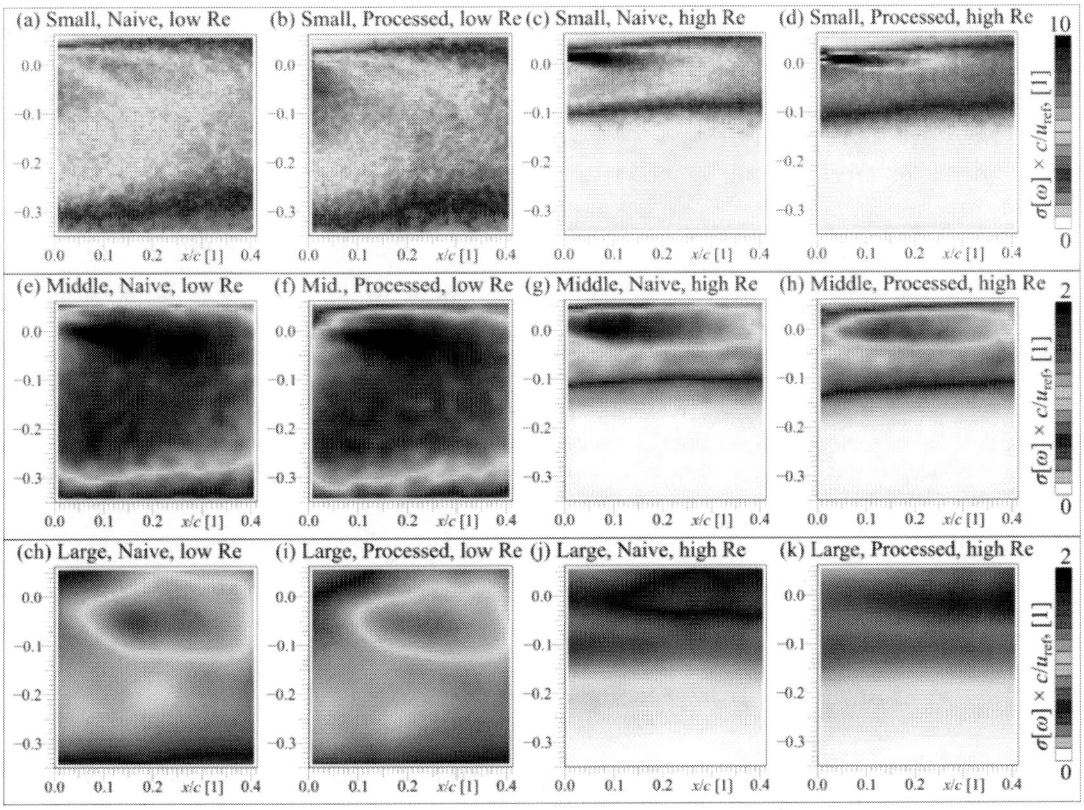

Fig. 8. Scale-dependent standard deviation of vorticity. Top row displays the smallest scales, middle row the middle scale (i.e. 0.020 – 0.027 c) and bottom row shows the vorticity fluctuations at larger scales. The colormap range keeps within each row, while it changes among rows, hence look to Fig. 7 for comparing values.

4 Conclusion

A pair of airfoil realizations is compared, the first one is created by "naive" approach: download coordinates and print, while the "processed" variant gains from the errors observed in the first case, thus its trailing edge is slightly remodeled and the surface is smoother by using sandpaper. Both airfoils produce massive separation at Reynolds number $1.6 \cdot 10^4$, while at higher Re $1.6 \cdot 10^5$ the flow is attached and the wake is more concentrated. The processed airfoil produces smaller vorticity fluctuations, mainly at small scales. The skewness and flatness of vorticity originate at the boundary of the wake. The flatness is produced by small-scale fluctuations, while this source changes with Reynolds number for skewness.

Acknowledgments

The work was supported from ERDF under project "Research Cooperation for Higher Efficiency and Reliability of Blade Machines (LoStr)" No. CZ.02.1.01/0.0/0.0/16_026/0008389.

References

[1] J.-Z. Wu, H.-Y. Ma and M.-D. Zhou, Vorticity and Vortex dynamics, Berlin Heidelberg New York: Springer, 2006.

[2] "Airfoiltools.com," [Online]. Available: http://www.airfoiltools.com/airfoil/details?airfoil=naca643618-il.

[3] S. Inkinen, M. Hakkarainen, A. Albertsson and A. Södergård, "From lactic acid to poly(lactic acid) (PLA): Characterization and analysis of PLA and Its precursors," *Biomacromolecules,* vol. 12, no. 3, pp. 523-532, 2011.

[4] S. Zhang, Handbook of 3D Machine Vision: Optical Metrology and Imaging, Boca Raton: CRC Press, 2013.

[5] J. Vagovský, I. Buranský and A. Görög, "Evaluation of measuring capability of the optical 3D scanner," *Procedia Engineering,* vol. 100, pp. 1198-1206, 2015.

[6] V. Yanovych, D. Duda, V. Uruba and T. Tomášková, "Hot-Wire Investigation of Turbulence Topology behind Blades at Different Shape Qualities," *Processes,* vol. 10, no. 3, p. 522, 2022.

[7] C. Tropea, A. Yarin and J. Foss, Springer Handbook of Experimental Fluid Mechanics, Berlin: Springer, 2007.

[8] D. Duda, V. Yanovych, V. Tsymbalyuk and V. Uruba, "Effect of Manufacturing Inaccuracies on the Wake Past Asymmetric Airfoil by PIV," *Energies,* vol. 15, no. 3, p. 1227, 2022.

[9] D. Duda, V. Uruba and V. Yanovych, "Wake Width: Discussion of Several Methods How to Estimate It by Using Measured Experimental Data," *Energies,* vol. 14, no. 15, p. 4712, 2021.

[10] M. La Mantia, P. Švančara, D. Duda and L. Skrbek, "Small-scale universality of particle dynamics in quantum turbulence," *Physical Review B,* vol. 94, p. 184512, 2016.

[11] D. Duda, V. Yanovych and V. Uruba, "An Experimental Study of Turbulent Mixing in Channel Flow Past a Grid," *Processes,* vol. 8, no. 11, p. 1355, 2021.

[12] A. Alexakis and L. Biferale, "Cascades and transitions in turbulent flows," *Physics Reports,* Vols. 767-769, no. 12, pp. 1-101, 2018.

[13] C. Williamson, "Three-dimensional wake transition," *Journal of fluid mechanics,* vol. 328, pp. 345-407, 1996.

[14] M. Thompson, K. Hourigan and J. Sheridan, "Three-dimensional instabilities in the wake of a circular cylinder," *Experimental Thermal and Fluid Science,* vol. 12, no. 2, pp. 190-196, 1996.

[15] I. Solís-Gallego, A. Meana-Fernández, J. Fernández Oro, K. Argüelles Díaz and S. Velarde-Suárez, "Turbulence Structure around an Asymmetric High-," *Journal of Applied Fluid Mechanics,* vol. 10, no. 4, pp. 1013-1027, 2017.

[16] V. Yanovych, D. Duda, V. Uruba and P. Antoš, "Anisotropy of turbulent flow behind an asymmetric airfoil," *SN Applied Sciences,* vol. 3, p. 885, 2021.

[17] M. Izakov, *Solar System Research,* vol. 47, no. 3, pp. 170-181, 2013.

[18] D. Duda and V. Uruba, "Spatial spectrum from particle image velocimetry data," *Journal of Nuclear Engineering and Radiation Science,* vol. 5, no. 3, pp. 030912-1-030912-7, 2019.

MATEC Web of Conferences **367**, 00018 (2022) https://doi.org/10.1051/matecconf/202236700018
Power System Engineering 2022

Numerical Simulations in SCR systems with Emphasis on Uniform Flow at the Catalyst Inlet

Martin Novák[1*], *Richard* Matas[2]

[1]Department of Power System Engineering, Faculty of Mechanical Engineering, University of West Bohemia, Univerzitní 2762/22, 301 00 Pilsen, Czech Republic
[2]Modelling and Simulations in Technical systems, New Technologies – Research Centre, University of West Bohemia, Univerzitní 2732/8, 301 00 Pilsen, Czech Republic

Abstract. Selective Catalytic Reduction (SCR) is one of the most widely used technologies for reducing emissions from diesel engines, which are widely used in the energy world. This technology is very complex, and it is not currently possible to address all aspects of it comprehensively. This study focuses on trying to improve the distribution of the exhaust stream at the catalyst inlet using models which represent the most common geometric distributions. Four types of geometries are selected, complemented by a fifth design, which is used as a reference, to find the maximum achievable values of the parameters under study. In this case, mainly the uniformity index is addressed, which indicates how uniformly the selected variable is distributed over the particular area. The study presents possible modifications of the flue gas duct which will be generally applied.

1 Motivation

It is essential to design an SCR system well, to meet requirements such as sufficient residence time of the injected AdBlue® and subsequent even distribution of ammonia at the catalyst inlet. Very often the system is not well designed, and geometry changes are needed to improve its functionality. As far as possible, efforts are made to minimise interference with the system design. A successful example of a simple geometry change was made by Li et al. who achieved the desired change by only slightly modifying the blades already in use. [1] Another example of a simple modification to the flue was made by Mi et al. who added vanes to the duct to provide better ammonia distribution. [2] Hong et al. also improved the flow in the duct by a similar change in geometry. [3]

Another commonly used component is a Static Axial Rectifier. It increases the effectiveness of the SCR system as reported by Mehdi et al. [4] They increased the effectiveness by increasing the dwell time of the droplets in the manifold, which results in higher evaporation and therefore higher effectiveness. Another reason for the higher effectiveness is that the heat transfer to the droplets is also increased due to increased turbulence, therefore the droplets evaporate faster. These claims are confirmed by Tan et al. in their publication. [5] Several interesting designs of rectifier modules are described by Xu

* Corresponding author: novakm42@ntc.zcu.cz

et al. It should also be pointed out that the distance to evaporation in their SCR system is very small, so one cannot expect the injection fluid to completely evaporate before entering the catalyst. However, the effectiveness of the system is still increased when using the modules. [6] There are other articles also confirming the improvement of the system function, see [7] [8] [9].

Modelling a catalyst is a complex matter as it consists of a huge number of small channels, so it is common to simplify this part of the model in CFD simulation by using a porous medium. This simplification has no significantly negative effects on the results. A comparison between an ANSYS Fluent simulation and an experiment is described by Holtz et al. [10] and indicates that the results are comparable.

Experimental validation of the calculations has been carried out, for example, by Wardana et al. [11] Their research involved the chemical reactions and evaporation process of a solution of water and urea. In their study, they investigated mixers for better mixing and increased conversion of the solution to ammonia. 3D simulations supported by their own experimental results were also reported by Luo et al [12]. They used the simplest reduction equations in the flue gas stream in their research, but they also considered the chemical processes in the catalyst, which they simplified by using a porous medium. They compared stationary and non-stationary approaches to the calculations and concluded that the results are similar. A comprehensive approach to 3D CFD simulation of an SCR system was taken by Lee [13] in his study. He included the evaporation process, thermolysis and hydrolysis of urea (water and urea solution), the aqueous film and the effect of the mixer on the overall system in his calculations. He compared the calculations with experiments and found very good agreement. He also investigated changing the position of the solution injection and the insertion of mixers into the flue gas stream. Zhang et al. [14] proposed very interesting mixers in the exhaust duct which included droplet evaporation, a water film on the mixer walls and chemical processes related to SCR. They verified their results experimentally.

The application of an SCR system in ship engines was investigated by Mehdi et al. [4], who also concluded that it is advisable to use mixers in the exhaust duct to improve the flow. This is only a numerical simulation without experimental verification. SCR systems can be applied on different devices, for example, Gao et al. [15] performed numerical simulations and an experiment on a flue downstream of a gas turbine. They used the three most common equations describing the chemical reactions of selective catalytic reduction and concluded that the results were equivalent.

All these studies consider a uniform input profile to the SCR technology. In addition, different inlet profiles are not used, which would establish the sensitivity of the system to different inlet parameters (velocity profiles). For most SCR systems, it can be assumed that the exhaust gas passes through a turbocharger that does not have a uniform exit profile at the outlet. Therefore, a small part of the study presented here will also point out the fact that different inlet profiles play a role. Because then one could argue whether the internal inserts, which are very popular for improving the effectiveness of these systems, are generally applicable or only applicable to the one particular case.

2 Geometry

It can be seen from the previous chapter that the design of an SCR device, whether it is a CHP unit or a large engine, can be divided into several basic designs. These are illustrated in Figure 1. In the following Figures 1, 2 and 3 are inlets displayed by blue arrows and outlets by red arrows.

The first is to place the catalyst behind the engine via two 90° elbows. The second solution has only one elbow. Variants one and two have the same length of pipe in front of the catalytic converter. In the first case, the pipe is divided by means of elbows into three equal length units. In the second design, the pipe is split in half. The next option for the location of the SCR catalyst is similar to the second option, except that the elbow is straight into the catalyst, i.e. there is no straight pipe beforehand to further direct the flow after the elbow. The fourth geometric solution, which was used in some cases mentioned in the previous chapter, is shown at the bottom right. This geometry is mainly chosen in designs where a compact design of the whole system is required. The fifth variant, shown here, serves as a reference. It is not a geometry that would be applicable in a real design, because building requirements usually do not allow a straight supply pipe. This variant serves as the basis for the set of variants shown in Figure 1. These variants are shown with the same type of catalyst, which has a circular cross section. This type of catalyst has already been used in [16], therefore it is considered to be a realistic model.

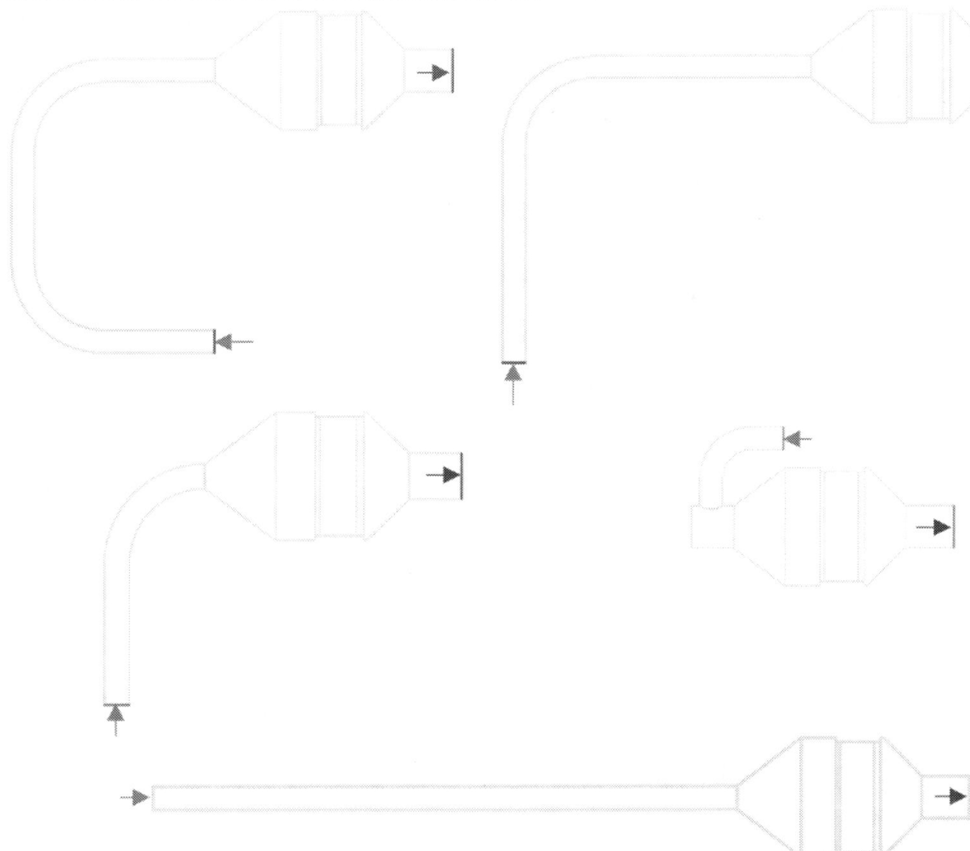

Fig. 1 a) upper left – variant "double_knee", b) upper right – variant "long_knee", c) middle left – variant "short_knee", d) middle right – variant "tdm", e) down – variant "straight"

Based on the reference variant, the set of variants represented here by only three geometries aims to determine the appropriate shape of the inlet chamber into the catalyst are shown in Figure 2. This set of variants includes changing the angle of the inlet chamber in 5° increments, up to 80°, and then changing the angle in 1° increments. The first variant has an angle of attack of 0°, the second geometry has an angle of attack of 30°, and the last model has an angle of attack of 85°.

Fig. 2 a) top – variant "straight_0deg", b) middle – variant "straight_30deg", c) down – variant "straight_0deg",

This set of variants essentially modifies the inlet to the catalyst and turns it into a diffuser where the pressure/velocity ratio is changed. This geometric modification can be used for all the other variants as well, since the results from the straight pipe indicate that changing the geometry in this direction is desirable, as it results in a more uniform flow and a better distribution of the flow field at the catalyst inlet.

A large number of geometric variations have been developed; therefore, only representative samples are shown here. In Figure 3 at the top left, a modified variant with two elbows is shown, but the pipe expansion starts from the inlet. At the top right, in the same figure, a similar variant is shown, i.e., the pipe diameter expands from the inlet, but the first elbow is rotated by 180°. The bottom left shows the same variant as the two previous ones, with two elbows, but the pipe widens at the end of the second elbow. The last of the variants in this figure is the 'TDM' variant with the inlet modified so that the pipes are connected by a T-piece rather than an elbow.

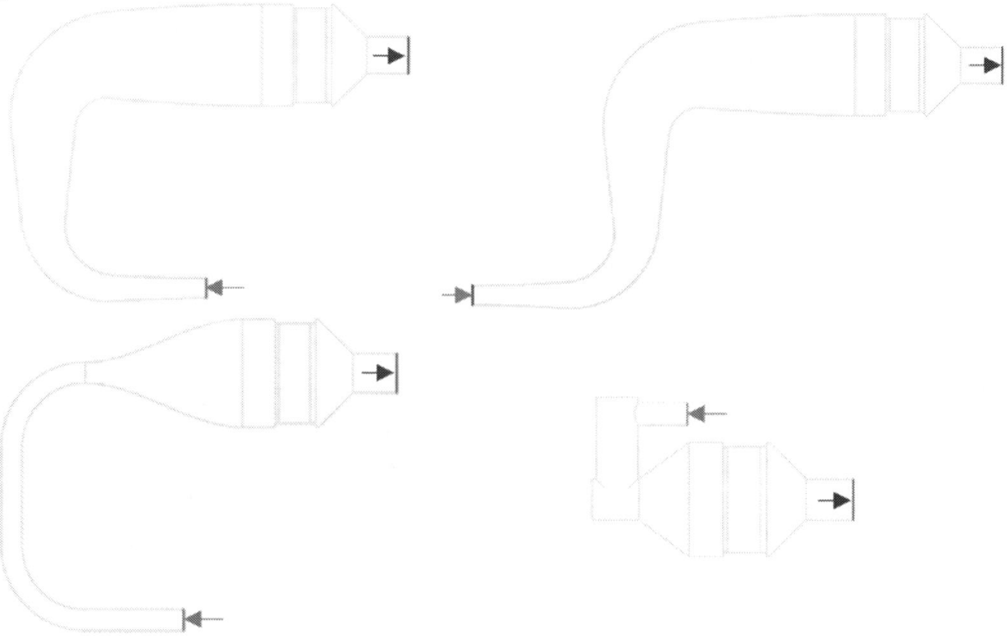

Fig. 3 a) upper left – variant "double_knee-cone", b) upper right – variant "double_knee-cone-rot", c) lower left – variant "double_knee-third-cone", d) lower right – variant "tdm-tupo-bigst"

As mentioned above, more than fifty geometric variations were designed. The variants with “cone” suffix were designed because ANSYS Fluent and its "Adjoint Solver" optimization function started to change the diameter of the inlet pipe. However, the program did not achieve any major changes in the flow with these modifications, as the computation time was considerable, and the rate of change was small. Therefore, the pipe diameter was changed manually for the straight inlet pipe variant and the other geometric variants were further modified.

The other variants are derived from the “double_knee” and “long_knee” types where the pipe diameter is changed. In addition, we designed variants of these two types with different elbow radiuses to investigate the effect of the parameters on the flow in the catalyst inlet chamber.

3 Simulation description

Fluid mechanics tries to find the values of four basic quantities. These are pressure p, velocity w, temperature T and density ρ. Three physical laws are used to solve these unknowns. The first is the law of conservation of momentum, the second is the law of conservation of mass, and the third is the law of conservation of energy. Density is considered to be constant, so the fourth equation is not necessary to solve the system of equations

The law of conservation of momentum is represented by the Navier-Stokes equations, which are analytically solvable only in exceptional cases and therefore we solve them numerically. Momentum is a vector, so it is broken it down into three components. [17]

$$\frac{\partial u}{\partial t}+\frac{\partial(uu)}{\partial x}+\frac{\partial(uv)}{\partial y}+\frac{\partial(uw)}{\partial z}=-\frac{1}{\rho}\frac{\partial p}{\partial x}+\nu\left(\frac{\partial^2 u}{\partial x^2}+\frac{\partial^2 u}{\partial y^2}+\frac{\partial^2 u}{\partial z^2}\right)+f_x \tag{1}$$

$$\frac{\partial v}{\partial t}+\frac{\partial(vu)}{\partial x}+\frac{\partial(vv)}{\partial y}+\frac{\partial(vw)}{\partial z}=-\frac{1}{\rho}\frac{\partial p}{\partial y}+\nu\left(\frac{\partial^2 v}{\partial x^2}+\frac{\partial^2 v}{\partial y^2}+\frac{\partial^2 v}{\partial z^2}\right)+f_y \tag{2}$$

$$\frac{\partial w}{\partial t}+\frac{\partial(wu)}{\partial x}+\frac{\partial(wv)}{\partial y}+\frac{\partial(ww)}{\partial z}=-\frac{1}{\rho}\frac{\partial p}{\partial z}+\nu\left(\frac{\partial^2 w}{\partial x^2}+\frac{\partial^2 w}{\partial y^2}+\frac{\partial^2 w}{\partial z^2}\right)+f_z \tag{3}$$

In equations (1) - (3), u, v, w represent the individual velocity components in the x, y, z direction. t is time, p is pressure and ρ is the density of the fluid. The kinematic viscosity is denoted by ν . Since this is an equation based on the laws of physics, it is possible to determine what each term means. The first product represents the acceleration of the fluid (local), and the next three terms represent the nonlinear internal inertial acceleration. These terms cause instabilities at higher Reynolds numbers. The first term on the right side represents the pressure gradient (inertial acceleration due to pressure). The next term is the acceleration from the shear stresses of the fluid caused by its viscosity. The last terms f_x, f_y, f_z are the components of the external volumetric accelerations. All terms have a dimension of acceleration. If it is necessary to write the equation in terms of forces, both sides can just be multiplied by the mass. [17] [18]

The law of conservation of mass describes the continuity equations. When considering density as invariable, the equation takes the following form:

$$\frac{\partial u}{\partial x}+\frac{\partial v}{\partial y}+\frac{\partial w}{\partial z}=0 \tag{4}$$

As in the Navier-Stokes equation, the velocity components u, v, w are in the three x, y, z directions. What is interesting about this equation is that the right side of the equation is 0. The left side of the equation must be zero because the amount of fluid entering the control volume is assumed to be the same as the amount that flows out of that volume. [17]

The law of conservation of energy is defined by the energy equation. This equation, for a constant density ρ, is discussed in the following lines. For clarity, substitution by ψ has been used, otherwise the equation would be long, and not entirely clear.

$$\frac{\partial T}{\partial t}+\frac{\partial (uT)}{\partial x}+\frac{\partial (vT)}{\partial y}+\frac{\partial (wT)}{\partial z}=\alpha\left(\frac{\partial^2 T}{\partial x^2}+\frac{\partial^2 T}{\partial y^2}+\frac{\partial^2 T}{\partial z^2}\right)+\mu\psi \tag{5}$$

$$\psi = 2\left(\left(\frac{\partial u}{\partial x}\right)^2+\left(\frac{\partial v}{\partial y}\right)^2+\left(\frac{\partial w}{\partial z}\right)^2\right)+\left(\left(\frac{\partial u}{\partial y}+\frac{\partial v}{\partial x}\right)^2+\left(\frac{\partial u}{\partial z}+\frac{\partial w}{\partial x}\right)^2+\left(\frac{\partial v}{\partial z}+\frac{\partial w}{\partial y}\right)^2\right) \tag{6}$$

$$\alpha = \frac{\lambda}{\rho \cdot c_p} \tag{7}$$

In equations (5) - (7), T represents the thermodynamic temperature; t is time; u, v, w are the velocity components; α is the thermal conductivity coefficient; λ is the thermal conductivity coefficient; ρ is the density; c_p is the specific heat capacity at constant pressure; μ is the dynamic viscosity. The first term in Equation 5 describes the heat accumulation in the liquid. The remaining terms on the left side of the equation represent convection caused by flow. The first term on the right side represents heat conduction. The last term, which is subsequently broken down, represents the heat from the deformation and movement of the fluid. [17] [19]

The calculation program was ANSYS Fluent [20]. The simulations were performed as stationary, the turbulence model was *SST k-ω* [21]. As mentioned above, the input velocity profile is not uniform. The input profile has already been described in a previous paper, which is the logical predecessor of this paper [16]. In the introduction of this text, it is mentioned that modelling the catalyst as a geometrically complex component is possible by using a porous zone that can replace the behaviour of the real component, so this simplification was also adopted in this case. In order to achieve this simplification, it was necessary to create a submodel of a realistic part of the catalyst. The pressure drop was determined from it and then the corresponding values were set in the porous zone accordingly.

4 Results

The Uniformity Index (UI) is one of the most commonly used parameters in evaluating the uniformity of the input field to the catalyst. This index was first introduced by Weltens et al. in 1993 in their paper dealing specifically with the distribution of the flow field upstream of an SCR catalyst inlet. [22] This index has proven to be an important variable for the quantitative assessment of uniform velocity field distribution. [23] Bressler et al. state that a series of experiments confirm the need to ensure a homogeneous velocity field distribution at the catalyst inlet, because this will ensure a longer lifetime and lower pressure losses. [24] In addition, the NOx conversion effectiveness will be increased, and the ammonia slip, which is also an emission, will be reduced. [25] [26] UI can be used to evaluate variables other than velocity. For example, Capetillo et al. used it to evaluate the uniformity of ammonia distribution, and Vedagiri et al. used it when evaluating concentrations. [27] [28]

The calculation of the UI index is given in equation (8) as reported by Girard et al., or Johansson et al., [29] [30]

$$UI = 1 - \frac{1}{2}\cdot\frac{\sum_{i=1}^{n}(|v_i - v_{mean}|\cdot A_i)}{A\cdot v_{mean}} \tag{8}$$

where v_i $[m/s]]$ is the velocity at the local location (cell), v_{mean} $[m/s]$ is the mean velocity over the evaluation area, A_i $[m^2]$ is the area at the local location (cell area) and A $[m^2]$ is the total size of the evaluation area.

The value of this index is between 0 and 1. At a value of 1 the field is completely uniform, which means the value of the variable is the same in each location (cell). [29] Weltens et al. state that for a fully developed flow in a tube of $Re = 64000\ [-]$, the value of $UI = 0.94\ [-]$. [22] Lu et al. state that the UI of the velocity field needs to be at least $0.98\ [-]$ to ensure maximum catalyst effectiveness. [25] Kurzydym et al. state that a UI value of $\geq 0.94\ [-]$ can be considered as a good result. [9] However, all the authors of publications related to this topic aim for the highest possible value of the uniformity index in their studies.

The uniformity index can be weighted by the mass flux in addition to weighted by the area, see equation (8). Both options are implemented in ANSYS Fluent using equations (9) to (12). The area-weighted UI is denoted as γ_a for the computed variable ϕ.

$$\gamma_a = 1 - \frac{\sum_{i=1}^{n}[(|\phi_i - \overline{\phi_a}|)A_i]}{2|\overline{\phi_a}|\sum_{i=1}^{n} A_i} \tag{9}$$

$$\overline{\phi_a} = \frac{\sum_{i=1}^{n} \phi_i A_i}{\sum_{i=1}^{n} A_i} \tag{10}$$

Where i is the area index of an area composed of n faces and ϕ_a is the area-weighted average value of the variable. In the following lines, γ_m is the area-weighted UI and ϕ_m is the average value of the variable weighted by the mass flux.

$$\gamma_m = 1 - \frac{\sum_{i=1}^{n}[(|\phi_i - \overline{\phi_m}|)(|\rho_i \vec{v_i} A_i|)]}{2|\overline{\phi_m}|\sum_{i=1}^{n}[|\rho_i \vec{v_i} A_i|]} \tag{11}$$

$$\overline{\phi_m} = \frac{\sum_{i=1}^{n}[\phi_i(|\rho_i \vec{v_i} A_i|)]}{\sum_{i=1}^{n}[|\rho_i \vec{v_i} A_i|]} \tag{12}$$

The uniformity index is an important parameter in this study and is one of the most important indicators for evaluating the quality of the design variant at this stage of the calculations, as it eliminates clearly unsuitable options.

The first results presented here are the straight pipe variants where only the angle (size) of the catalyst inlet chamber is changed, see Figure 4. The graph in Figure 4 contains two sets of results, which are composed of four different velocities - $70\ [m/s], 120\ [m/s], 171\ [m/s] and\ 210\ [m/s]$. One result shows the UI mean drifting velocity (only the velocity towards the catalyst) weighted by mass flow (UI MW - Uniformity Index Mass flow Weighted) and the other results are area weighted (UI AW - Uniformity Index Area Weighted). At first sight, the UI values vary depending on what they are averaged over. There are no consistent variations until above 80°, until then it can be clearly said that the more conservative UI values are represented by the area weighting, so those values will continue to be considered.

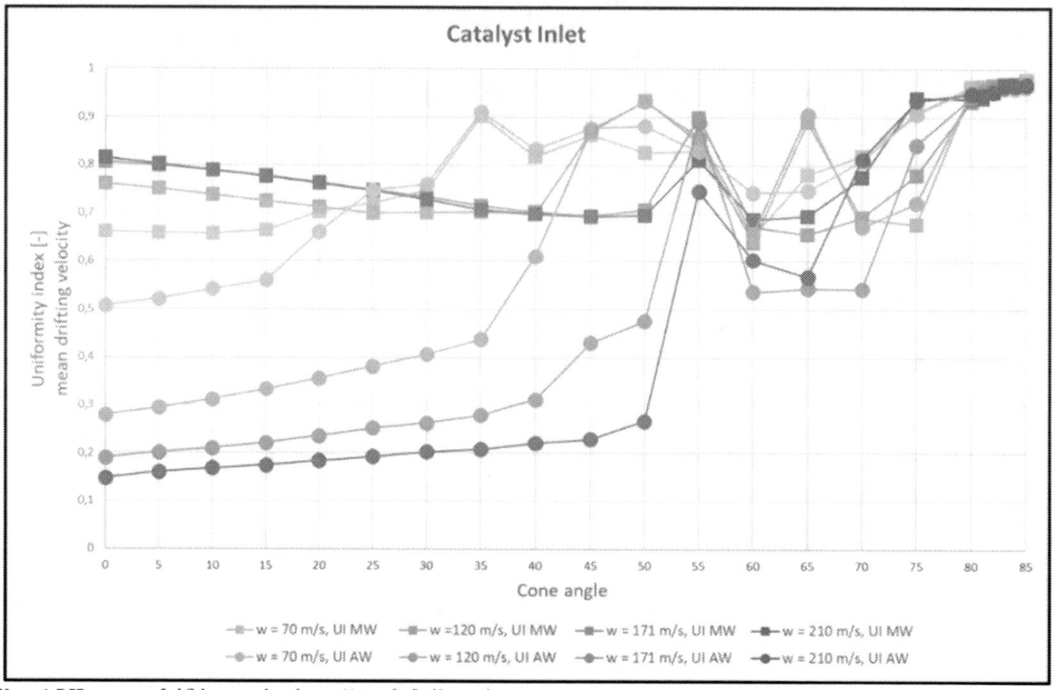

Fig. 4 UI mean drifting velocity –"straight" variant

It is mentioned above that it is necessary to achieve the highest possible UI value, ideally values above 0.94 (better above 0.98). In this case, these values are achieved above an angle of 80°, with the UI MW and UI AW values being identical only at angles greater than and including 83°.

Another important factor to observe is the ratio of the area of the drift velocity (in the direction of flow) to the total area. This parameter tells how much flow is in that direction or how much is not flowing there (backflow, stalling, perpendicular flow, etc.). The values of this variable are shown in Figure 5. It can be seen that at lower velocities the flow is stable, and everything flows in one direction. At increasing velocities and small angles, the flow is concentrated only in one part (the middle part) of the catalyst and in other parts the flow is rotating in the inlet chamber. Therefore in some situations ($w = 210\ [m/s]$) the value of the area ratio is less than 0.5, which means that the flow direction is mostly towards the catalyst. At higher angles, as in the previous case above 80°, or 83°, the area ratio is equal to one, meaning that there is flow in one direction (into the catalyst) throughout the channel.

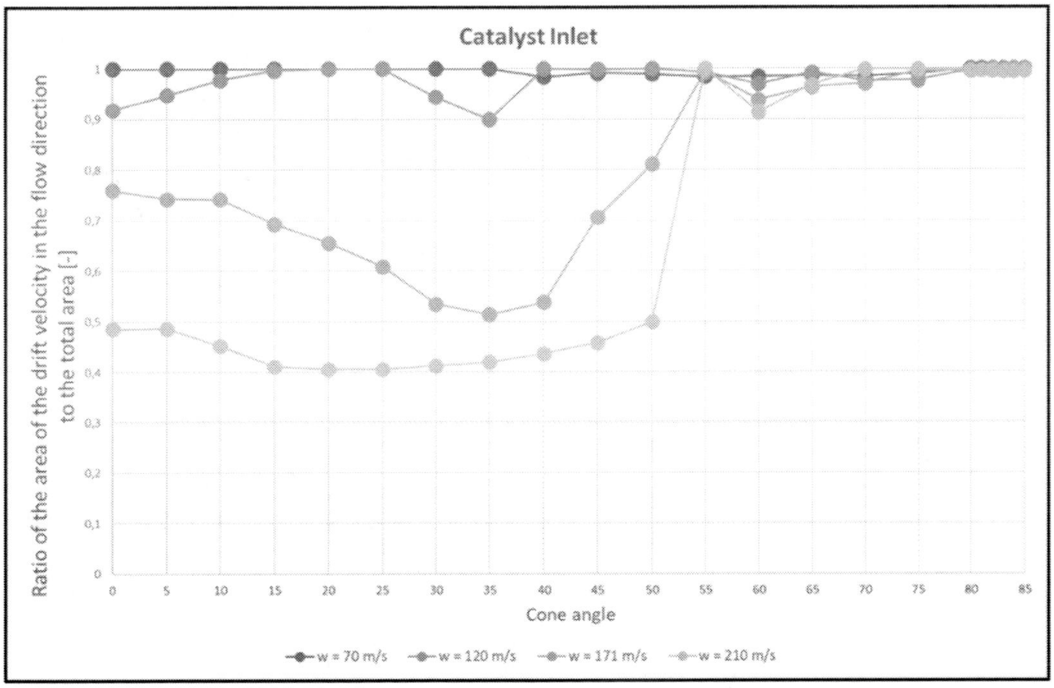

Fig. 5 Area ratio, "straight" variant

The uniformity index is supposed to take values between 0 and 1. However, in some special cases, according to all available data, it has other values. See Figure 6, where it can be seen that the UI values are not only negative but also greater than one (if absolute values are taken). The negative values are due to the fact that the flow is complex and it flows in more directions than just the catalyst axis direction (drift direction), or there is also backflow in a large part of the channel, which is undesirable. These negative values are recorded on a surface that is in the middle of the straight part of the inlet chamber, so that it is equidistant from the catalyst inlet for all variants. It can be seen that in this section the backflow is significant and affects the UI evaluation. But looking at the area in the catalyst inlet, the UI is already evaluated correctly and does not take on values that can be claimed as erroneous.

Because of the way UI behaves, this factor should only be used in areas (locations/ cross-sections) where there is a clear direction of flow and problematic backflow is not occurring to any significant degree. Therefore, below we only deal with the evaluation of the UI at the catalyst inlet, where there is no introduction of inappropriate values into the evaluation.

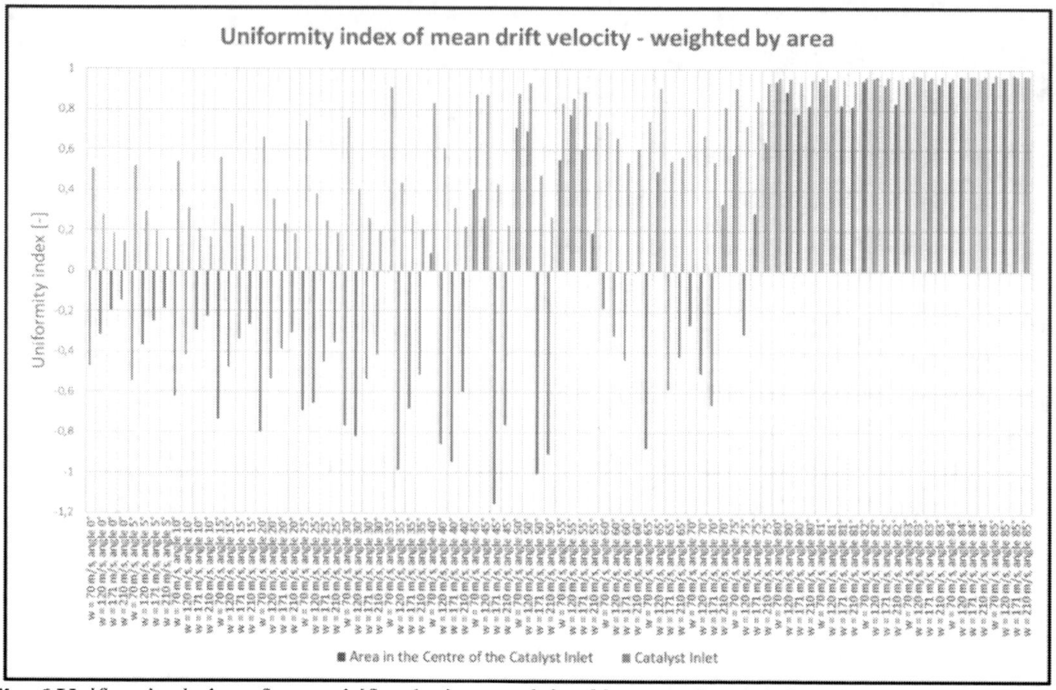

Fig. 6 Uniformity index of mean drift velocity – weighted by area, “straight” variant

The above results are for the reference variant and are used to assess the behaviour of the system and to prepare a suitable apparatus for comparison. It is also used to determine the potentially best possible value of the area to UI ratio, which can be expected to be the best possible for a straight pipe, as the geometry is not affected by any negative effects in the form of elbow (drift behind these elements).

Figure 7 contains the results from the "other" variants. These variants have already been roughly outlined in the introduction of this chapter, but here it can be clearly seen that there are many more variants solved than mentioned above. However, all the variants are based on the four basic ones - "double_knee", "long_knee", "short_knee" and "tdm", plus some results for the "straight" variant where certain parameters are changed.

The results in the figure show the ratio of the drift velocity area to the total area at the catalyst inlet. Here, it is necessary to point out the range of the area ratio, which does not start at 0, but with a value of 0.5. However, results with an area ratio below 0.85 can be evaluated as inappropriate since, even at this value, 15% of the area is backflowing, or 15% of the catalyst area is not used.

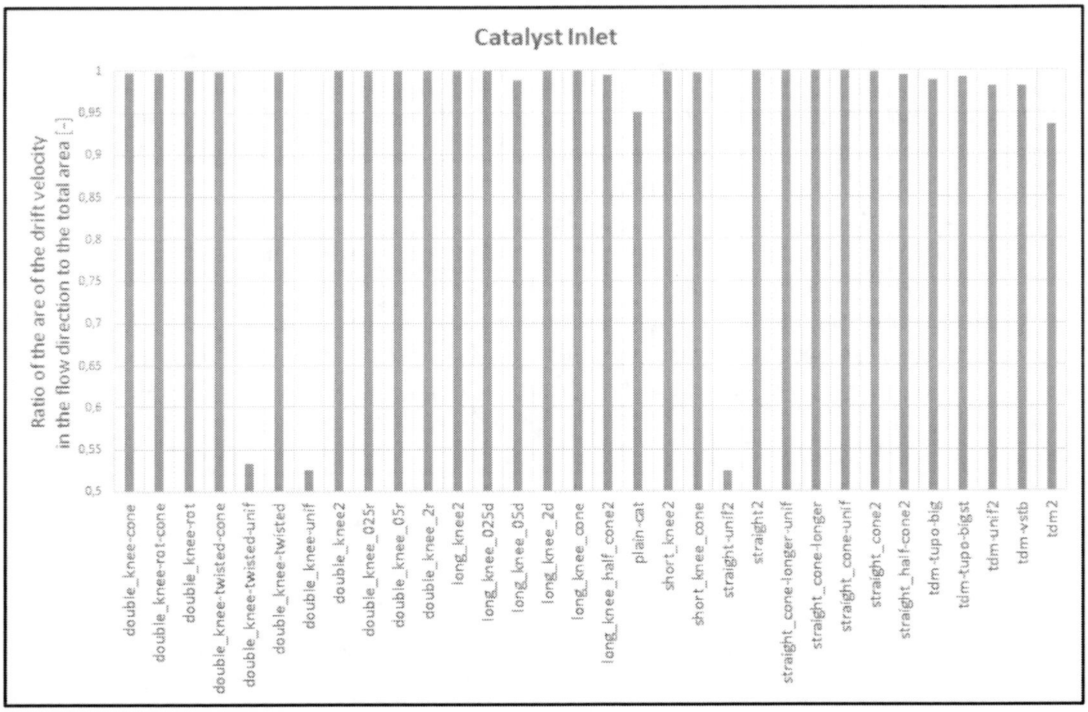

Fig. 7 Area ratio, remaining variants

The second criterion for selecting a suitable solution is the UI. The variants mentioned above are evaluated with respect to UI in Figure 8.

The result of the basic variant "tdm" is "tdm2". It does not achieve the highest UI values, which can be clearly seen from the results. Several variants of geometry modifications are designed and it can be seen that the variant "tdm-tupo-bisgt", which has a 34% higher UI value over the basic variant. The second best geometric design for this option is "tdm-vstb". However, it contains an insert that may be susceptible to damage, and these kinds of modifications will be further addressed in follow-up research. When comparing the "tdm2" and "tdm-unif2" variants, which differ only in their inlet velocity profiles, it can be seen that the velocity distribution at the catalyst inlet needs to be addressed as this also has a significant impact on the UI.

The "straight" variants have already been largely covered in the previous section, which dealt with changing the angle of the inlet chamber to the catalyst. These variants are here referred to as "straight" + "cone". However, here it is useful to compare two variants, namely, "straight2" and "straight-unif2". These differ only in the distribution of the inlet velocity (velocity profile), where the "straight-unif2" variant has a uniform inlet. Here a significant dependence on the inlet profile can be seen, as the UI differs by 42% in this case, so the dramatic change in velocity distribution has a significant effect on how uniform the flow is at the catalyst inlet.

There are only two examples of "short_knee" variants given here. The difference between this variant and the "long_knee" variant can be seen at the beginning of this text in the section describing the geometry. Comparing the two results, it can be clearly seen that by changing the geometry of the pipe to an expanding pipe (cone/cone shape), there is a 32% change in UI.

The "long_knee" set of variants also has many derivatives. Variants with the "d" suffix contain a different elbow radius from the original radius (it varies proportionally according to the values given before the "d" suffix). The change to a larger radius is positive, the change to a smaller radius is negative. The original "long_knee" variant achieved a significant improvement in UI of 30% after the geometry change, called the "long_knee_cone" variant, by using an expanding pipe diameter.

The original variant "double_knee2" is the default variant for the last set of geometries shown in the following figure, reaching the highest UI value. This seems to be a unique issue, as any intervention in the geometry has a negative effect on the flow, so a stable solution to the problem needs to be found. Variants containing the "unif" suffix are those that contain a uniform inlet profile and at first glance appear to achieve the lowest values. This is the last piece of evidence that the inlet profile fundamentally affects the UI. It is clear from the results that the highest values of uniformity are achieved when an expanding pipe, the "cone" extension, is used. It does not matter whether both elbows are bent in the same direction, as the variants containing the "rot" have one original elbow and the other rotated 180° when the pipe is expanded. So, when comparing the "double_knee_cone" and "double_knee_rot_cone" variants, it can be seen that the results are almost identical. Some variants are calculated with only the shorter extension section of the pipe, but these do not achieve as high UI values as those with the full extension pipe.

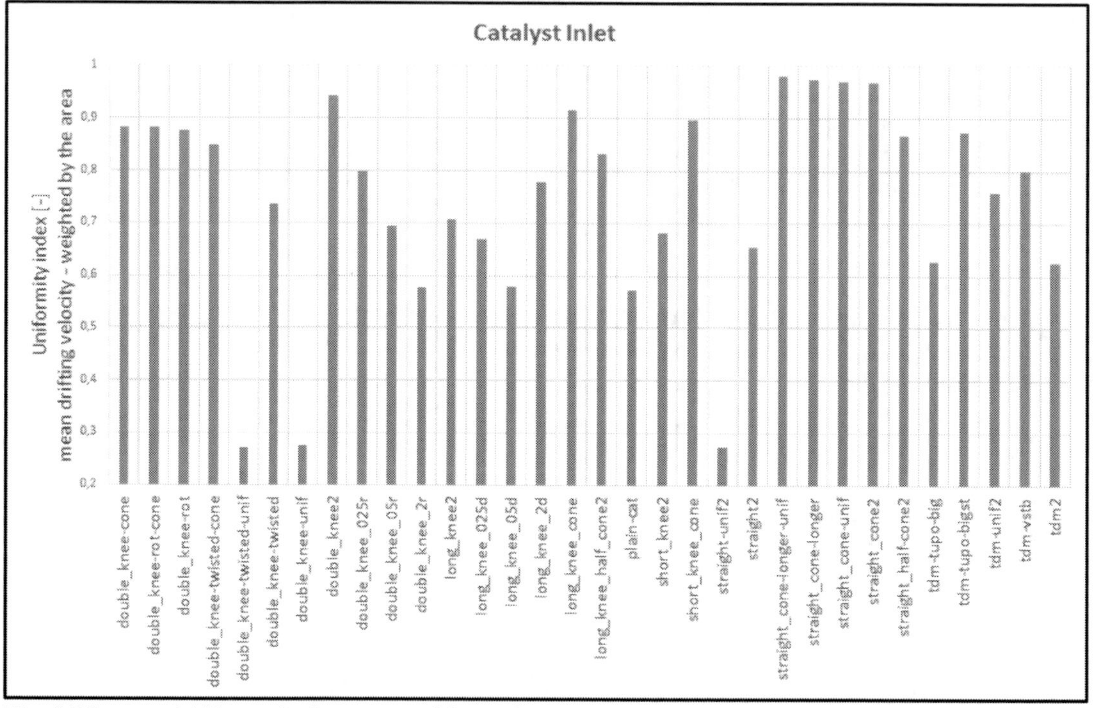

Fig. 8 UI mean drifting velocity – rest of the variants

Another variable that can indirectly affect UI is the diameter of the pipe and the radius of the elbows. The radius of the elbows has been mentioned above, and the following figure shows the results from these calculations. The results are for the larger pipe diameter, which in this case is $106.5\ [mm]$ (the original pipe diameter is $71\ [mm]$). From these results at low velocities ($70\ [m/s]$) the UI is minimally affected, when the radius of the elbows is changed. In the other cases, a clear trend can be seen that the UI increases with increasing radius. These values can be compared with those in the previous figure and a larger pipe diameter in this case negatively affects the UI.

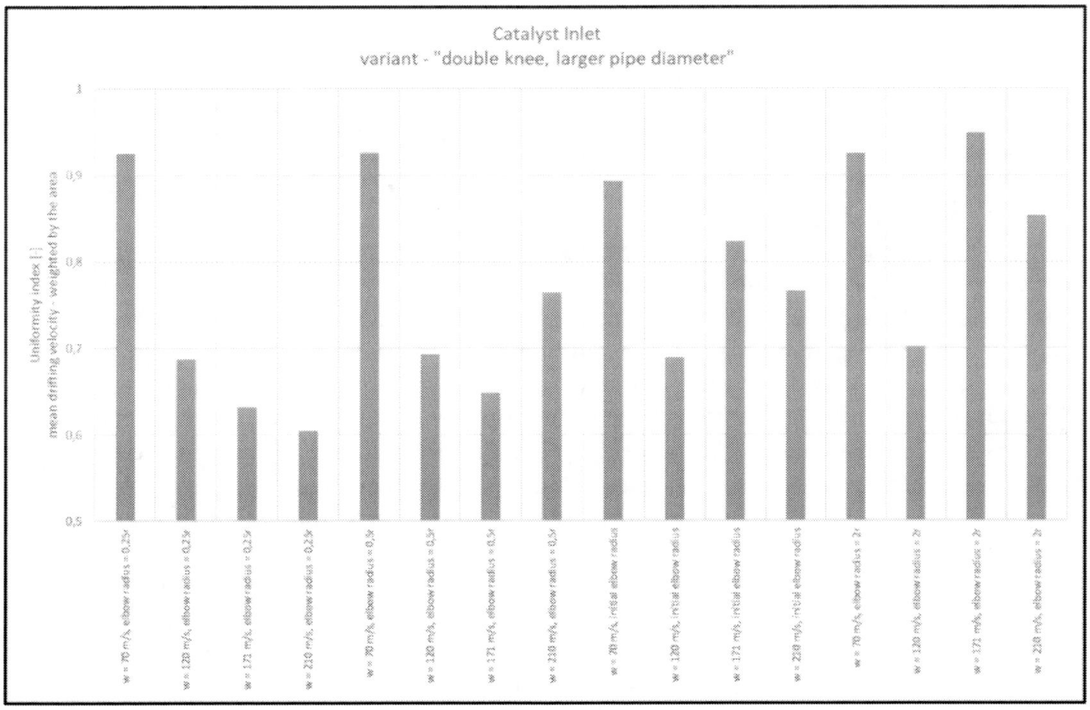

Fig. 9 UI, variant “double_knee”

The "long_knee" variant was modified in a similar way to the "double_knee" variant (in the previous figure), i.e., the diameter of the pipe was increased and the radius of the elbow was changed. It is not possible to clearly see what trend is occurring here. For the highest and lowest velocities, it can be seen that the UI value increases as the radius of the elbows increases, however, for the middle velocities ($120\ [m/s]$ and $171\ [m/s]$) this is not the case, and the UI decreases as the radius of the elbow increases. It is clear that the larger diameter has a positive effect on the UI value in this case.

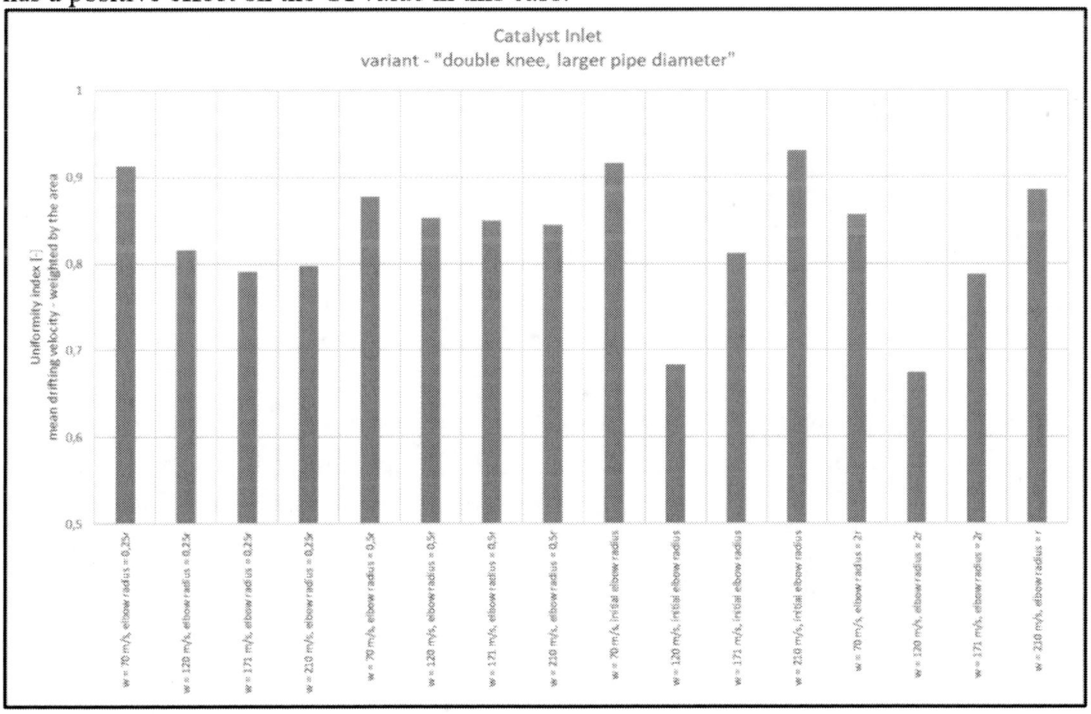

Fig. 10 UI, variant “long_knee”

5 Conclusion

Selective catalytic reduction is an important topic which is relevant, as reducing emissions is an important aspect for the survival of society and the planet. Therefore, we need to keep improving these systems to achieve the best results and highest effectiveness using the best available technologies, which are becoming more sophisticated every year. CFD simulations includes a lot of physical phenomena and since these are simulations it is always necessary to think about whether the results make sense and are physical. Experience in fluid mechanics gained through previous projects is needed to assess whether the results are meaningful. And even that it is always great to make experiments which proves the CFD results, but it is not possible to do it in this case because there is a plenty of geometrical modifications and only the best one deserves to be tested since manufacturing the pipeline costs time and money.

The present work deals with the modification of the catalyst inlet duct in order to increase the utilized area of the catalyst and at the same time to make the distribution of the exhaust stream on it as uniform as possible. For this purpose, two parameters are presented - the ratio of the drift velocity area to the total area, and the uniformity index. However, this index is only applicable in cases where the flow is in one direction without significant eddies and backflow effects.

The best options for solving the issue seem to be those where the pipe gradually increases its cross-section up to the diameter of the catalyst inlet chamber. These variants are "double_knee-cone", "short_knee_cone" "long_knee_cone" and "tdm_tupo_bigst". However, even derivatives of these variants, which have a shorter cone section, reach quite good values. So, according to the available data, it is clear that these changes also improve the situation at the catalyst inlet.

The next line of research will focus on further increasing the UI by using the internal inserts which are briefly mentioned in this paper. The next steps of research in this area will look at variants into which droplets of aqueous urea solution fluid will be injected, along with evaporation. The distribution (UI) of this evaporated second phase at the catalyst inlet will be monitored. The next step will involve solving the basic chemical equations that are integral to SCR technology. Once the simulations show a significant improvement to a real-life SCR system then the experiment will be conducted on that particular engine.

Acknowledgments

This work was financially supported by student project SGS-2022-023 (Research and development of power machines and equipment).

References

1. LI, Mao, Hongjie YAN a Jiemin ZHOU. Numerical Simulation and Optimization of Flow Field in the SCR Denitrification System. In: *2009 International Conference on Energy and Environment Technology* [online]. IEEE, 2009, s. 415-418 [cit. 2022-04-26]. ISBN 978-0-7695-3819-8. Available from: doi:10.1109/ICEET.2009.565
2. MI, Jia, Dan A. PITSKO a Tim HASKEW. CFD Applications on Selective Catalytic NOx Reduction (SCR) Systems. In: *Volume 1:*

Fora, Parts A, B, C, and D [online]. ASMEDC, 2003, s. 2139-2146 [cit. 2022-04-26]. ISBN 0-7918-3696-7. Available from: doi:10.1115/FEDSM2003-45436

3. HONG, Wen Peng, Hao Shu DING a Yue HUANG. Numerical Simulation and Optimization of the Flow Field in the SCR System. *Advanced Materials Research* [online]. 2013, **860-863**, 1464-1469 [cit. 2022-04-26]. ISSN 1662-8985. Available from: doi:10.4028/www.scientific.net/AMR.860-863.1464
4. MEHDI, Ghazanfar, Song ZHOU, Yuanqing ZHU, Shafquat HUSSAIN a Arif HUSSAIN. A CFD analysis of static mixer to study its impacts on SCR performance in marine diesel engine. In: *2019 2nd International Conference on Computing, Mathematics and Engineering Technologies (iCoMET)* [online]. IEEE, 2019, s. 1-6 [cit. 2022-04-26]. ISBN 978-1-5386-9509-8. Available from: doi:10.1109/ICOMET.2019.8673421
5. TAN, Ligang, Pengfei FENG, Shubao YANG, Yage GUO, Shaochun LIU a Ziwen LI. CFD studies on effects of SCR mixers on the performance of urea conversion and mixing of the reducing agent. *Chemical Engineering and Processing - Process Intensification* [online]. 2018, **123**, 82-88 [cit. 2022-04-26]. ISSN 02552701. Available from: doi:10.1016/j.cep.2017.11.003
6. XU, Zhengxin, Jingping LIU a Jianqin FU. Experimental investigation on the urea injection and mixing module for improving the performance of urea-SCR in diesel engines. *The Canadian Journal of Chemical Engineering* [online]. 2018, **96**(6), 1417-1429 [cit. 2022-04-26]. ISSN 00084034. Available from: doi:10.1002/cjce.23082
7. QU, Litao, Chao LI, Fengyuan HE, Xiaohui QI, Honghai YU, Jia DU a Dexin WANG. Study on Flow Field Optimization of SCR Denitrification Reactor Based on CFD Numerical Simulation Technology. *IOP Conference Series: Earth and Environmental Science* [online]. 2019, **242** [cit. 2022-04-26]. ISSN 1755-1315. Available from: doi:10.1088/1755-1315/242/4/042009
8. MEHDI, Ghazanfar, Song ZHOU, Yuanqing ZHU, Ahmer SHAH a Kishore CHAND. Numerical Investigation of SCR Mixer Design Optimization for Improved Performance. *Processes* [online]. 2019, **7**(3) [cit. 2022-04-26]. ISSN 2227-9717. Available from: doi:10.3390/pr7030168
9. KURZYDYM, Damian, Adam KLIMANEK a Zbigniew ŻMUDKA. Experimental research and CFD analysis of flow parameters in a SCR system for the original part and WALKER's replacement. *Combustion*

Engines [online]. 2019, **179**(4), 13-20 [cit. 2022-04-26]. ISSN 2300-9896. Available from: doi:10.19206/CE-2019-402

10. HOLTZ, Dorian, Conrad GIEROW, Robert BANK, Dirk KADAU a Flavio SOPPELSA. CFD simulation of particle deposition in exhaust gas treatment systems. SIEBENPFEIFFER, Wolfgang, ed. *Heavy-Duty-, On- und Off-Highway-Motoren 2019* [online]. Wiesbaden: Springer Fachmedien Wiesbaden, 2020, s. 69-82 [cit. 2022-04-26]. Proceedings. ISBN 978-3-658-31370-8. Available from: doi:10.1007/978-3-658-31371-5_6
11. WARDANA, Muhammad Khristamto Aditya, G. M. Hasan SHAHARIAR, Kwangchul OH a Ocktaeck LIM. Ammonia Uniformity to Predict NOx Reduction Efficiency in an SCR System. *International Journal of Automotive Technology* [online]. 2019, **20**(2), 313-325 [cit. 2022-04-26]. ISSN 1229-9138. Available from: doi:10.1007/s12239-019-0031-x
12. LUO, Zhaoyu, Parvez SUKHESWALLA, Scott A. DRENNAN, Mingjie WANG a P. K. SENECAL. 3D Numerical Simulations of Selective Catalytic Reduction of NOx With Detailed Surface Chemistry. In: *Volume 2: Emissions Control Systems; Instrumentation, Controls, and Hybrids; Numerical Simulation; Engine Design and Mechanical Development* [online]. American Society of Mechanical Engineers, 2017, s. - [cit. 2022-04-26]. ISBN 978-0-7918-5832-5. Available from: doi:10.1115/ICEF2017-3658
13. LEE, Changhee. Numerical and experimental investigation of evaporation and mixture uniformity of urea–water solution in selective catalytic reduction system. *Transportation Research Part D: Transport and Environment* [online]. 2018, **60**, 210-224 [cit. 2022-04-26]. ISSN 13619209. Available from: doi:10.1016/j.trd.2017.04.015
14. ZHANG, Chaofeng, Chuan SUN, Meiping WU a Kai LU. Optimisation design of SCR mixer for improving deposit performance at low temperatures. *Fuel* [online]. 2019, **237**, 465-474 [cit. 2022-04-26]. ISSN 00162361. Available from: doi:10.1016/j.fuel.2018.10.025
15. GAO, Xiang, Ben WANG, Xudong YUAN, Siyuan LEI, Qinggong QU, Chuan MA a Lushi SUN. Optimal design of selective catalyst reduction denitrification system using numerical simulation. *Journal of Environmental Management* [online]. 2019, **231**, 909-918 [cit. 2022-04-26]. ISSN 03014797. Available from: doi:10.1016/j.jenvman.2018.10.060
16. NOVÁK, Martin a Richard MATAS. *Numerical simulation of flue gas in selective catalytic reduction system* [online]. In: . s. 040003- [cit. 2022-04-26]. Available from: doi:10.1063/5.0041560

17. BLEJCHAŘ, Tomáš. *Turbulence modelového proudění - CFX: učební text*. Vyd. 1. Ostrava: Vysoká škola báňská - Technická univerzita Ostrava, 2012. ISBN 978-80-248-2606-6.
18. KOZUBKOVÁ, Milada. *Modelování proudění tekutin FLUENT, CFX* [online]. 1. vyd. Ostrava: Vysoká škola báňská - Technická univerzita, 2008 [cit. 2022-04-26]. ISBN 978-80-248-1913-6.
19. PINKUS, Oscar a Beno STERNLICHT. *Theory of hydrodynamic lubrication*. New York: McGraw-Hill, 1961.
20. *ANSYS Fluent User's Guide 2021 R1* [PDF]. 2021. U.S.A., 2021 [cit. 2022-04-26].
21. MENTER, F. Zonal Two Equation k-w Turbulence Models For Aerodynamic Flows. In: *23rd Fluid Dynamics, Plasmadynamics, and Lasers Conference* [online]. Reston, Virigina: American Institute of Aeronautics and Astronautics, 1993, s. - [cit. 2022-04-26]. Available from: doi:10.2514/6.1993-2906
22. WELTENS, Herman, Harald BRESSLER, Frank TERRES, Hubert NEUMAIER a Detlev RAMMOSER. *Optimisation of Catalytic Converter Gas Flow Distribution by CFD Prediction* [online]. In: . s. - [cit. 2021-12-20]. Available from: doi:10.4271/930780
23. SUN, Ke, Haiyang ZHAO, Kui ZHAO, Da LI a Shuzhan BAI. Optimization of SCR inflow uniformity based on CFD simulation. *Open Physics* [online]. 2020, **18**(1), 1168-1177 [cit. 2021-12-20]. ISSN 2391-5471. Available from: doi:10.1515/phys-2020-0221
24. BRESSLER, H., D. RAMMOSER, H. NEUMAIER a F. TERRES. *Experimental and Predictive Investigation of a Close Coupled Catalytic Converter with Pulsating Flow* [online]. In: . s. - [cit. 2021-12-20]. Available from: doi:10.4271/960564
25. LU, Kai, Dewen LIU, Yan WU, Shusen LIU a Shuzhan BAI. A study on the high-efficiency mixer of the SCR system. *Mechanics & Industry* [online]. 2021, **22** [cit. 2021-12-20]. ISSN 2257-7777. Available from: doi:10.1051/meca/2020101
26. MCKINLEY, Thomas L., Andrew G. ALLEYNE a Chia-Fon LEE. Mixture Non-Uniformity in SCR Systems: Modeling and Uniformity Index Requirements for Steady-State and Transient Operation. *SAE International Journal of Fuels and Lubricants* [online]. 2010, **3**(1), 486-499 [cit. 2021-12-20]. ISSN 1946-3960. Available from: doi:10.4271/2010-01-0883
27. CAPETILLO, Azael J., Fernando IBARRA, Dominik STEPNIEWSKI a Jo VANKAN. Multiphase Modelling of SCR Systems: Using the Taguchi Method for Mixer Optimisation. *SAE International Journal of*

Engines [online]. 2017, **10**(1), 61-71 [cit. 2021-12-20]. ISSN 1946-3944. Available from: doi:10.4271/2017-26-0113

28. VEDAGIRI, Praveena, Leenus Jesu MARTIN a Edwin Geo VARUVEL. Characterization of urea SCR using Taguchi technique and computational methods. *Environmental Science and Pollution Research* [online]. 2021, **28**(10), 11988-11999 [cit. 2021-12-03]. ISSN 0944-1344. Available from: doi:10.1007/s11356-020-08743-y
29. GIRARD, James W., Figen LACIN, Charles J. HASS a Joseph HODONSKY. *Flow Uniformity Optimization for Diesel Aftertreatment Systems* [online]. In: . s. - [cit. 2021-12-20]. Available from: doi:10.4271/2006-01-1092
30. JOHANSSON, Åsa, Ulf WALLIN, Mikael KARLSSON, Annika ISAKSSON a Phillip BUSH. *Investigation on Uniformity Indices Used for Diesel Exhaust Aftertreatment Systems* [online]. In: . s. - [cit. 2021-12-20]. Available from: doi:10.4271/2008-01-0613

Optimization of pressure drop in curved conduits using Ansys DesignXplorer

Aleš Palkovič[1*], *Petr* Pečený[1], and *Jiří* Teichman[1]

[1]TechSoft Engineering, spol. s r. o., Na Pankráci 322/26, 140 00 Praha 4-Nusle, Czech Republic

Abstract. This paper describes the effects of guide vanes installed in curved conduits with the main emphasis on pressure loss reduction. In curved parts of conduits, eddy zones are created due to the flow separation resulting in an increased value of pressure drop. Installation of guide vanes can prevent the creation of large eddy zones. On the other hand, the cross-section of the conduit is decreased by placing the vanes, leading to the growth of the pressure drop. That is why the properties of guide vanes needed to be selected carefully. The goal of this paper is to describe the process of finding the optimal number of vanes in order to obtain the lowest value of pressure drop in the curved conduit. This is achieved by a parametric case study in Ansys DesignXplorer combined with CFD simulations in Ansys Fluent. The case study parameters are the number and thickness of vanes, dimensions of the conduit, and the inlet velocity. Our results are compared with recommendations on the optimal number of vanes from Idelchik [1].

1 Introduction

With the development of software and hardware, numerical simulations became a commonly used tool when creating new products or when improving the current ones. Nowadays, numerical simulations are employed in a wide range of fields including fluid, structural and thermal analysis, optics, electronics and others. The main advantage of numerical simulations is that the performance of proposed designs is evaluated in a short time without the need of manufacturing and testing the physical model. The common goal is to make the product better and cheaper at the same time i.e. to optimize the product.

Parametric optimization is a method of finding the optimum value of output parameters when changing a set of input parameters in defined ranges. A simple tool for such parametric studies is Ansys DesignXplorer which is suitable for data coming either from simulations, inhouse codes or even experiments. Thanks to the concept of Response Surface, DesignXplorer provides the resulting optimum with a significantly reduced number of executed simulations.

DesignXplorer is used to optimize the flow in a rectangular pipe elbow by installing guide vanes. The idea is that there is an existing elbow and the goal is to find the optimal design for the pressure drop to be minimal and compare the results with experimental results published in the literature. The number of guide vanes and their thickness are defined as the input parameters and pressure drop with velocity distribution are monitored

* Corresponding author: palkovic@techsoft-eng.cz

as the criteria of optimality. These output parameters are evaluated by a numerical simulation in Ansys Fluent.

2 Theoretical background

Centrifugal forces acting on the fluid passing through a curved conduit causing a decrease in the pressure at the inner wall and an increase in the pressure at the outer wall. These changes in pressure are related with the corresponding changes in velocity. As described for example in [1], we speak about the diffuser effect at the outer wall and the bellmouth effect at the inner wall. Due to described effects, the flow is separated from the inner wall and an eddy zone is created leading to a reduction of the main stream section and resulting in increased resistance of the pipe i.e. the pressure drop.

The pressure drop in the curved conduit can be decreased by installing guide vanes. For elbows, it is possible to use simple vanes with constant diameter or airfoils. On the other hand, concentric vanes are recommended for bends. Thanks to the vanes the flow is deflected towards the inner wall and the eddy zone is minimized, therefore the pressure drop is decreased and the velocity distribution behind the turn is improved [1]. In order to accomplish this, the properties of guide vanes such as dimensions, number, shape and angle need to be selected carefully.

Several experimental studies on different guide vanes setups were presented in the literature [2, 3, 4] resulting in guidelines for specific vanes setup and main parameters of guide vanes to consider. Idelchik [1] claims that the most efficient way to decrease the elbow resistance (i. e. the pressure drop) and equalize the velocity distribution is to eliminate the eddy zone at the inner wall and presents a general formula for an optimal number of guide vanes (n) in a 90° elbow:

$$n = 2.13 \left(\frac{r}{D}\right)^{-1} - 1 , \tag{1}$$

where r and D stand for the inner curvature radius of the elbow and the diameter of the pipe.

3 Model and simulation

In order to execute a parametric study, a parametric model of a 90° rectangular elbow was created in Creo Parametrics. The number of vanes and their thickness are defined as parameters. The overall dimensions of the elbow are kept constant (Fig. 1). This geometry is coupled with Ansys Workbench and serves as an input for meshing in Fluent. In the CFD simulation is set up, that air enters the pipe through the velocity inlet at 10 *m/s* and leaves at the pressure-outlet. The turbulence is modelled by k-ω SST model. As a result, the elbow pressure drop and velocity uniformity after the turn are monitored. The velocity uniformity is monitored in order to quantify the velocity distribution described by Idelchik and it is computed by Fluent function called Uniformity Index – Area Weighted (γ_a) [5]:

$$\gamma_a = 1 - \frac{\sum_{i=1}^{n}\left[\left(\left|\phi_i - \phi_{avg}\right|\right)A_i\right]}{2\left|\phi_{avg}\right|\sum_{i=1}^{n} A_i}, \quad \phi_{avg} = \frac{\sum_{i=1}^{n} \phi_i A_i}{\sum_{i=1}^{n} A_i}, \tag{2}$$

where ϕ is the specified field variable, in our case the velocity magnitude, i is the facet index of a surface with n facets and A_i is the area of a facet.

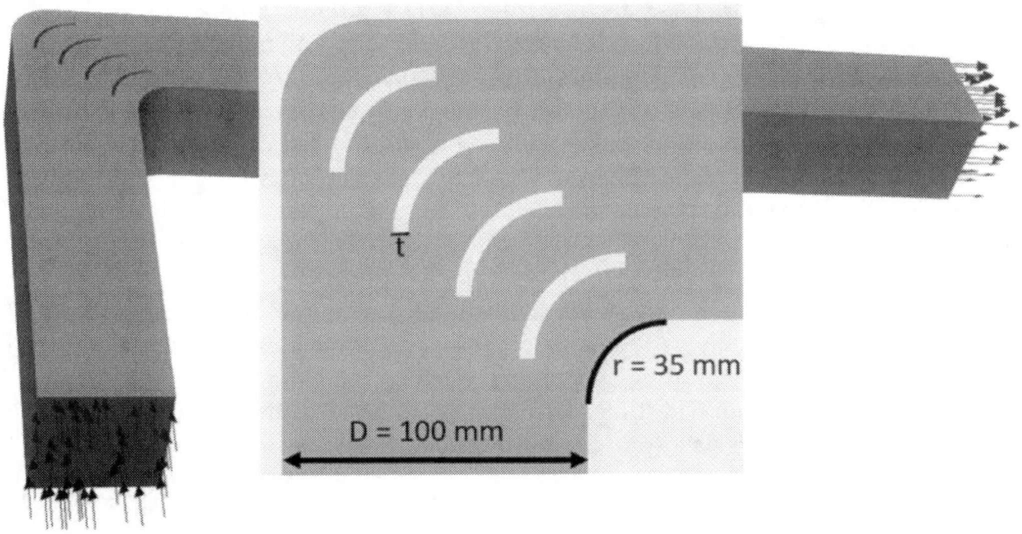

Figure 1: The example of elbow geometry with four vanes defined by characteristic length, radius and thickness (*t*).

In addition to this standard setup, features of DesignXplorer are added to the Workbench thus completing the scheme necessary for running the parametric study and optimization (Fig. 2).

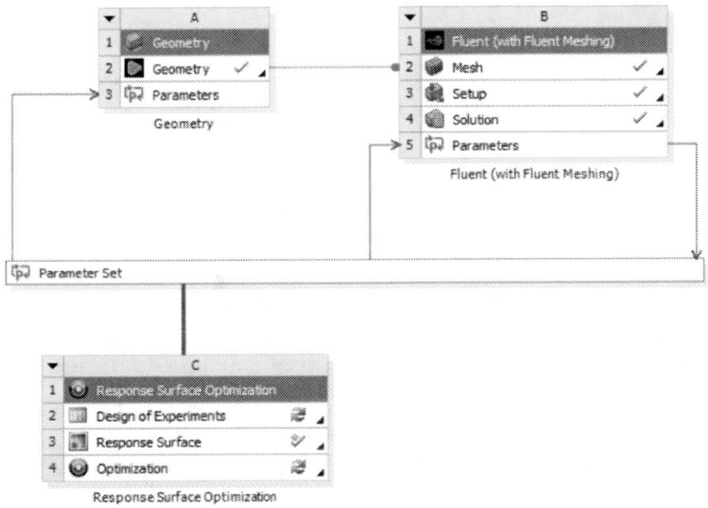

Figure 2: Scheme of the case in Workbench consisting of 3 blocks: geometry, solution and optimization.

4 DesignXplorer

The optimization in DX (DesignXplorer) is divided into three steps. The first step is called Design of Experiments where the range of input parameters is defined by the user. Based on that, DX proposes a "smart" set of Design Points with varying parameters to be calculated by the solver (Fluent or others), in order to sufficiently describe the trends within the defined range.

Secondly, the calculated results from Design of Experiments are approximated by a surface in the step called Response Surface (Fig. 3). DX offers a number of methods for approximating the calculated data. For this case, Second Order Polynomial method is used.

Finally, the Response Surface Optimization is executed based on the Response Surface and the defined objectives. In our case, the objectives are to minimize the pressure drop and maximize the velocity uniformity and the Screening method based on Shifted Hammersley Sampling is used [6, 7]. As results, DX provides the best candidates for defined objectives and in the case of multiple objectives, it also shows the graph of Pareto fronts. Pareto fronts graph is a helpful result that describes the trade-off between conflicting objectives.

In a general optimization algorithm, a large amount of output parameters evaluations is necessary to find the optimum. In order to evaluate the output parameters, the CFD simulation needs to be converged therefore thousands of evaluations for finding the optimum would be excessively time-consuming. That is why the concept the Response Surface Optimization is used. The time consumption is reduced due to the fact, that the optimization algorithm draws data from the Response Surface.

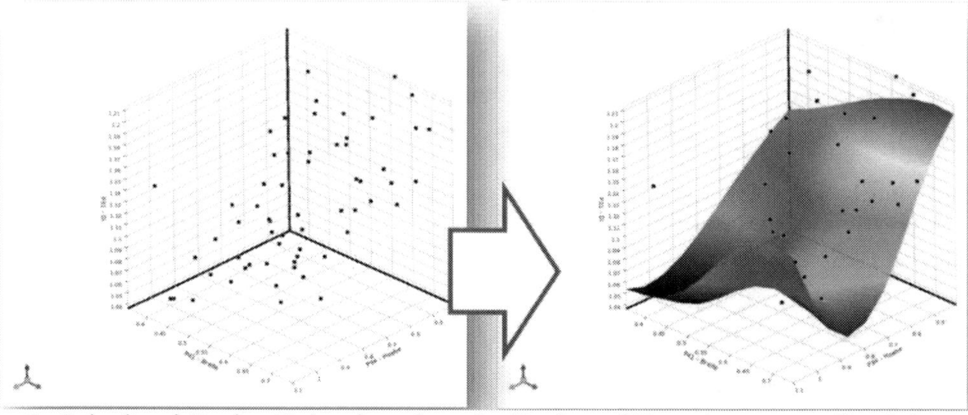

Figure 3: Calculated Design Points based on the Design of Experiments are fitted through with a Response Surface.

5 Results

Based on the Design of Experiments, five simulations with different vanes thickness were computed in Fluent for each number of installed vanes resulting in 30 so-called Design Points. Through these Design Points, a Response Surface is fitted in order to obtain dependency of pressure drop and velocity uniformity on the input parameters (Fig. 4). Computation of the Design Points is the most time-consuming step of the whole optimization in the process of Response Surface Optimization since it requires converged CFD simulations. Our whole optimization case was computed in under 24 hours.

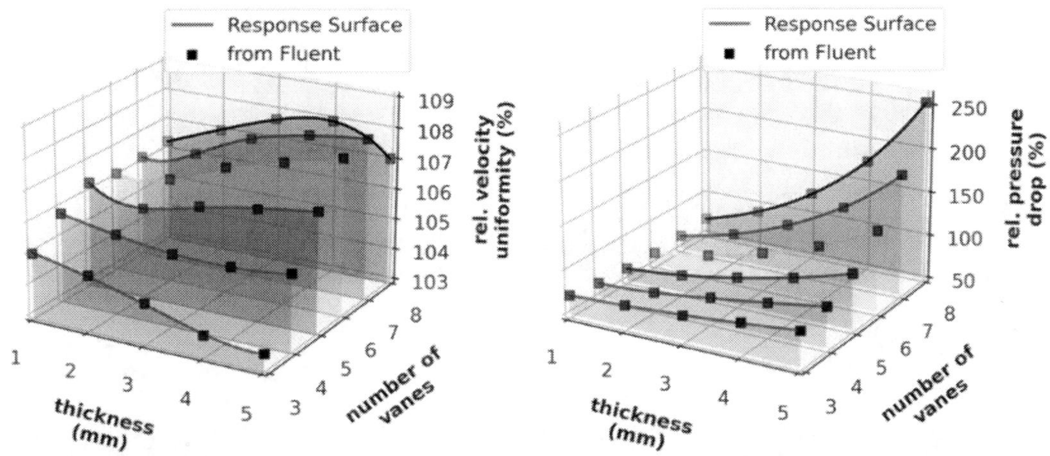

Figure 4: Dependency of the relative velocity uniformity (left) and the relative pressure drop (right) on the input parameters as a result of the Response Surface.

Due to the fact that the number of vanes is defined with discrete values the resulting Response Surface is separated to those discrete values. The use of discrete values also causes the number of Design Points calculated in Fluent to be higher in comparison to cases where the input parameters are defined continuously. For discrete parameter values, DX is not able to approximate for example the values of pressure loss with 5 installed vanes from Design Points computed for 6 and 4 vanes.

Different representation of the result from Figure 4 are displayed in Figure 5. It confirms the fact that vanes decrease the pressure drop when setup correctly. Naturally, the pressure drop increases with the growing thickness of vanes.

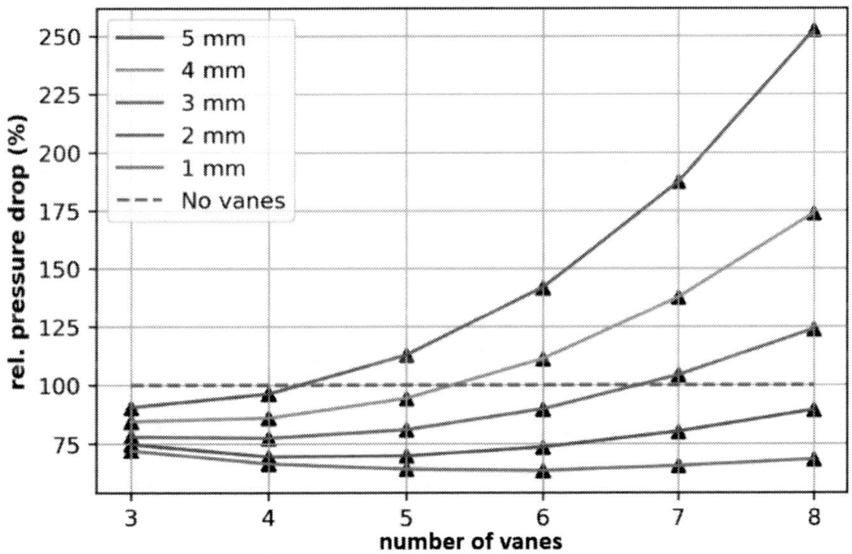

Figure 5: Graph of relative pressure drop depending on the number of vanes for different vane thicknesses.

In the process of finding the optimal vanes parameters, several optimizations were conducted. Firstly, the objective was to minimize the pressure drop. The optimal number of vanes is 6 with thickness of 1 mm (see Fig. 5). This is caused mainly by the fact that with growing thickness and number of vanes the pipe cross-section decreases significantly.

For a graphical comparison of the flow, a case with 4 vanes that are 3 mm thick is compared with the case without any vanes (Fig. 6). From this picture, it is evident that the eddy zone near the inner wall is significantly reduced, and the flow is more equalized when using vanes. This graphic result follows our expectations and agrees with historical theoretical studies.

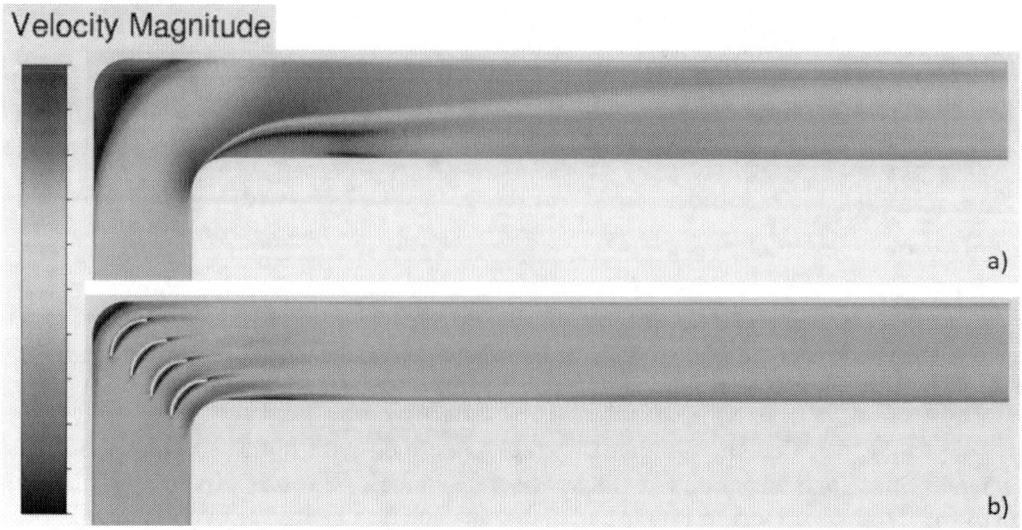

Figure 6: Contours of velocity magnitude for geometry without vanes (a) and with 4 vanes that are 3 mm thick (b).

In addition to the two input parameters previously used, a set of simulations was computed with different inlet velocities. In Figures 7 and 8, examples representing the minimum and the maximum vane thickness are presented. For the vane thickness of 1 mm, the relative pressure drop for different inlet velocities reaches similar values. On the other hand, the relative pressure tends to differ with the growing number of vanes due to the coefficient of local resistance generally depends on the flow velocity. This behaviour is the most noticeable for the case of vane thickness of 5 mm and inlet velocity 35 m/s (see Fig. 8).

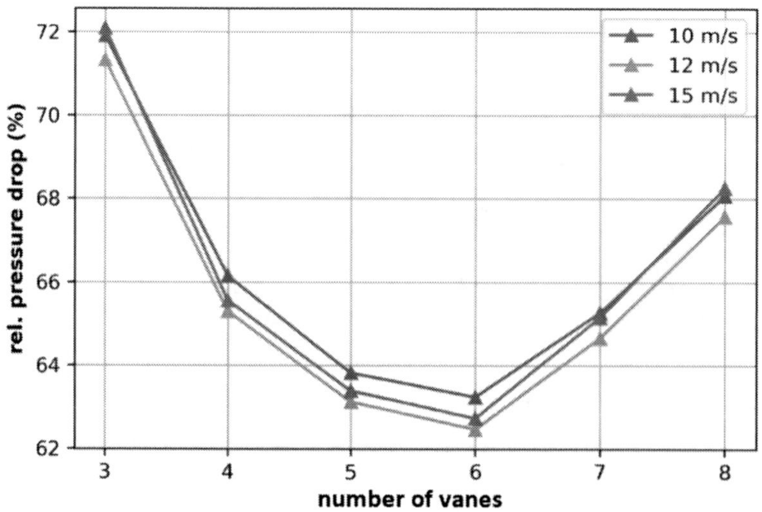

Figure 7: Dependency of relative pressure drop on the number of vanes for different inlet velocities. The vane thickness is 1 mm.

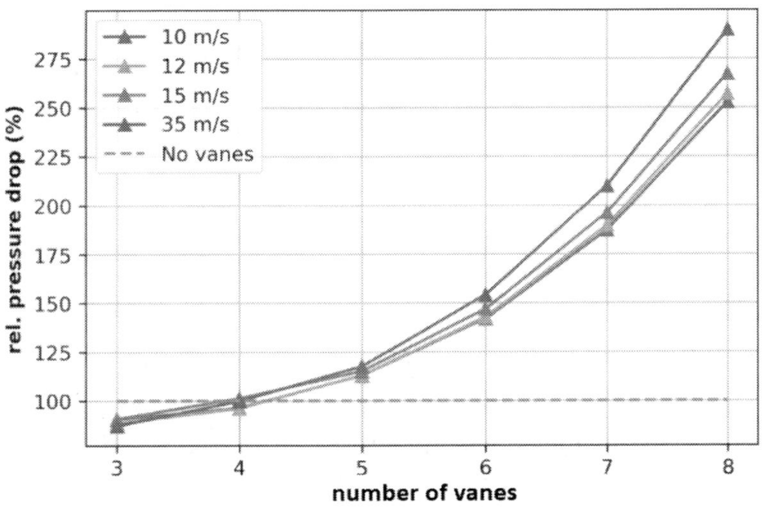

Figure 8: Dependency of relative pressure drop on the number of vanes for different inlet velocities. The vane thickness is 5 mm.

Idelchik claims that the resistance of the elbow is decreased by equalizing the velocity distribution, therefore an optimization with the objective to maximize the uniformity of velocity was calculated. However, this result is different from the previous optimization as the maximum velocity uniformity is obtained with 8 installed vanes that are 5 mm thick.

That is why the third optimization was conducted with both previous objectives used together. It is tricky to find the optimum when multiple objectives are defined, that is why Pareto fronts are used (Fig. 9). Pareto fronts show how the resulting objectives change when varying input parameters. In multi-objective optimization, it is usually necessary to make a compromise between the two or more goals as moving closer to one goal causes a drop in the performance of the other objective.

This chart shows that a design with 5 vanes and 1 mm thickness (green line) is the best candidate for the defined objectives as we are looking for the lowest pressure drop and highest uniformity. This result corresponds with the general formula presented by Idelchik when applied to our pipe dimensions.

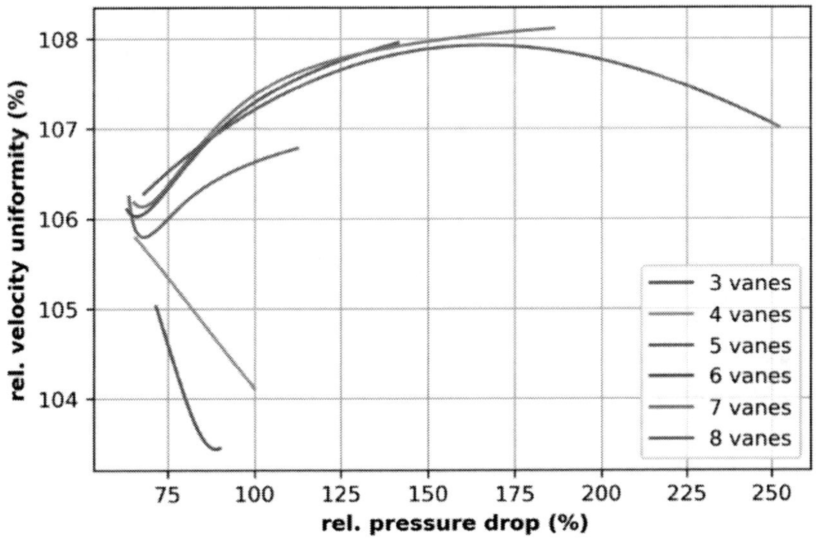

Figure 9: Graph of Pareto fronts used for finding the optimum number of vanes.

6 Conclusion

This paper deals with finding an optimal number of guide vanes and optimal thickness in order to decrease the resistance of a curved conduit, more specifically a 90° elbow. To achieve this a parametric optimization tool called Ansys DesignXplorer is used as well as a CFD solver Fluent. DesignXplorer proved to be a suitable tool for this type of problem as it was possible to quickly execute several optimizations with different objectives when the Response Surface was once obtained. The concept of Response Surface simply increases the efficiency of finding the optimal design when dealing with simulations.

When comparing results from optimization with literature, our assumption that the optimal vane parameters would be the same for minimizing the pressure drop as for maximizing the velocity uniformity behind the bend did not prove to be correct as the optimal vane parameters differ for these objectives. That is why a multi-objective optimization was conducted with both previously mentioned objectives. The resulting Pareto front chart shows that the optimal number of vanes agrees with the general formula presented by Idelchik.

Our findings show that the general formula is certainly useful as a guideline when dealing with such problems, but does not necessarily provide the result that some users expect, e. g. achieving the most uniform flow does not imply the lowest possible pressure drop. Moreover, it is possible to expect that the optimal vane parameters will differ in conduits with various shapes before and after the elbow (multiple bends, change in diameter). That is why a parametric study is a helpful and efficient tool when dealing with such cases.

In future work, the optimal design from our calculations could be enhanced by using the Adjoint Solver [8] in Ansys Fluent which is able to adjust the shape of vanes based on defined criteria through gradient based optimization. Based on comparison of our methodology and literature results optimization of more complex conduit systems could follow.

References

[1] Idelchik, Isaak E. "Handbook of hydraulic resistance." *Washington* (1986).

[2] Lindgren, Björn, Jens Österlund, and Arne V. Johansson. "Measurement and calculation of guide vane performance in expanding bends for wind-tunnels." *Experiments in fluids* 24.3 (1998): 265-272.

[3] Liou, Tong-Miin, Hsin-Li Lee, and Chin-Chun Liao. "Effects of guide-vane number in a three-dimensional 60-deg curved side-dump combustor inlet." *J. Fluids Eng.* 123.2 (2001): 211-218.

[4] Sahlin, Alexander, and Arne V. Johansson. "Design of guide vanes for minimizing the pressure loss in sharp bends." *Physics of Fluids A: Fluid Dynamics* 3.8 (1991): 1934-1940.

[5] ANSYS, ANSYS Fluent Theory Guide (2022).

[6] ANSYS, DesignXplorer User's Guide (2022).

[7] Diwekar, Urmila M., and Saadet Ulas. "Sampling techniques." *Kirk-Othmer Encyclopedia of Chemical Technology* (2000).

[8] ANSYS, ANSYS Fluent User's Guide (2022).

Development of a micro ORC power unit using digital prototyping methods

Jan Pejchal[1], *Petr* Kolář*[2]

[1]TechSim Engineering, Innovation Centre, Purkyňova 648125, 612 00 Brno, Czech Republic
[2]TechSim Engineering, Headquarters, Budějovická 1550/15a, 140 00 Praha 4, Czech Republic

Abstract. The development of the prototype of the ORC micro turbine includes already in the design phase the creation of a digital prototype based on the use of computational simulations, especially the CFD flow program Simcenter STAR CCM +, strength and rotor-dynamic FEA analyses of system behaviour - Simcenter 3D. This is followed by the optimization of the device by increasing its efficiency and improving the mechanical properties of the system. For this purpose, we used the optimization program HEEDS and the topological optimization method Simcenter 3D. This was followed by the production of a prototype, where 3D printing technology was used to produce turbine housing moulds and other components. A huge attention of this contribution is devoted to these additive production processes.

1 Introduction

The principle of an ORC power unit is a steam engine cycle where the working fluid is an organic medium that receives heat from an external source into a closed loop. The device converts the thermal energy into mechanical work, which can be used to generate electricity or to drive other equipment (e.g. a refrigeration compressor). The working fluid undergoes phase changes during the cycle. The duty cycle of the Organic Rankine Cycle (ORC)unit is shown in Fig. 1.

Refrigerants are most commonly used as the working medium to allow the use of low potential heat, from any energy source, both from the primary source and waste heat (WHR). Then it is a renewable source with zero emissions and universal use.

The aim of our project is the development of a micro ORC micro combine heat and power (μCHP) unit with a capacity of 1.5 and 5 kW.

1.1 Use of micro ORC units

In the case of micro cogeneration, this is up to 50 kWe. The principle of cogeneration is the production of both thermal and electrical energy from both primary and secondary sources. In the case of primary sources, it will allow us to cover local electricity consumption, reduce transmission losses and allow operation in an independent island system. It is a highly efficient process for using primary energy, with an efficiency of up to 95%.

In the case of waste heat recovery (WHR), it is possible to process residual heat from the flue gases of biomass boilers, process heat from food, wood, glass and chemical plants or conventional combustion engines used to drive compressors or energy centres.

* Corresponding author: kolar@techsim.cz

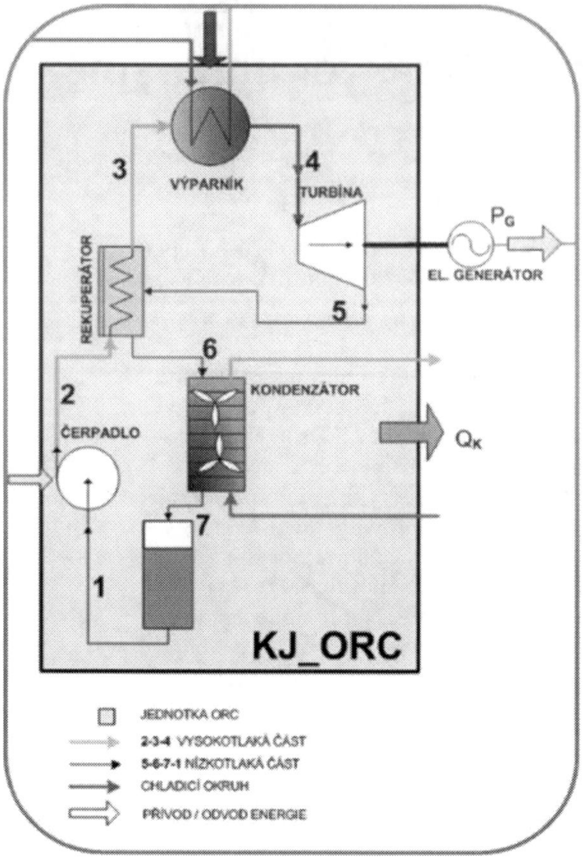

Fig. 1. Demonstration of the ORC cycle principle
1-2: the working medium in liquid state is supplied and compressed from the reservoir by a circulating pump
2-3: in the recuperator the medium is preheated by steam from the turbine outlet
3-4: in the evaporator the medium is further heated, evaporated and superheated (dry steam)
4-5: steam expands in a turbine, which does mechanical work and drives e.g.an electric generator
5-6:in the recuperator, the outlet steam is partially cooled by heat exchange with the liquid medium
6-7:in the condenser, the medium is cooled by the cooling circuit (air, water...) and condense
7-1: condensate is collected in the reservoir and then supplied by pump = the cycle is closed

2 Specifics of the proposed solution

The main criterion of our approach to the design of the μCHP ORC unit was to achieve the highest possible efficiency of the whole plant and low production costs. This is based on the economic viability of these small power plants, where the return on the investment is at the break-even point. Thus, there was a need to design a device that would have high efficiency for power generation and yet be inexpensive.

The main components we focused on were the choice of the expand machine used and the choice of the appropriate working medium.

The aim was to develop a μCHP unit with a capacity of 1.5 and 5 kW. The design and performance parameters of the 1.5 kW unit are shown in Table 1.

Table 1. Input and power parameters of the unit

Parameter	Nominal value	Unit
Inlet pressure	10.4	bar
Inlet temperature	100	°C
Heat output of the source Q_s	21	kW_t
Electrical power at terminals	2,0	kW_e
Net electrical power	1,5	kW_e
Inlet temperature of cooling water	30	°C
Outlet temperature of cooling water	50	°C
Usable heat Q_c	18	kW_t
Total efficiency of the ORC circuit (Q_c/Q_s)	85	%

2.1 Micro turbine

Most attention was paid to the choice and design of the expander. We needed to achieve the highest possible efficiency over the widest possible operating range of temperatures and steam flow rates, and at the same time we needed the most compact dimensions so that the entire ORC unit could be supplied in combination with 25 kW biomass boilers. In the end, we chose a single-stage micro turbine with an axial arrangement, partial admission and action impeller stage.

We designed the micro turbine unit in a unique arrangement subject to a utility model. The unit includes the turbine itself with inlet spiral, partial spray nozzles, impeller with balancing holes and outlet diffuser.

The turbine shaft is connected to the generator or alternator gearbox by an electromagnetic coupling. The electromagnetic coupling contains a baffle which hermetically separates the refrigerant circuit from the surrounding environment and predisposes the entire ORC unit to automatic and maintenance-free operation. The ball bearings are lubricated with refrigerant and no servicing is required. An example of the ORC micro turbine arrangement with gearbox and generator is shown in Fig.2.

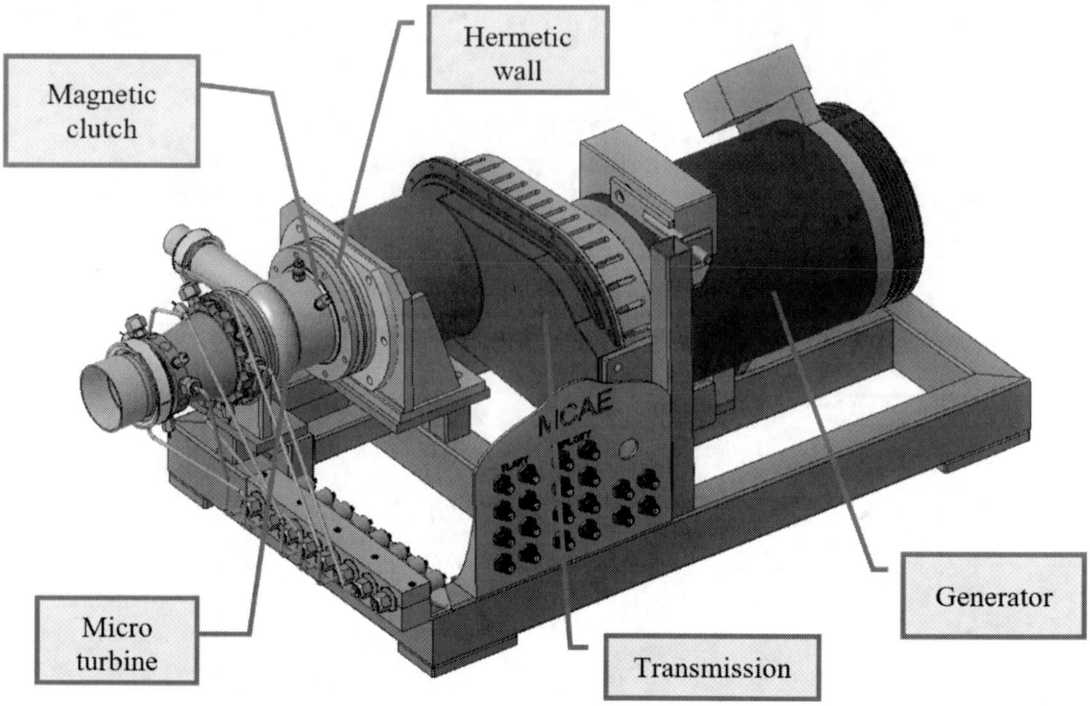

Fig. 2. Example of a 1.5 kW micro turbine with magnetic clutch, gearbox and generator.

2.2 Choosing a suitable refrigerant

The thermodynamic properties of the refrigerant can strongly influence the efficiency of the entire cycle. The thermodynamic efficiency depends on the interaction of thermodynamic properties of the medium such as critical point, specific heat, density, etc. A high vapour density will give us a smaller volume and therefore a smaller device size, a low viscosity means better heat transfer and lower frictional losses, and a lower evaporation pressure will make the device design easier.

Stability at higher temperatures (eliminating degradation) and the safety and toxicity of the medium and flammability are also important properties of the medium. Recently, there has also been a lot of attention paid to the environmental impact of GWP media (Kyoto, Montreal Protocol and Paris Agreements). Last but not least, the availability and cost of the media are equally important. An illustration of the characteristics of different media is shown in Fig. 3 together with the design of the ORC cycle of a μCHP unit in a T-S diagram. For the temperature range we used, we chose the refrigerant R245fa which was eventually replaced by the environmentally friendly version R1233 Zd(E).

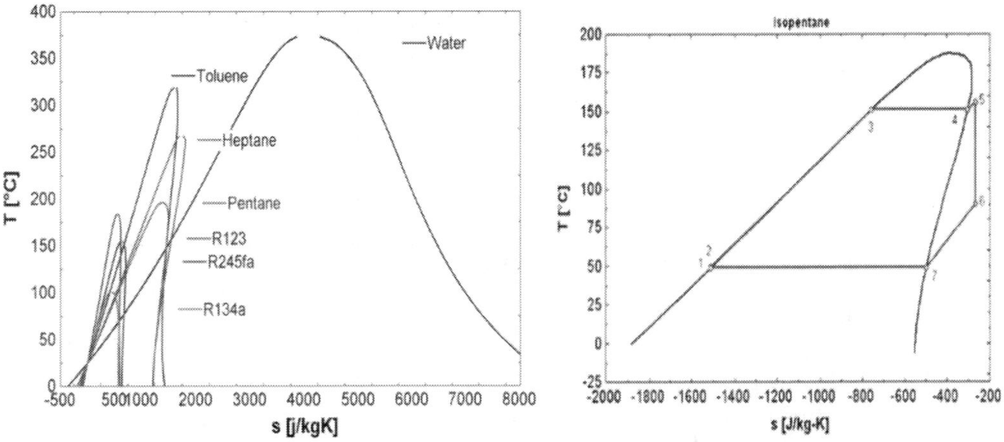

Fig. 3. On the left is an example of the behaviour of different types of refrigerants in a T-S diagram and on the right an example of the designed ORC cycle of a micro CHP unit.

2.3 Description of the μORC unit test circuit

The 1.5kWe μCHP test plant was operated in combination with an 18kWe electric boiler, which supplied 1000°C water to the steam super heater. The steam, which was subsequently processed in the micro turbine, condensed in an air-cooled condenser and from there was again supplied to the steam super heater by a pump via a recuperator.

A schematic of the ORC test circuit is shown in Fig. 4.

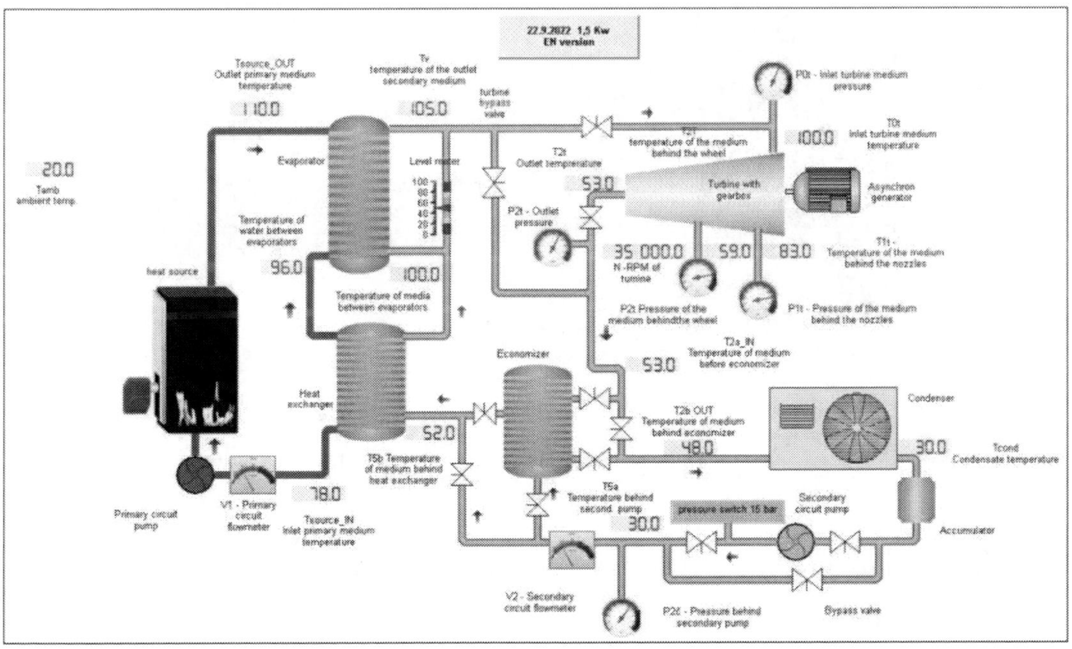

Fig. 4. Schematic of the 1.5 kW μORC test circuit.

3 Digital prototype of micro turbine with magnetic coupling

The development of the entire μORC micro turbine-generator unit was carried out using rigorous digital prototyping methods. The input values for the design of the current part of the micro turbine are shown in Table 2.

Table 2. Design parameters of the flow section of the microturbine

Parameter	Value	Unit
Working medium	R1233zd (E)	
Input temperature at the turbine casing inlet t_0	100	°C
Input pressure at the turbine casing inlet p_0	10,42	bar
Outlet pressure behind the turbine p_2	1,3	bar
Mass flow rate m_T	0,366	kg.s^{-1}
Shaft speed n_T	35 000	min^{-1}

The complete nozzle design was performed using the STAR CCM+ CFD system and the Simcenter HEEDS optimization engine using the Sherpa assistant system to select the most appropriate combination of optimization methods. An example of the optimized flow path is shown in Fig. 5.

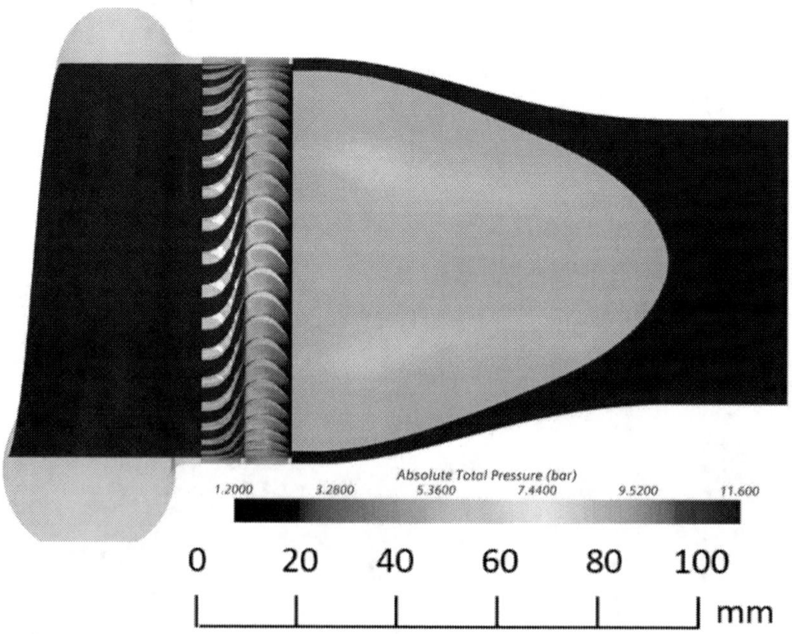

Fig. 5. Flow design of input volute and output diffuser using Simcenter STAR CCM+

Figure 6 shows the fluid flow parameters determined by CFD simulation in STAR CCM+ and the optimized shape of the input volute.

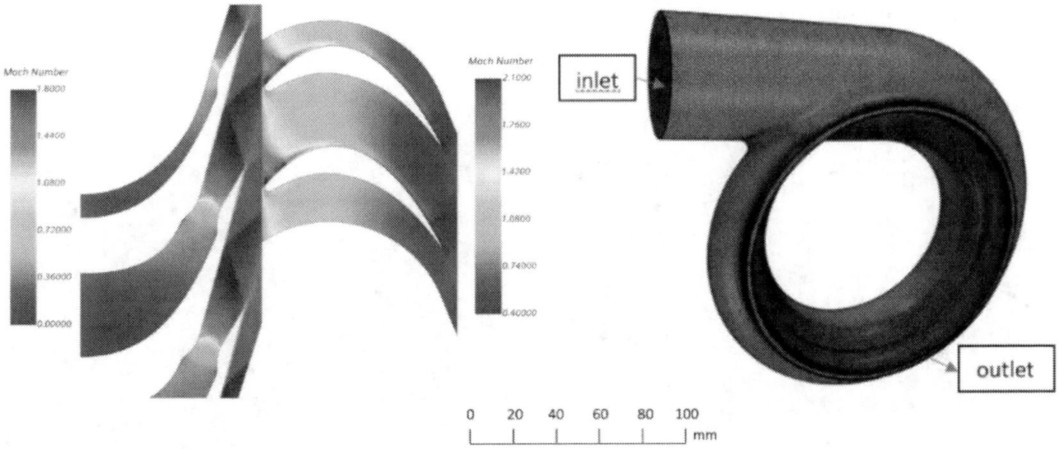

Fig. 6. Computational simulation of the turbine axial action stage and input volute using the CFD program Simcenter STAR CCM+

Furthermore, static and dynamic computation of the impeller and magnetic clutch was performed. The design of the magnetic clutch for high operating speeds of up to 35,000/min proved critical during development. It was necessary to use high-strength materials for the design of the clutch bodies and special magnet mountings due to the high stresses from centrifugal forces.

All static calculations were performed using the FEA method in Simcenter 3D. NX Nastran solvers were used to check the dynamics of the two mass system of the impeller and clutch.

4 Prototype production

The production of the components of the ORC micro turbine unit was carried out with the consistent use of 3D printing methods for the creation of working mock-ups of the unit and also functional parts of the turbine casing moulds. The flow-through parts of the turbine were final machined using 5-axis machining. An example of the printed parts and the final impeller shape is shown in Fig. 7.

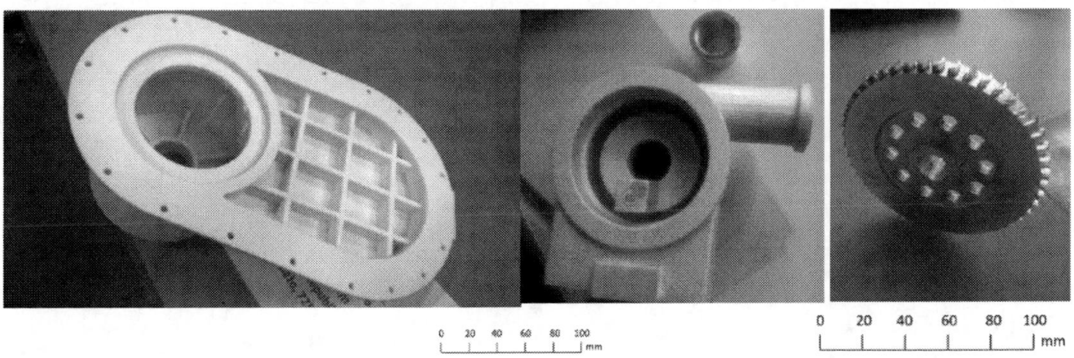

Fig. 7. On the left the gear cover made by 3D printing, in the middle the printed mould for the casing casting and on the right the impeller of the micro turbine after final modification

5 Experimental measurements

The measurements were carried out in a test room where a hot water electric boiler with an output of 18 kW was used as a waste heat source. The ORC micro turbine unit itself and the entire ORC unit at the test site are shown in Fig. 8.

Fig. 8. On the left the micro turbine unit with generator, on the right the complete ORC unit in experimental operation.

The prototype was equipped with more than 30 measuring points each to monitor physical quantities such as pressures, temperatures, turbine speeds, working medium flows, cooling water, evaporator levels, etc. Figure 9 shows the installation of both turbines with sensors and the output of signals to a sub panel of connectors for connection to the central control system (control, data acquisition).

The ORC unit was completely developed, mostly manufactured in cooperation with MCAE Systems. Completion of the circuit, piping connection and filling with the working substance (refrigerant R1233zd(E)) including the supply of the measuring panel and the connection of the sensors was carried out in cooperation with an external partner qualified for working with refrigerants.

In order to measure the power characteristics, MCAE Systems has also developed an electrical load (power resistor) with proportional power control to cover the required power range. The load in Fig. 4 is designed as a set of power resistors that are switched in parallel to the generator output according to the output current (power) requirement. It is designed to be universal for both power variants of ORC units. The regulation is discrete, in 12 steps:

For the ORC 1.5 kWe version from 0W to 1265 W, step approx. 90 W, at an operating voltage of 14 V DC. An example of the measurement record is shown in Fig. 9.

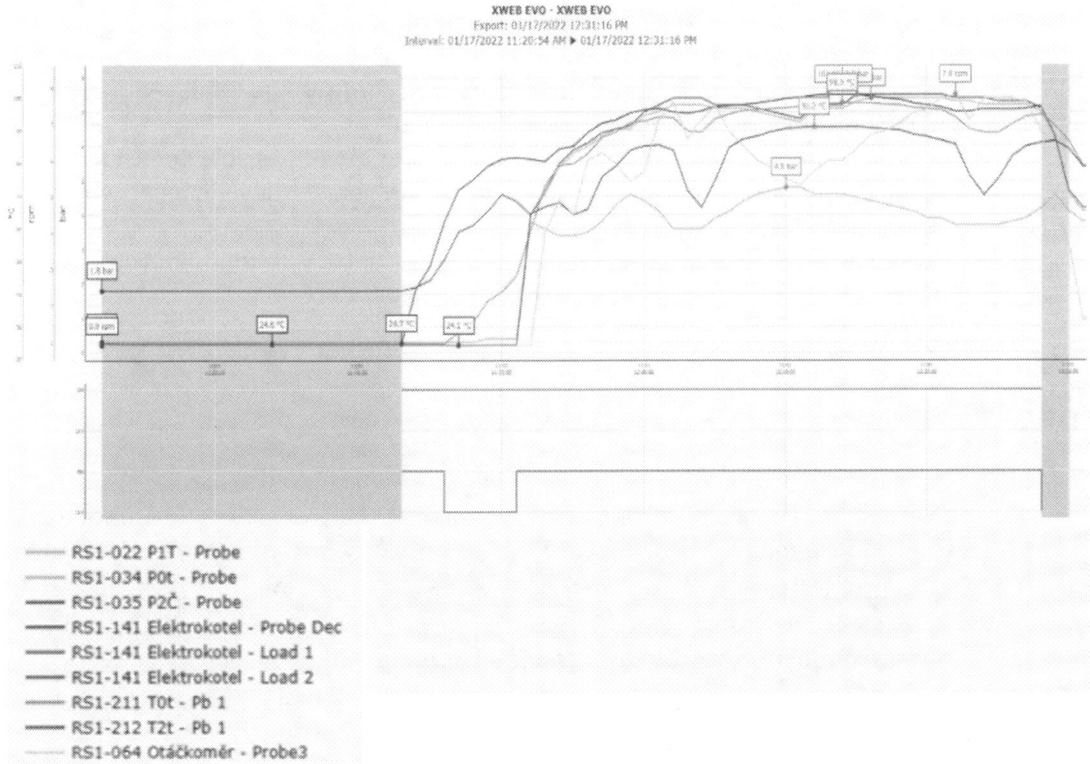

Fig. 9. The measurement record of the ORC unit shows the temperature, pressure and rotational speed.

6 Conclusion

The results of the tests confirmed that the chosen path of the ORC unit concept (we followed up on our utility model registered with the Industrial Property Office of the Czech Republic under no 34240) is correct. In this context, it is necessary to continue the established trend and to further develop the necessary technical and personnel infrastructure, to plan an extensive test programme using the existing very valuable knowledge as well as new opportunities in the field of rapid prototyping, e.g. metal printing, which is available to us, etc.

The tests have verified performance parameters, turbine efficiencies, overall circuit efficiencies, etc. under various anticipated operating modes.

The problems I encountered were of a mechanical nature at the magnetic coupling of the hermetic bulkhead. The magnetic coupling was experiencing defomation in the screw connections holding the magnets and the bulkhead was experiencing increased deflection due to the high temperatures and low temperature resistance of the material used. This has now been replaced by a more temperature and strength resistant type of GRP. The temperature field around the magnetic coupling was unproblematic and did not even reach 1000C. There was no problem with eddy currents due to the non-conductive material of the baffle.

3D printing played a major role in the development of the ORC unit, especially in the preparation of the moulds for the turbine casing casting. Using the mouldable model method. Furthermore, the gearbox body, including the housing, the panel for mounting the connectors of the measuring sensors and other small components of the miroturbine were made by 3D printing.

Further, long-term durability and reliability tests need to be carried out with monitoring of wear and tear of the unit components to gain insights to further improve the functionality and durability of the unit for full commercial use.

The overall behavioural results of the prototypes confirm the achievement of the predicted parameters in the field of μCHP.

For future development of the ORC-based unit, the variety of possible applications and thus the range of variations in unit layout and use and associated development work also needs to be taken into account.

The new product will be offered to end customers through TechSim Engineering's own business activities as well as in cooperation with business partners.

References

1. M. Bianchi, L. Branchini, A. Suman, *Experimental Performance of a Micro-ORC Energy System for Low Grade Heat Recovery*, Energy Procedia (September 2017)
2. M. Bianchi, L. Branchini, N. Torricelli, *Performance and Total Warming Impact Assesment of Pure Fluids and Mixtures Replacing HFCs in Micro-ORC Energy Systems*, Applied Thermal Engineering, 9 December 2021 (2021)
3. G. Qui, Y. Shao, S. B. Riffat, *Experimental Investigation of a Biomass-fired ORC-based Micro-CHP for Domestic Applications*, Fuel (June 2012)
4. B. Saleh, G. Koglbauer, J. Fischer, *Working Fluids for Low-temperature Organic Rankine Cycles*, Energy (July 2007)
5. S. Quolin, M. Van den Broek, V. Lemort, *Techno-economic Survey of Organic Rankine Cycle (ORC) Systems*, Renewable and Sustainable Energy Reviews (June 2013)
6. T. Deethayat, T. Kiatsiriroat, Ch. Thawonngamyingsakul, *Performance Analysis of an Organic Rankine Cycle with Internal Heat Exchanger Having Zeotropic Working Fluid*, Case Studies in Thermal Engineering (September 2015)
7. I. Hernandez-Carrillo, Ch. Wood, H, Liu, *Development of a 1000 W Organic Rankine Cycle Microturbine-generator Using Polymeric Structural Materials and its Performance Test with Compressed Air*, Energy Conversion and Management 6 April 2019 (2019)
8. P. Iora, G. Di Marcoberardino, R. Bini, *Dynamic Analysis of Off-grid Systems with ORC Plants Adopting Various Solution for the Thermal Storage*, Energy Procedia (September 2017)

Experimental indirect evaporative air conditioning system - a possible implementation

Norbert Szaszák[1*], *Attila* Juhász[2]

[1]Institute of Energy Engineering and Chemical Machinery, Department of Fluid and Heat Engineering, University of Miskolc, H-3515 Miskolc, Hungary
[2]University of Miskolc, H-3515 Miskolc, Hungary

Abstract. This article presents the principle of operation of an experimental indirect evaporative cooling system which applies liquid desiccant solution as a drying agent. This mobile system is going to be built for the investigation of the effects of different working parameters (e.g. regenerating temperature of the desiccant salt-solution, air flow rates, solution flow rates, droplet size and mixing path length, etc.) on the produced cooled and dehumidified air, as well as the effects of the initial hot-air parameters (temperature and humidity) on the effectiveness (cooling and dehumidification rate) of the system. In this paper the basic mechanisms of both the direct and indirect evaporative cooling systems are presented with their advantages and disadvantages. It is shown how the solar energy by means of solar collector(s) can be utilized as an energy source of the regeneration of the diluted desiccant solution. Besides the 3D drawing and the parts of the experimental cooler and air dryer system will be presented and explained.

1 Introduction

Both in the industrial and residential buildings the appropriate conditions of the indoor air quality (humidity and temperature) are requirement to provide adequate comfort for the people. During hot seasons (especially in summer) air conditioning systems are applied to create indoor air conditions which meets the requirements of the comfort theory [1]. This consists of two basic physical processes: one is the cooling and the other is the dehumidification of the air. According to some sources for human beings, the most ideal air-condition in a room for low-power work (e.g. in an office) is the temperature of 25°C and relative humidity of 55% [2]. The relative humidity, φ is the ratio of the partial pressure of the water vapor ($p_{p,water\ vapor}$) in the air to the pressure of equilibrium water vapor $p^*_{water\ vapor}$ belongs to a certain air temperature and atmospheric pressure (p_a):

$$\varphi = \frac{p_{p,water\ vapor}}{p^*_{water\ vapor}(T_a, p_a)} \cdot 100\% \qquad (1)$$

* Corresponding author: norbert.szaszak@uni-miskolc.hu

If the air is totally dry, its relative humidity φ=0%. If φ=100% the wet air is saturated: more water can not be evaporated into the air (the cooling effect of sweating which is based on the evaporation does not occur), or by cooling the saturated air condensation occurs.
As an example, the results of a NASA-case study can give us a clear correlation between the effectiveness and the accuracy of the human work versus the room temperature (see Table 1).

Table 1. NASA case study: effectiveness and the accuracy of the human work vs. the room temperature. [3]

Temperature, °C	24	27	29	32	35	38	41
Loss of work output, %	3	8	18	29	45	62	79
Error rate increase, %	-	5	40	300	700	-	-

Based on the data of Table 1 it is clear that the temperature has to be kept within limits to get maximal productivity and ideal comfort. In the point of view of the human comfort, besides the optimal temperature range (20°C $\leq T_a \leq$ 26°C), the humidity of the air must be in the range of 40% $\leq \varphi \leq$ 60% based on the literature [4]. This comfort zone can be observed on a psychometric diagram where the properties of the wet-air is presented as a function of the dry-bulb temperature and several parameters (absolute humidity ratio, relative humidity φ, specific enthalpy, specific volume), see Figure 1.

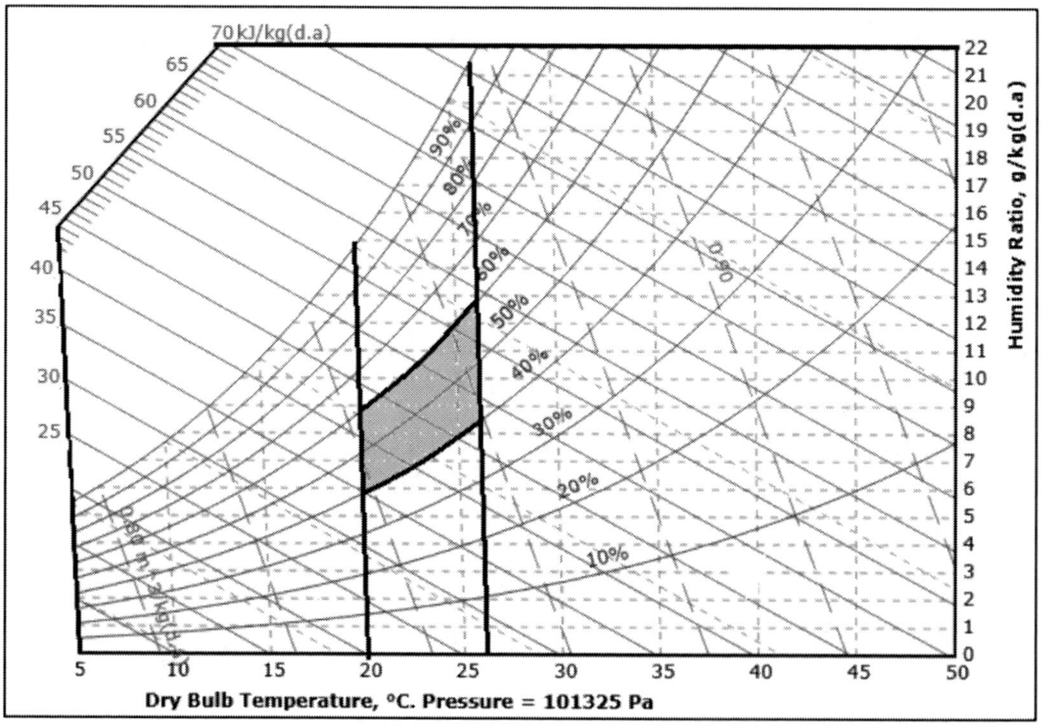

Fig. 1. The interval of the comfort zone (green) on a psychrometric chart (edited figure, original [5]).

Based on the diagram it can be seen that the human comfort zone is only a narrow region of the possible conditions: therefore – especially during hot seasons in continental and Mediterranean or the tropic climate – not only the cooling but also the dehumidification of the indoor air is necessary.
For this reason, the widespread split-air conditioning systems are applied which working principle is based on the compression of the vapor of the working medium (refrigerant). All

these kinds of A/C units require a huge amount of electrical power, primarily for the compressors driven by electric motors (a smaller portion of power is required for the control unit and the fan of the evaporator in the conditioned room and the fan of the condenser unit outside). The most residential use units have an average coefficient of performance (COP) in the range of 3÷4 W/W which means, each 1W of electrical power carries 3 or 4W of heat from the conditioned space to outside environment. This may seem a good value, however the production of the electricity and the total energy chain has relatively low efficiency which compensates the optimistic COP value of such an A/C device.
An alternative of the compressor A/C unit is the evaporative air cooling (in some cases air conditioning) system. Both direct and indirect evaporative devices are available. The direct term means that the temperature of the cooled air is decreased by the energy demand of the evaporation of water in the treated air itself, which goes together increased relative humidity. This kind of direct evaporative coolers hence are called swamp coolers (the air of the swamps is very humid). A mathematic model for the direct evaporative cooling was presented by Camargo et al [6], and later the theory was proven by their experimental results [7]. Following them, a numerical model was developed and validated by experimental results by Kovacevic and Sourbron [8]. Mathematical calculations and experiments were performed in topic of evaporative cooling in a greenhouse by Franco et al. [9]. The review of the environmental effects and the potential in energy saving of the evaporative cooling was presented by Xuan et al. [10,11]. Evaporative cooling based on absorbent material was reviewed by Rafique et al. [12], while the effectiveness of modern materials was compared to materials used in traditional systems. In case of the indirect evaporative system the air which is humidified and therefore cooled is physically separated but is thermodynamically coupled with the indoor air which is hence cooled but not humidified. An indirect evaporative system with Maisotsenko-cycle was investigated on hot, Mediterranean climate in Greece by Rogdakis et al. [13]. It was found that based on the experimental data this kind of cooler is capable to satisfy the cooling demand at a high efficiency.

Based on the previous information it can be seen that both the direct and indirect evaporative cooler solutions can be applied, the appropriate type is strongly depended on the demand regarding the humidity of the cooled air. Where the high humidity is not an issue or preferred (e.g. mushroom-farms), direct cooling should be chosen, otherwise indirect system is preferred.

2 Direct evaporative cooling

The direct evaporative cooling is based on the vaporization of the water which is an endothermic process. It is well known that the boiling of the water requires a huge amount of heat while the temperature of the boiling water is constant. In this process the heat added to the water helps in the phase-change of the liquid to vapor, and it is called latent heat of vaporization or heat of transformation. This latent heat, l_v (energy per unit mass of water) is depended on the temperature of the boiling (saturation temperature, T_{sat}), which is a function of its pressure (during the boiling process it is called saturation pressure, p_{sat}). Table 2 shows some values of the latent heat as a function of the saturation temperature and pressure.

Table 2. Latent heat of vaporization of the water [14].

T_{sat} [°C]	20	40	60	80	100	120
p_{sat} [bar]	0.0234	0.0738	0.1994	0.4739	1.014	1.985
l_v [kJ/kg]	2454.1	2406.7	2358.5	2308.8	2257.0	2202.6

It can be seen that the specific latent heat is greater in the direction of lower temperatures, but it is more important that it has a huge value in the whole ambient temperature range while the cooling of the air can be necessary: at the temperature of 40 °C more than 2.4 MJ of heat is needed to evaporate 1 kg of water. Thus, in case of direct evaporative cooler the evaporation of water in the air removes a huge amount of heat energy which cause the decrease of the temperature of the moist air. The enthalpy of the moist air is the energy content relative to a datum of the dry air portion which is the function of its temperature and the energy content of the water vapor in the air at the temperature of the air and at the partial pressure of the vapor (thus the relative humidity, φ). If water (which temperature equals to the temperature of the air) is evaporated in the air, the total specific energy of the wet air (namely the enthalpy) remains constant while a part of its energy is used to evaporate the water. This can happen only if the temperature of the air decreases during the humidification process and it is called isenthalpic process. The process of the direct evaporative cooling can be observed in Fig. 2 where the initial conditions (**1**): T_1=35°C and φ_1=20% after the cooling (**2**): T_2≈20°C and φ_2=90%.

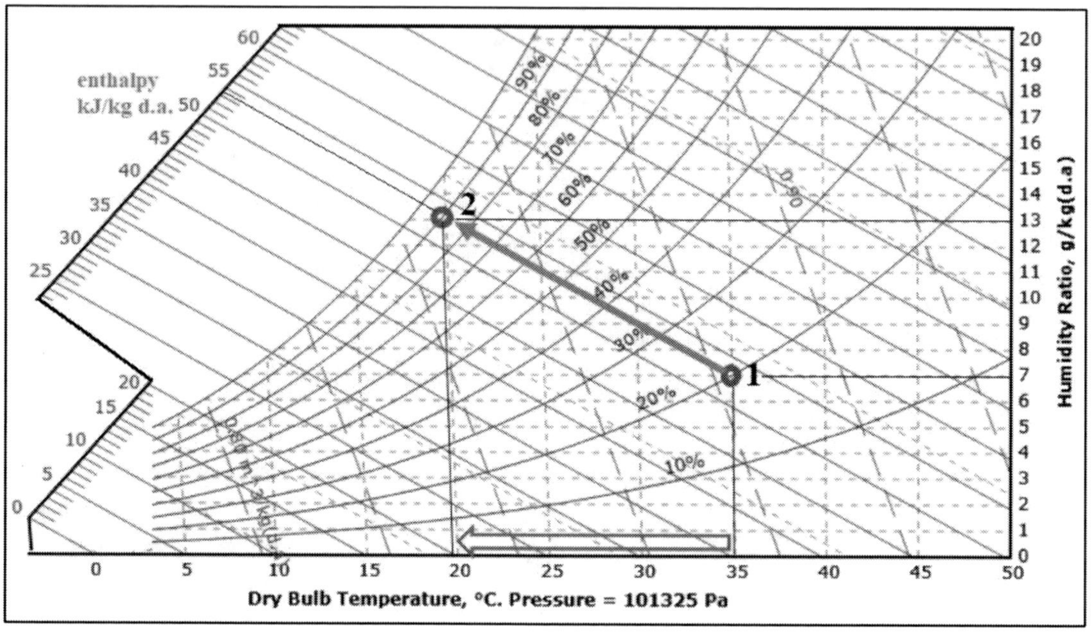

Fig. 2. Adiabatic, direct evaporative cooling on a psychrometric chart (edited figure, original [5]).

The arrow of the process is parallel to the neighbour constant enthalpy lines and points in the direction of the increasing relative humidity (and so the increasing absolute humidity, see vertical axis on the right). Based on the vertical axis of the diagram, in order to decrease the temperature of the hot air (**1**) to the condition (**2**), the evaporation of totally 13-7=6 g water/kg dry air is required. Moreover, it can be seen that the minimum temperature which can be achieved by evaporative cooling is theoretically given: the temperature which belongs to the φ=100%, which is called dew point. This limits the maximum achievable temperature drop and is depended on the initial properties of the air.

In order to evaporation occurs the water has to be in contact with the flowing air. One way is the atomization into the air stream by spraying the water at high velocity. The atomized water droplets are small therefore their sedimentation rate is negligible, so they have enough time to be evaporated before getting out from the air stream. On the other hand, the smaller the droplet size, the greater the specific area is which is one of the most important conditions of the evaporation. Such devices are the vapor gates with high pressure water sprayers or the swamp coolers which in the most cases use axial fan equipped with water sprayer.

Another solution to achieve great contact surface between air and water is the application of a porous evaporator media which is wetted by the water and has low flow resistance so the air can pass it while it evaporates the water by contacting the surface of the liquid film. The evaporator media can be made of wood shavings, cellulose or synthetic threads. Nowadays new, high-performance materials are used as the cellulose-honeycomb, coconut hair, melamine-paper, ceramics, or steel wool. Bishoyi et al. [15] proved experimentally that the honeycomb-structure is the most effective solution as an evaporative medium.
All of the previously presented direct coolers have the same disadvantage: the cooling goes together with the humidification, which may cause discomfort and in case of coolers with evaporative medias bacteria and fungi can settle causing odour and infections. To overcome these issues, so called indirect evaporative coolers can be chosen.

3 Indirect evaporative cooling

The difference between the direct and indirect systems is that the evaporation of the water does not increase the absolute humidity of the conditioned air, instead a secondary air stream is used to evaporate the water thus cool down. This cooled and humid air is then applied to cool down the air to be treated by using heat exchanger. Such a solution is the Maisotsenko cycle [16], which is capable to cool the air to the dew point temperature of the incoming secondary air. This cycle applies a special crossflow heat exchanger and evaporator unit in which a portion of the incoming air is used to cool down another part of the air while evaporating the water.
If the air with a certain vapor content is cooled e.g. in a surface heat exchanger, its absolute humidity ratio is constant until its dew point, where the relative humidity reaches the value of 100%. This cooling method can be seen in Fig. 3 where the hot air (state **1**) is cooled to condition (**2**).

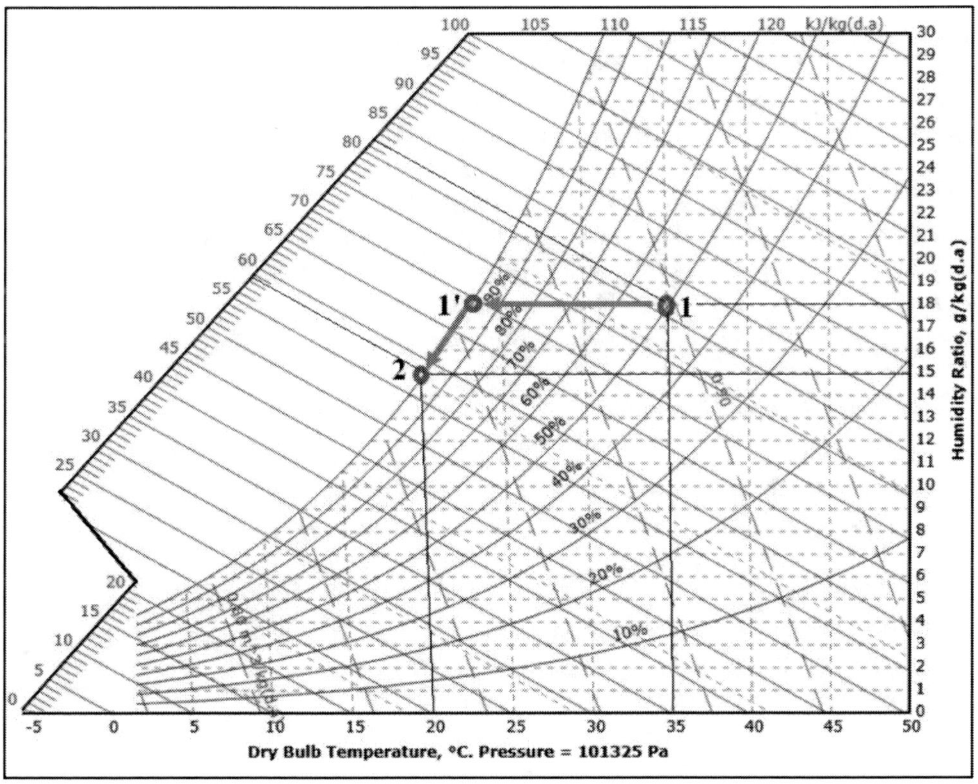

Fig. 3. Adiabatic, indirect evaporative cooling on a psychrometric chart (edited figure, original [5]).

During the indirect cooling of the initial state of the air is T_1=35°C and φ_1=50%. While cooling its temperature decreases along with its enthalpy, while its absolute humidity keeps constant until state **1'** while its relative humidity becomes 100%, which is called the dew point. From this point if more heat is subtracted by the heat exchanger the temperature and the enthalpy of the air decreases while its relative humidity keeps constant (**1'–2**). This means a decrease in the absolute humidity thus the difference of the absolute humidity (18-15=3 g/kg d.a., see diagram) condensates on the surface of the heat exchanger. This process happens in case of the indoor part of the conventional compressor driven A/C units. The reason why the relative humidity of the air in the room is not high is that the air in state **2** is mixing with the air in the conditioned room while its temperature increases and its relative humidity decreases. This process finally decreases both the temperature and the humidity since both energy subtraction and condensation occurs.
In the previous case condensation occurred so the processed air had less humidity. But this is not the case if the initial humidity is lower or the temperature drop is not enough to reach the dew point, which can happen when indirect evaporative cooling is used with smaller temperature drop. In this case there is no change in the absolute humidity, only the temperature decreases. This means a cooled air with higher relative humidity which is may not suitable for the people. To solve this problem, it is necessary to dry the air in a separated process or, as it is going to be presented a complex process can be applied where both cooling and dehumidification occurs. This process will be applied in our experimental evaporative air conditioning system.

4 Experimental indirect evaporative cooler and dehumidifier unit

As it was shown, due to the small temperature drop of the evaporative cooling condensation may not occur in case of indirect cooling which results cooled but humid air. To solve this problem a hygroscopic material, so called desiccant has to be applied. The desiccant may be a solid substrate with high specific surface area, e.g. several metal-salts or silica gel which adsorbs a huge amount of water on its surface, or liquid desiccant, e.g. salt-water solution. The cooled air has to be in contact with the desiccant material in order to give some of its humidity to the desiccant and hence it becomes dryer. This process is exothermic, so it increases the temperature of the desiccant and the air too, so it has to be taken into account.
Our experimental device was inspired by Robert Heron by his indirect-evaporative cooler video on the "Tech Ingredients" YouTube-channel [17]. This device applies solution of calcium-chloride ($CaCl_2$) and water which is pumped into a sprinkler in a desiccant and cooling-tower where both cooling and drying happens. During the process of drying the initially concentrated desiccant solution becomes more and more diluted so it has to be regenerated. This can be done by heating up and then mixing with ambient air which becomes more humid while the desiccant becomes more concentrated.
The 3D model of the experimental system can be observed in Figure 4.

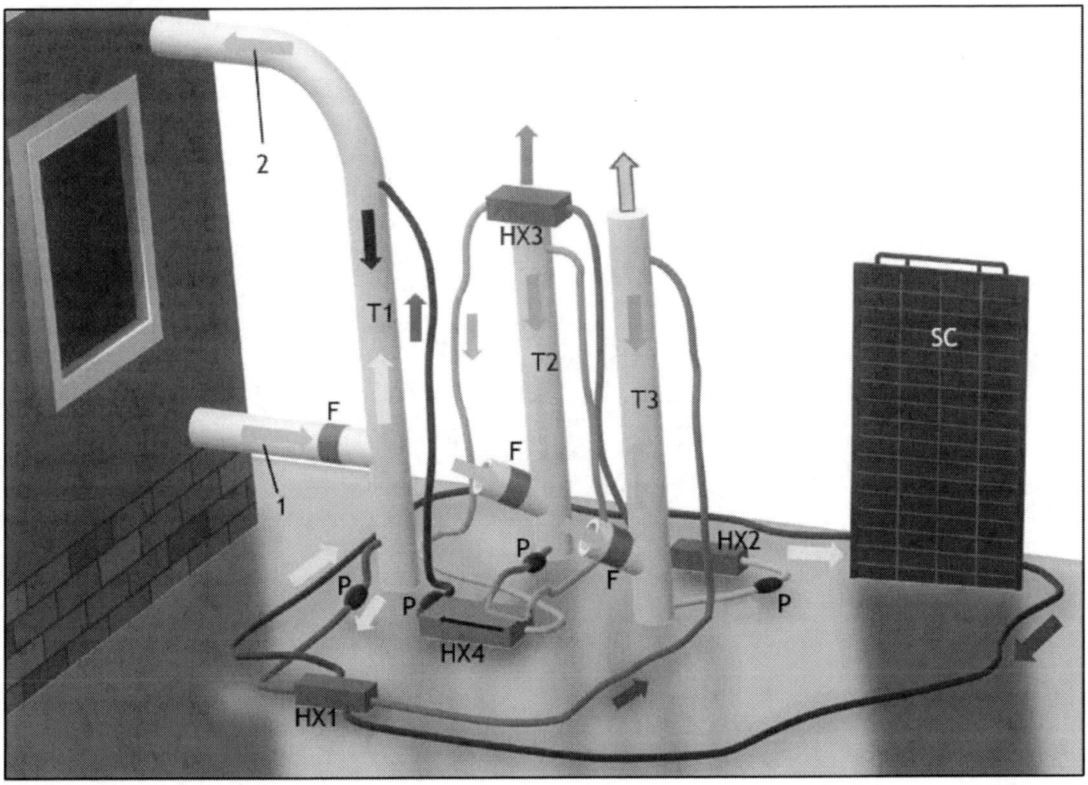

Fig. 4. 3D drawing of the experimental indirect evaporating A/C system. Notations: 1: room air to be treated, 2: conditioned air, F: fan, HX: heat exchanger, T: tower, P: pump, SC: solar collector.

The operation of this system is the following. The warm and humid room-air (1) is circulated by an axial fan (F) through the cooler and dehumidifier tower (T1) and is re-circulated to the air-conditioned space through pipe 2. In tower T1 a sprinkler is used to produce desiccant droplets of cooled and concentrated $CaCl_2$ solution that falls downward, in counterflow direction. Here the desiccant absorbs moisture from the air while it cools the air down, too. As a result, the desiccant solution warms up and become diluted, so it must be regenerated. Following this diluted solution is circulated by a pump (P) and it passes through the liquid-liquid heat exchanger (HX1) in which it is heated up by the hot heat transfer fluid coming from the solar collector (SC). The heat which is required to heat up the diluted desiccant solution can be produced by using several kind of energy sources, like electrical energy, natural gas or, in our case it utilizes the energy captured in solar collector. In terms of the whole A/C system this energy input is the dominant to maintain the system itself which makes it clear that this energy must be provided by a cheap and or a sustainable source. Since thermal energy is needed, one of the most obvious solution is the solar energy utilizing solar collectors.

The hot and diluted solution is then introduced into the regenerator tower (T3) where a sprinkler makes droplets which free fall downwards, against the flowing outside air which is blown through the tower by an axial fan (F). As a result, the hot diluted solution gives a part of its heat and water content to the flowing air, so a warm and concentrated solution leaves the tower at its bottom. A pump (P) pumps the warm solution through an air-liquid heat exchanger (HX2) where it gives off some of its heat thus it cools down a little bit higher temperature than that of the ambient air. Following this solution is pumped through another air-liquid heat exchanger (HX3) where it is cooled down below the temperature of the ambient air. To do so, evaporative cooling is applied in the tower 2 (T2).

In T2 tower distilled water is circulated in counterflow direction to the ambient air. A sprinkler is used at the top of the tower to make water droplets which fall downwards in the upwards flowing air. As a result, the droplets partially evaporate (need to be refilled) and hence cool down while the flowing air is cooled as well (see Fig. 2 for the process of the direct evaporative cooling). This cooled and moist air is therefore capable to cool down the desiccant solution below the temperature of the ambient air. Since in tower 2 the water is recirculated in a closed system by a pump (P), its temperature – with good approximation – decreases to the dew point of the ambient air. This cooled water is then used to further cooling of the diluted solution coming from HX3 by using a counterflow coaxial liquid-liquid heat exchanger (HX4). Finally, the concentrated solution has almost the same temperature as the dew point temperature of the ambient air, and it is pumped back to the tower 1 (T1) to cool and dry the indoor air so the cycle starts again.
For better understanding of the cycle, the heat transfer and the temperature-changes of the medias involved in the process can be observed on a temperature as a function of transferred heat (*Q*-*T*) diagram in Figure 5. Note that nor the temperatures nor the transferred heats are to scale.

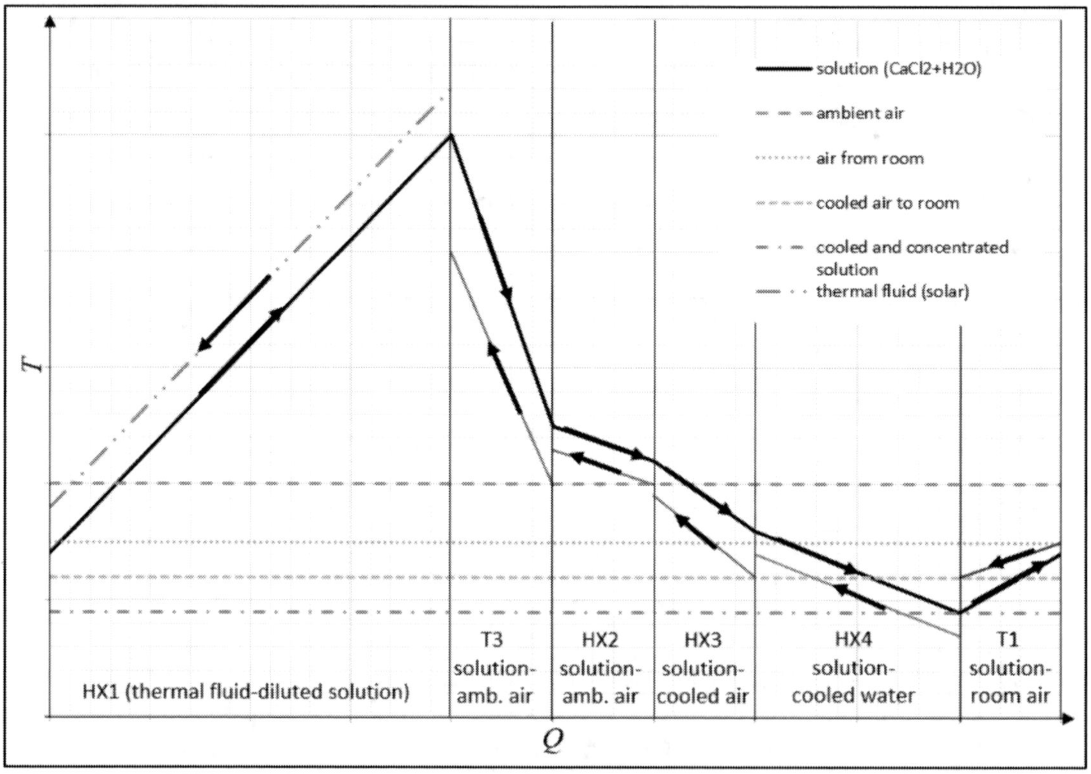

Fig. 5. The *T*-*Q* diagram of the heat transfers occur in the medias of the cycle. Ascending arrows denote heating while descending arrows denote cooling of the media.

The previously introduced system is going to built in the year 2022. As a first step the heat required for regeneration will be provided by electric power – instead of solar power – which can be controlled well. The system will be equipped with several thermocouples and humidity-sensors as well as mass flow meters and electric power meters to measure the power consumption of the fans and the pumps. The laboratory of the department has air conditioning chamber with 41 m^3 of airspace which in one case will be used as the room to be air conditioned so the experimental system will be placed outside of the chamber. During experiments target the investigation of the effects of the outer air properties the system will be placed in the chamber and only the air inlet (1) and outlet (2) will be connected to the outer air (the air of the laboratory). The purpose of the experimental system and its investigation is to characterize its working parameters and to optimize the parts of the system.

5 Conclusion

In this paper the fundamental processes applied in evaporative coolers were presented with a possible implementation of an experimental indirect evaporative air cooling and dehumidification system which applies hydrophilic salt solution as a desiccant and cooling agent. It was presented how the direct evaporative systems increase the humidity of the cooled air which has to be avoided ins several applications. Besides the disadvantage of the indirect evaporative cooling was presented as well, namely the increase of the relative humidity while the absolute humidity remains constant. This is due to the relatively low achievable temperature drop by the evaporative cooling which is not sufficient to develop condensation which would decrease the water content of the treated air.
Both leaded to the indirect evaporative system which is capable to decrease the absolute humidity while cooling the treated moist and wet air. The power consumption of the presented system is mainly due to the heating of the desiccant solution for regeneration. This heat input can be provided by utilizing solar energy (e.g. solar collector). In this case only the power demand of the fans, pumps and control should be provided by electric power, which makes such a system both economical and sustainable.

References

1. D. Bishoyi, K. Sudhakar, *Experimental performance of a direct evaporative cooler in composite climate of India*, Energy and Buildings, **153**, pp. 190-200 (2017)
2. ANSI/ASHRAE standard 55–2010: *Thermal Environmental Conditions for Human Occupancy*, American Society for Heating, Refrigerating and Air Conditioning Engineers Inc, Tullie Circle, NE, Atlanta (1791)
3. T. Bottomley, E. Roth, *Compendium of human responses to the aerospace environment. 6. Thermal environment.* NASA CR-1205(1). NASA Contractor report, United States, NASA (1968)
4. M. N. Shaharon, J. Jalaludin, *Thermal comfort assessment-A study toward workers' satisfaction in a low energy office building.* American Journal of Applied Sciences, **9**,7, 1037. (2012).
5. Online Interactive Psychrometric Chart. http://www.flycarpet.net/en/psyonline; access: 05.02.2022.
6. J.R. Camargo, C.D. Ebinuma, S. Cardoso, *A mathematical model for direct evaporative cooling air conditioning system,* Engenharia Térmica **4**, pp. 30–34. (2003)
7. J.R. Camargo, C.D. Ebinuma, J.L. Silveira, *Experimental performance of a direct evaporative cooler operating during summer in a Brazilian city*, Int. J. Refriger. **28**, pp. 1124–1132. (2005)
8. I. Kovacevic, M. Sourbron, *The numerical model for direct evaporative cooler*, Appl. Therm. Eng. **113**, pp. 8–19. (2017)
9. A. Franco, D.L. Valera, A. Pena, *Energy efficiency in greenhouse evaporative cooling techniques: cooling boxes versus cellulose pads*, Energies **7**, pp. 1427–1447. (2014)

10. Y.M. Xuan, F. Xiao, X.F. Niu, X. Huang, S.W. Wang, *Research and application of evaporative cooling in China: a review (I)−research*, Renew. Sustain. Energy Rev. **16**, pp. 3535–3546. (2012)
11. Y.M. Xuan, F. Xiao, X.F. Niu, X. Huang, S.W. Wang, *Research and application of evaporative cooling in China: a review (II) −system and equipment*, Renew. Sustain. Energy Rev. **16**, pp. 3523–3534. (2012)
12. M. M. Rafique, P. Gandhidasan, S. Rehman, L. M. Al-Hadhrami, *A review on desiccant based evaporative cooling systems*, Renew. Sustain. Energy Rev. **45** pp. 145–159 (2015)
13. E. D. Rogdakis, I. P. Koronaki, D. N. Tertipis, *Experimental and computational evaluation of a Maisotsenko evaporative cooler at Greek climate*, Energy Build. **70** pp. 497–506. (2014)
14. J. M. Moran, H. N. Shapiro, D. D. Boettner, M. B. Bailey, *Fundamentals of Engineering Thermodynamics*. John Wiley & Sons Inc. (8th issue), (2014)
15. D. Bishoyi, K. Sudhakar, *Experimental performance of a direct evaporative cooler in composite climate of India*, Energy and Buildings **153** pp. 190-200. (2017)
16. H. Caliskan, A. Hepbasli, I. Dincer, V. Maisotsenko, *Thermodynamic performance assessment of a novel air cooling cycle: Maisotsenko cycle*, I. J. Refrigeration, **34**, 4 pp. 980-990, (2011)
17. R. Heron: Tech Ingredients channel (https://www.youtube.com/c/TechIngredients); access: 07.13.2022.

MATEC Web of Conferences 367, 00022 (2022) https://doi.org/10.1051/matecconf/202236700022
Power System Engineering 2022

Experimental Investigation of 3D Dynamical Effects in a Wake behind Circular Cylinder

Václav Uruba[1,2*], *Pavel* Procházka[1]

[1]Department of Fluid Dynamics, Institute of Thermomechanics, Czech Academy of Sciences, Dolejškova 5, 182 00, Prague, Czech Republic.
[2]Power System Engineering, Faculty of Mechanical Engineering, University of West Bohemia in Pilsen, Univerzitní 22, 306 14 Pilsen, Czech Republic.

Abstract. A circular cylinder in crossflow is subjected to the study of 3D dynamical structure of its wake. The typical dynamics is characterized by quasi-periodic behaviour called Kármán – Bénard vortex street with the typical frequency in dimensionless form known as Strouhal number. The experimental study relies on stereo Particle Image Velocimetry method, the plane of measurement is perpendicular to the flow in the distance 3.8 cylinder diameters in streamwise direction. Reynolds number was around 5 thousand. The structure and dynamical behaviour of the wake along the cylinder axis is studied in details. On the top of the known streamwise velocity deficit definition of the wake, the streamwise oriented dynamically evolving vortices are detected. For the detailed dynamics examination, the Oscillation Pattern Decomposition method was used. Waves of various structures travelling along the cylinder axis as well as some pulsations were identified. The wake dynamics is characterised by the streamwise velocity component deficit and the streamwise vorticity concentration. Both structures are moving periodically in transversal direction with the same frequency, the velocity deficit is shifted by a quarter of period behind the vorticity concentration in the shedding process.

1 Introduction

A circular cylinder in crossflow is a typical engineering problem, appearing in practice very often in various forms. In fluid mechanics, this case is considered as a typical canonical case, with relatively simple and straightforward boundary conditions, but complex structure of the flow. The wake is highly nonstationary and 3D in the same time.

In literature, extensive information resources could be found, starting from dawn of fluid mechanics until now. The first systematic work by von Kármán [1] followed by overview paper by Roshko [2], and Williamson [3] to mention the most cited. Interest of researches is not changing a lot during the time, however the technical means applied in the research, involving both experimental approach and mathematical modelling, develops substantially.

* Corresponding author: uruba@it.cas.cz

Recently we have appropriate means available for addressing even the most complex aspects of the phenomenon.

The wake dynamics has been simplified very often to a single-frequency, periodical process with 2D topology. The Kármán vortex street is defined and characterized by the single Strouhal frequency. The wake structure along the cylinder is considered to be synchronized, and thus 2D. In reality, this approach could be accepted for the fully laminar case only characterized by a very low Reynolds number (*Re*), bellow 150. For higher *Re* values the turbulence is present in the wake, the flow should be a broad-band process with random component and fully 3D topology. Neglecting 3D aspects could lead to fully unphysical conclusions, e.g. zero drag, so called d'Alembert's paradox, which could be resolved only by considering the 3D reality, see e.g. Hoffman, Johnson [4]. Some more ideas could be found e.g. in [5].

In the presented paper we are about to address the 3D aspects of the flow in the turbulent wake behind a long cylinder.

The general conclusions are valid not only for circular cylinders, but for cylinders with the different cross-sections as well, e.g. rectangular, the situation is qualitatively similar, see e.g. [6].

2 Description of experiments and analysis methods

The experiments were carried out in the Laboratory of Turbulent Shear Flows in the Institute of Thermomechanics ASCR, Prague. The existing flow-down facility with test section 250 x 250 mm^2. The generated flow was low turbulent level and regular. The facility is described in more details e.g. in [7].

2.1 Experimental setup

As a typical case we have chosen the circular cylinder with diameter $D = 15$ mm and the incoming velocity $U_e = 5$ m/s. The Reynolds number was thus around 5 thousand. The cylinder was sufficiently long (about 250 mm) to consider 2D boundary condition, its aspect ratio was about 16. The cylinder (in grey) and plane of measurement (in red) are shown in Figure 1.

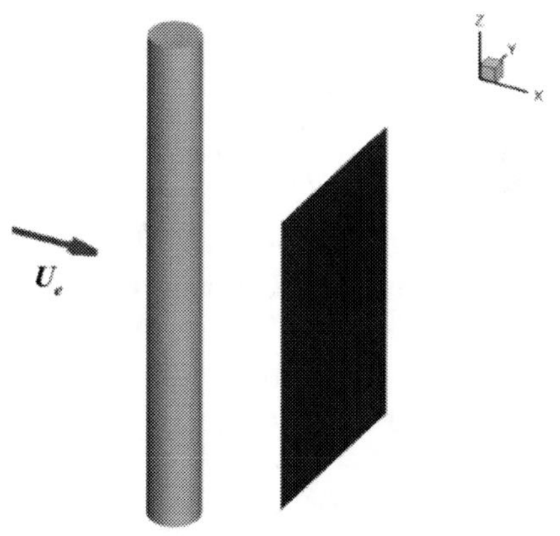

Fig. 1. Cylinder and plain of measurement.

The plane of measurement was chosen in the position 3.83 *D* downstream the cylinder axis. It was parallel to the cylinder axis and perpendicular to the incoming flow. The distance from the cylinder has been chosen just behind the "near wake", where the back-flow appears temporarily. The situation could be compared with results of measurement in the streamwise plane of measurement, results were presented in [7].

2.2 Method of measurements

The Particle Image Velocimetry (PIV) method was applied in its stereo version. This method allows for evaluation of all 3 velocity components in the plane of measurement.

The measuring apparatus consists of laser and 2 CMOS cameras by Dantec Company. The laser is New Wave Pegasus, Nd:YLF double head with wavelength of 527 nm, with maximal frequency 10 kHz and shot energy of 10 mJ (for 1 kHz), thus the corresponding power is 10 W per one head. The 2 cameras VEO 410 with resolution of 1 280 x 800 pixels are able to acquire double-snaps with frequency up to 2 500 Hz (full resolution) and they use internal memory 16 GB each. The Scheinpflug mounting were used for the cameras lenses to get the focus planes identical with the laser-sheet plane. The situation is depicted in Figure 2.

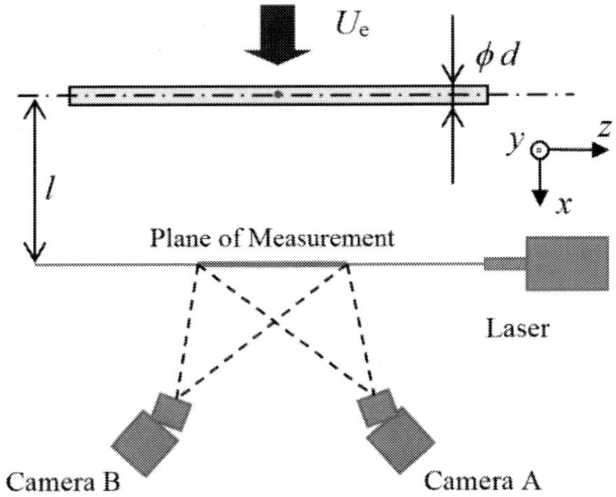

Fig. 2. Stereo PIV configuration.

Please note the Cartesian coordinate system in Figure 2 with origin on the cylinder axis (red dot), x in the streamwise direction and z is identical to the cylinder axis and y is perpendicular to both flow and cylinder axis.

The measurement records consist of 4 000 measurements performed with frequency 2 kHz. The Safex generator of particles in the form of oil droplets was used, the mean diameter of the particles was 1 μm, as shown in experimental studies by the apparatus supplier (see [11]). The data were acquired and post-processed using the Dynamic Studio and Tecplot softwares. The resolution of the field of interest was 94 x 56 vectors.

More details on the measuring technique are e.g. in [8].

2.3 Methods of data analysis

The methods of raw data evaluation and analysis are to be presented now.

The used statistical methods are based on averaging of acquired data over time. The mean values of the velocity components and Turbulent Kinetic Energy (*TKE*) distributions are to be presented. The *TKE* is defined as usual:

$$TKE = \frac{1}{2}\left(\overline{u^2} + \overline{v^2} + \overline{w^2}\right), \tag{1}$$

here u, v, and w are fluctuating parts of velocity components in x, y and z directions, respectively, the bar denotes averaging.

The Oscillation Pattern Decomposition (OPD) method has been used to study the flow-field dynamics, see [9]. The OPD method provides series of OPD modes, each OPD mode is characterized by the fluctuations topology in complex form (consisting of real and imaginary parts), frequency and attenuation of the pseudo-periodic (oscillating) behaviour. The fluctuations are expected in the form of cyclostationary process, when the real and imaginary topology appear in frequency of a mode quarter period, first positive and then negative velocity distributions. The Attenuation or amplitude decay is described by so called e-folding time representing the mean time period of the mode amplitude decay by factor "*e*". The other decay characteristic is the dimensionless "periodicity" which expresses the e-folding time in multiples of periods of the OPD mode defined by its frequency:

$$p = \tau_e \times f\,, \tag{2}$$

here p is periodicity, τ_e is e-folding time and f is the mode frequency.

The periodicity value could be considered as a measure of "relevance" of a given OPD mode. The frequency will be represented in dimensionless form as Strouhal number Sr:

$$Sr = \frac{f \times D}{U_i} \tag{3}$$

More details on the OPD method could be found in [9] and [10].

3 Results

The results of experiments are to be presented in 3 parts. First, analysis of the instantaneous topologies are to be shown, then the statistics, and finely the OPD analysis of the spatio-temporal data will be presented.

The plane of measurement (yz) will be shown, the cylinder axis is the line $y = 0$, in all graphs also the cylinder silhouette will be indicated by the broken lines $y = \pm 0.5$.

Please note that all coordinates are dimensionless, expressing numerically multiples of the cylinder diameter D. The velocities and TKE are normalised by U_e, while vorticity by the combination of D and U_e.

3.1 Acquired data

The quasi-periodical behaviour of the wake is characterized by the Strouhal frequency with corresponding Strouhal number $Sr = 0.21$. The Strouhal frequency was identified about 71 Hz. In the time-series of snapshots were selected situations corresponding to limiting positions of the wake.

Some of these situations are to be shown graphically in two forms. In both graphics the in-plane velocity vectors are to be shown. The first "velocity" form (a) shows the third, streamwise instantaneous velocity component u by colour map, this representation indicates the low-velocity fluid in the wake. In the second "vorticity" form (b), the colour represents the streamwise vorticity component, negative in blue and positive in red. This form shows presence the streamwise vortices in the instantaneous wake. To facilitate this identification, the vector-lines were added arbitrarily.

The 4 situations were selected to be presented, characterizing one period of the vortex shedding process. The wake is moving periodically from position y positive values to negative and back. We will show the situation in the following order: $y > 0$, $y = 0$, $y < 0$, $y = 0$. Those situations are shifted by a quarter of the shedding period.

The first example, snapshot No. 193, represents situation when the wake is located in top dead centre, close to the upper cylinder contour, y is positive. The situation is shown in Figure 3, the velocity form is on left (a) and the vorticity form on right (b). The vector lines constructed over the spanwise velocity components are used to make the streamwise vortical structures clearer. Please note strong streamwise vortices of both orientations distributed symmetrically around the cylinder axis, while the streamwise velocity deficit is shifted significantly upwards.

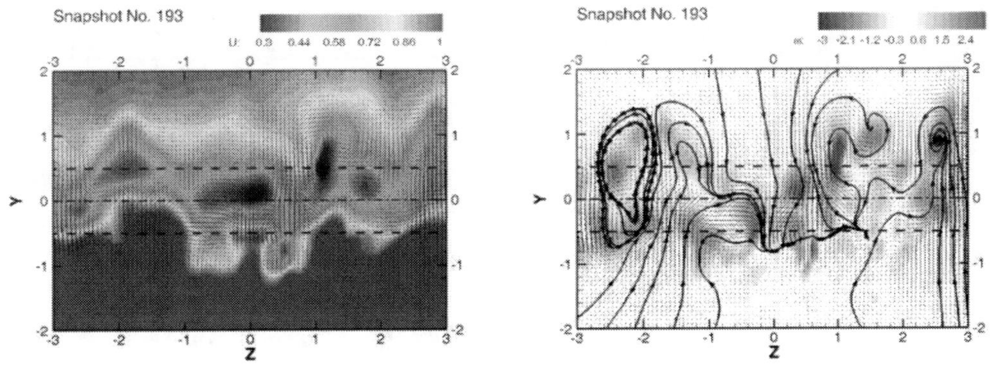

(a) (b)

Fig. 3. Instantaneous snapshot, wake shifted to the upper position. Streamwise velocity component distribution (a), streamwise vorticity component distribution (b).

Next stage is the central wake position, the wake is moving from up to down. In Figure 4a there is a snapshot No. 200 showing symmetrical wake in vertical direction, y is close to 0. However the vorticity distribution in Figure 4b suggests position of streamwise vortices on the level of lower cylinder contour $y = -0.5$, the upper part is nearly vortex-free.

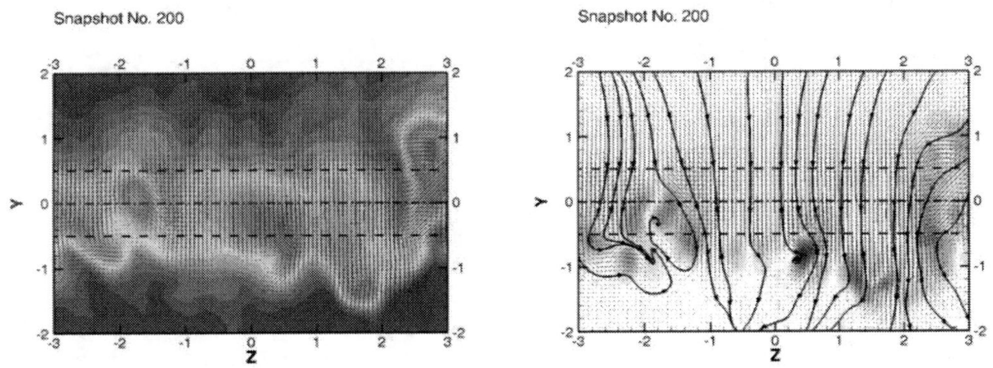

(a) (b)

Fig. 4. Instantaneous snapshot, wake in the middle position directed downwards. Streamwise velocity component distribution (a), streamwise vorticity component distribution (b).

The next position is represented by the snapshot No. 207 showing the situation when the wake is located close to the bottom dead centre, y is negative, see Figure 5a. The vorticity is distributed more-or-less symmetrically, similar to the situation when the wake is shifted to the upper position, compare to Figure 3b.

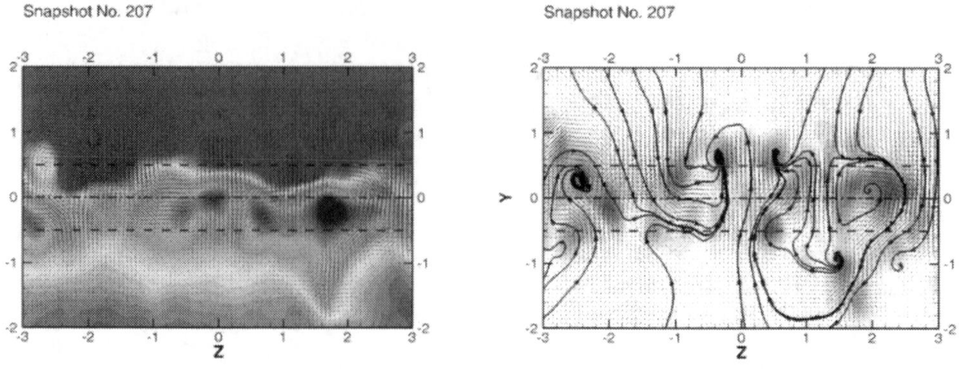

(a) (b)

Fig. 5. Instantaneous snapshot, wake shifted to the lower position. Streamwise velocity component distribution (a), streamwise vorticity component distribution (b).

The last example in Figure 6 shows again the situation, when the wake is in the middle position, y is 0, but moving from down to up (snapshot No. 214), see Figure 6a. In this case the streamwise vortices are concentrated in the upper part above the cylinder, while the lower part is vortex-free, compare Figures 4b and 6b.

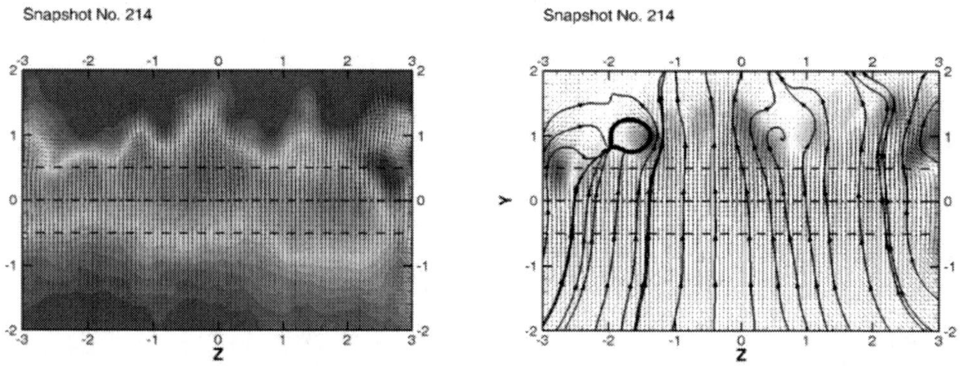

(a) (b)

Fig. 6. Instantaneous snapshot, wake in the middle position directed upwards. Streamwise velocity component distribution (a), streamwise vorticity component distribution (b).

From the above presented results we could derive an important information on the wake structure and its dynamics.

The wake could be defined in two ways. First, the wake is defined as a streamwise velocity component deficit. We could recognize the random velocity fluctuations along the cylinder axis z. Those fluctuations are strong in the dead ends positions (Figures 3a, 5a), while in central positions, they are weak (see Figures 4a, 6a).

The second definition could rely on appearance of the streamwise oriented vortices with random distribution along the cylinder axis z and random vorticity sign, negative or positive. The vorticity concentration is much stronger in central positions (see Figures 3b, 5b) than close to dead ends (see Figures 4b, 6b).

Both structures, streamwise velocity component deficit and streamwise vorticity concentration, are moving periodically in transversal direction y with the same frequency.

However, the vorticity concentration is shifted by a quarter of period before the velocity deficit in the shedding process.

3.2 Statistics

Statistical analysis relies on averaging of instantaneous data series.

In Figure 7 the mean streamwise velocity component U (perpendicular to the plane of measurement) distribution is shown in the left-hand-part (a). Legend of U indicates values in multiples of U_e. The vector U_e is shown as a reference in left-upper corner, to demonstrate that the in-plane velocity components V and W are negligible.

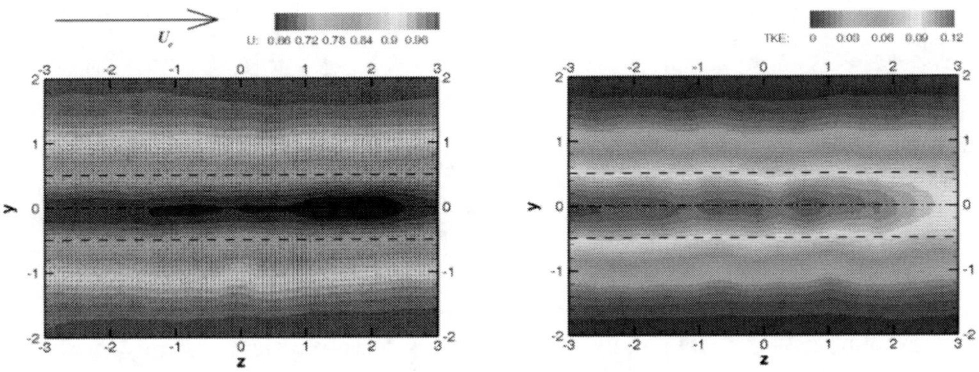

(a) (b)

Fig. 7. Mean streamwise velocity component (a), Turbulent Kinetic Energy (b) distributions.

The *TKE* distribution in Figure 7b shows the fluctuating activity maximum is located in the plane of symmetry $y = 0$ and it decays rapidly in transversal direction. In the position $y = 2$ the *TKE* approaches the free-stream value close to zero.

From Figure 7 it is clear, that the mean characteristics distributions, both U and *TKE*, are almost perfectly 2D, varying along the z axis of the cylinder insignificantly. The irregularities along the z-axis relate to experimental inaccuracy and limited statistical ensemble size (record of 2 s only).

3.3 Dynamics

The OPD analysis has been performed and 10 modes have been evaluated. Please note, that only fluctuations are analysed, the mean velocity field shown in Figure 7a is subtracted from the data set. This means, that the Reynolds decomposition into average and fluctuating parts of all quantities relevant to the problem is applied as a very first step.

In Figure 8 the OPD spectrum is shown in p-Sr plain. The dominant modes 1 and 2 are located around the shedding frequency $Sr_1 = 0.21$. The mode 3 is at a little higher frequency and the others higher order modes are of much lower frequencies.

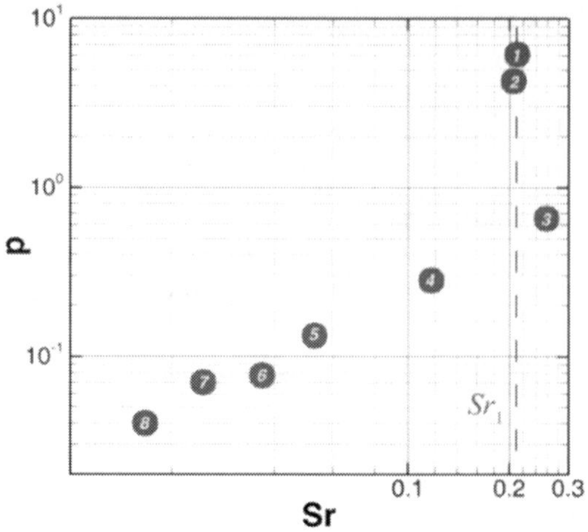

Fig. 8. OPD spectrum.

The first 5 OPD modes show not negligible values of periodicity, thus these modes are relevant as a part of cyclostationary process.

Next, the topologies of those relevant OPD modes are to be presented. The real part will be on the left (a) and the imaginary part on the right (b). The colour means here the out-of-plane velocity component U, positive in red and negative in blue. The in-plane velocity components are represented by arrows. Please note, that the absolute value of the fluctuating velocity components, both real and imaginary, has no direct physical meaning, as the modes are normalized. Only the topology is of relevance.

The first two OPD modes 1 and 2 are of the very same frequency and also the topology is very similar. It seems that these modes are in reality two versions of a single dominant mode. The real and imaginary parts are shown in Figures 9 and 10, respectively.

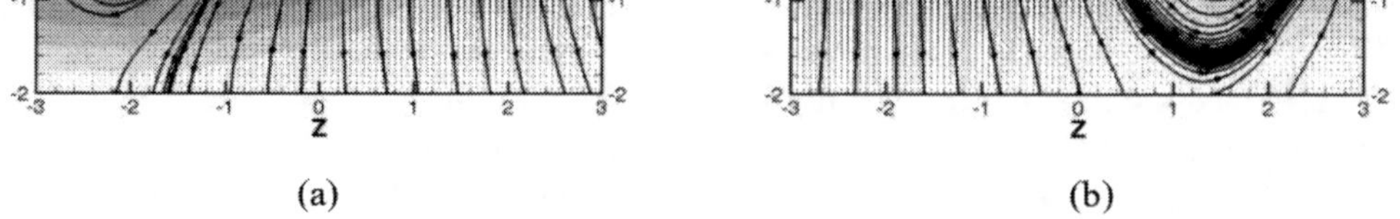

Fig. 9. OPD mode 1, real (a) and imaginary (b) parts.

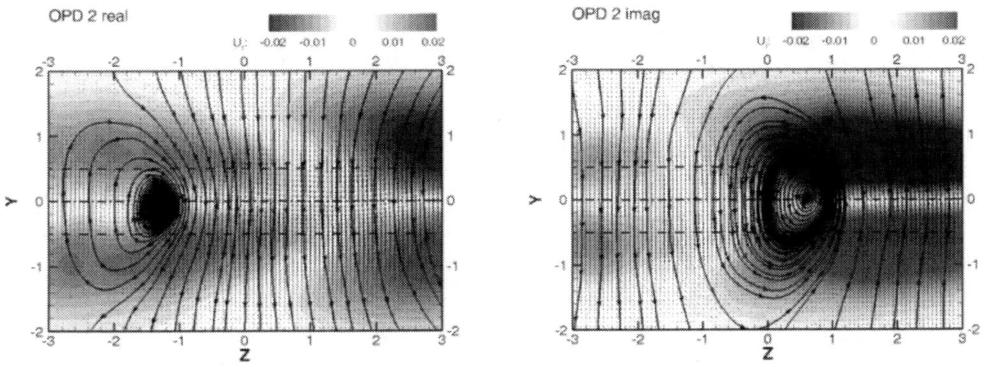

(a) (b)

Fig. 10. OPD mode 2, real (a) and imaginary (b) parts.

The dominant mode is formed by big velocity strips close to the cylinder silhouette, forming a chequered pattern, opposite orientation in upper – lower parts and right – left parts, respectively. On the top of that, the in-plane vortex is present, this means that the vortex has its axis in the streamwise direction x. Both the vortex and velocity strips are moving from the right to the left-hand-side forming the waves moving along the cylinder axis z. The frequency of this mode is close to the Strouhal frequency of the vortex street.

The OPD mod No. 3 is dominated by positive streamwise velocity component fluctuation close to the cylinder axis and negative fluctuations close to the cylinder silhouette. Those fluctuations form spots moving along the cylinder axis and changing the sign in the second half of the period. On the top of that, the weak flow in transversal y direction is present changing orientation during the period.

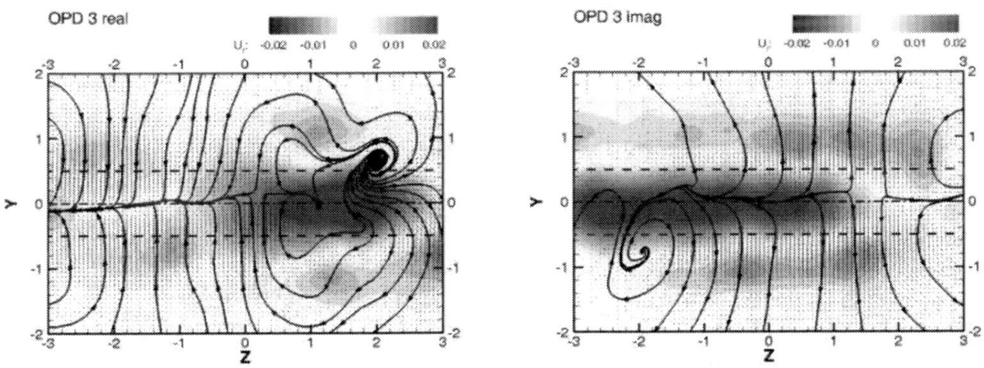

(a) (b)

Fig. 11. OPD mode 3, real (a) and imaginary (b) parts.

The topology of the next OPD mode No. 4 is shown in Figure 12. The 3 big vortices along the cylinder axis are moving along it in z direction. Those dominant structures are accompanied by system of pulsating jets on the cylinder's silhouette. The frequency is about one half of the Strouhal frequency.

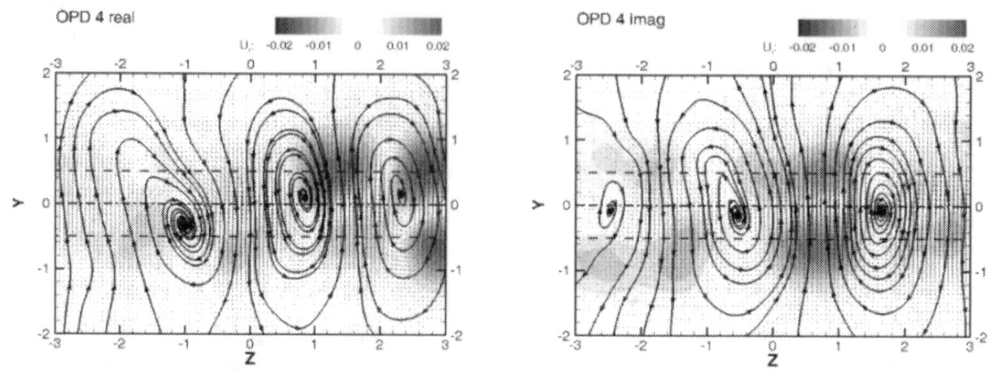

(a) (b)

Fig. 12. OPD mode 4, real (a) and imaginary (b) parts.

The last OPD mode presented here is the mode No. 5 in Figure 13. The mode is represented by the system of vortices and jets randomly distributed along the cylinder silhouette and changing theirs topology in time. There is no distinct motion along the axis *z*.

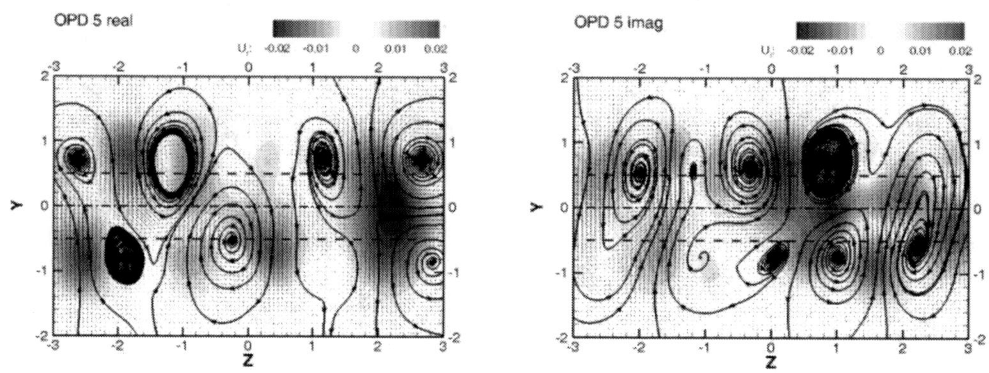

(a) (b)

Fig. 13. OPD mode 5, real (a) and imaginary (b) parts.

The higher OPD modes are significantly damped and thus not relevant to the overall dynamics.

4 Conclusions

The dynamical aspects of a fully turbulent wake behind a circular cylinder in cross-flow were studied experimentally using time resolved stereo PIV method. Reynolds number was about 5 thousand. Advanced methods of spatio-temporal data analysis have been used to gather a new information on the wake dynamics in 3D.

The results of the study show clearly, that despite the fact of presence 2D geometry, the flow in the wake is always 3D, at least its dynamical part. The time-mean structure could be 2D.

Typical dynamical patterns topology is presented in the paper. The wake is dominated by streamwise vortices appearing in distinct locations and phases of the quasi-periodical vortex shedding process randomly. Typical dynamics of those structures is wave motion along the cylinder axis. The other distinct dynamical behaviour is connected with appearance streamwise velocity component fluctuating spots, which are again moving along the cylinder axis in waves, or they pulsate. Both dynamical structures in the wake, the streamwise velocity component deficit and the streamwise vorticity concentration, are moving periodically in transversal direction with the same frequency, the vorticity concentration is shifted by a quarter of period before the velocity deficit in the shedding process.

Acknowledgements

This publication was supported by the project of University of West Bohemia in Pilsen SGS-2022-023.

References

1. T. von Kármán, Aerodynamics. Columbus: McGraw-Hill. (1954)
2. A. Roshko, On the development of turbulent wakes from vortex streets. NACA Rep. 1191, pp.801-825. (1955)
3. C.H.K. Williamson, Vortex Dynamics in the Cylinder Wake, Annu. Rev. Fluid. Mech. **28**, pp. 477-539. (1996)
4. J. Hoffman, C. Johnson, Resolution of d'Alembert's paradox, J. Math. Fluid Mech., **12** 3, pp.321-334. (2010)
5. V. Uruba, On 3D instability of wake behind a cylinder, AIP Conference Proceedings, 1745, art. no. 020062. (2016)
6. V. Yanovych, D. Duda, V. Uruba, Structure turbulent flow behind a square cylinder with an angle of incidence, European Journal of Mechanics, B/Fluids, **85**, pp. 110-123, (2021)
7. V. Uruba, and P. Prochazka, On Harmonic Contents of the Wake behind a Circular Cylinder, 26th International Conference on Engineering Mechanics (IM), Engineering Mechanics 2020 (IM2020) , pp.500-503, (2020)
8. V. Uruba, P. Procházka, V. Skála, On the 3D Dynamics of the Wake Behind a Circular Cylinder, In Proceedings Topical Problems of Fluid Mechanics 2020, Prague, Edited by D. Šimurda and T. Bodnár, pp. 240-248, (2020)
9. V. Uruba, Decomposition methods in turbulent research, EPJ Web of Conferences, **25** 01095. (2012)
10. V. Uruba, Near Wake Dynamics around a Vibrating Airfoil by Means of PIV and Oscillation Pattern Decomposition at Reynolds Number of 65 000, Journal of Fluids and Structures, **55**, pp. 372-383. (2015)
11. DynamicStudio 7.4, User's Guide, Dantec Dynamics (2021)

Turbulence topology behind different sections of the wind turbine blade

Vitalii Yanovych[1, 2, *], *Daniel* Duda[1], *Václav* Uruba[1, 2], *Pavlo* Kosiak[1, 2] and *Vít* Horáček[1]

[1]Power System Engineering, Faculty of Mechanical Engineering, University of West Bohemia in Pilsen, Univerzitní 22, 306 14 Pilsen, Czech Republic.
[2]Department of Fluid Dynamics, Institute of Thermomechanics, Czech Academy of Sciences, Dolejškova 5, 182 00, Prague, Czech Republic.

Abstract. This work aimed to investigate the turbulence characteristics behind different parts of the wind turbine blade. Airfoils S807, S813, S817, and S803, which characterize the shape of the blade at different lengths, were selected for estimation. During the experiment, the chord-based Reynolds number was 2.6×10^5, while the angle of attack was zero. Measuring cross-sections were placed behind the trailing edge at $x{\cdot}c^{-1}\approx0.2$, 0.4 and 1.0. For the determination flow topology, we used a Hotwire anemometry with a split fiber probe 55R55 and a miniature X-wire probe 55P64. The obtained data allowed us to determine and compare the evolution of the wake behind different types of airfoils in streamwise and spanwise directions. Thus, the largest and smallest velocity deficit located behind S817 and S803 airfoils, respectively. This trend is also evident in the Normalized Reynolds shear stress distributions. Finally, we determined the spectrum and calculated the Integral length scale, the Taylor and Kolmogorov microscale of turbulent flow. According to the results, profile S817 contributes to the formation of a flow with a large scale of turbulence, while the S803 is contrary.

1 Introduction

The energy efficiency of wind turbines is a key factor in the economic evaluation of this type of renewable energy. It is clear that the general characteristics of this parameter depend on many subsystems [1]: blades, gearbox, generator, and control. Among these, the shape of the blade deserves special attention. Because the efficiency of wind energy transformation into the rotational motion of the generator strongly depends on aerodynamic characteristics of the blade. It should be noted that its overall design consists of different types of profiles that differ in the shape and size of the chord. Thus, the flow formed around the general geometry of the blade is a very complex three-dimensional phenomenon.

Over the last decade, the airfoil families commonly used for horizontal axis wind turbines have become widespread, including the NACA 44, NACA 23, NACA 63, and NASA LS [2]. But the main problem for them is the significant deterioration in aerodynamic characteristics due to the effects of roughness caused by contamination of the leading edge. To avoid this disadvantage for wind turbine blades, the National Renewable Energy Laboratory and

* Corresponding author: yanovych@kke.zcu.cz

Airfoils Inc. developed a new airfoil family (NREL) [3]. According to the developers, these blades are providing significant increases in annual energy production, better lift-to-drag ratios, and, in the case of stall-regulated rotors, the use of the more swept area for a given generator size. Despite these positive qualities, the issue of noise reduction and structural vibration caused by flow turbulence remains open.

That is why the scope of this paper is to perform a detailed experimental investigation of the airfoil shape effect on the turbulence evolution at different parts of blade. The general geometry of which was formed by the NREL airfoils.

2 Experimental setup and equipment

For analysis of turbulent flow evolution in different parts of the wind turbine blade, we chose a model geometry that consists of NRL airfoil profiles. This blade uses different types of airfoils, but only four of them were selected for our study S807, S813, S817, and S803. They are kept along the span of the blade from root to tip. A sketch of the wind turbine blade and its cross-section profiles shows in Fig.1 a). The airfoils were produced of thermoplastic filament material PLA by the 3D printer PRUSA i3. Thus, the quality of their surface roughness according to [4] was Ra ≈ 21.8 μm. The manufactured airfoils had similar dimensions of chord length and span of $c \approx 100$ mm and of $l{\cdot}c^{-1} \approx 1$, respectively. But their thickness was somewhat different. Namely for S807 - $b{\cdot}c^{-1} \approx 0.19$, S813 - $b{\cdot}c^{-1} \approx 0.16$, S813 - $b{\cdot}c^{-1} \approx 0.17$ and S803 - $b{\cdot}c^{-1} \approx 0.15$.

The experiment was conducted in a low-speed open-type wind tunnel at the department of Power System Engineering, University of West Bohemia [5]. The internal dimensions of the test section were 0.3 m in height and 0.2 m in width, and its length was 0.75 m. The natural turbulence level amounted to about 0.1% in the working section input. Changing the position of the measuring point relative to the y and z axes was carried out using a traverse system.

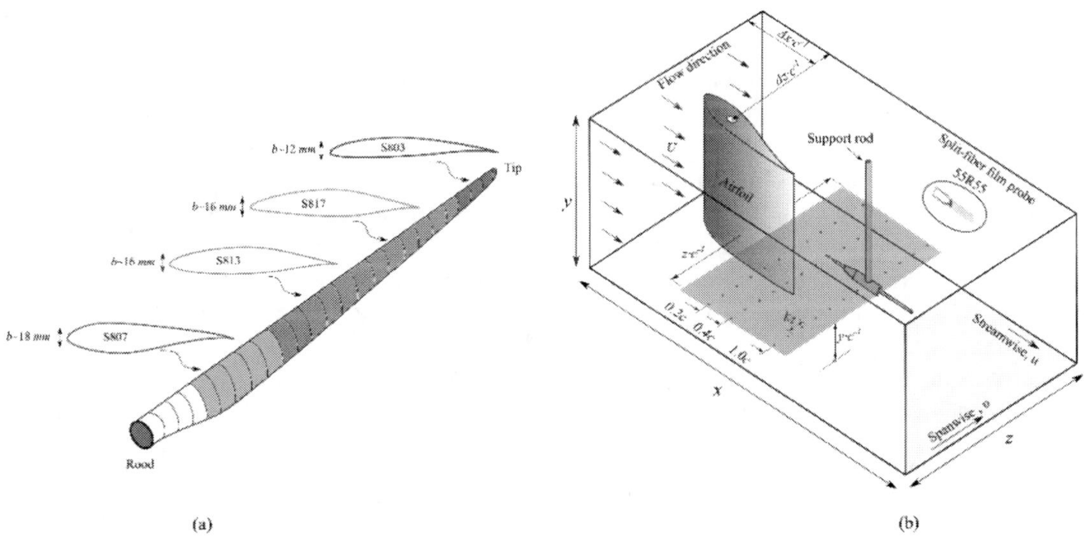

Fig. 1. a) Sketch of the wind turbine blade and its cross-section profiles. b) Airfoil model setup in the test-section and the coordinate system. The arrows in blue indicate the direction of air flow U. The green plane shows the Hot-wire measuring field. Where $z{\cdot}c^{-1}$, $x{\cdot}c^{-1}$ and $y{\cdot}c^{-1}$ the width, length and location height of the measuring fields, respectively. The red dots lines are a set of measuring points behind the airfoil at different cross-sections. While k_z is the distance between the measuring points. The coordinates $\Delta z{\cdot}c^{-1}$and $\Delta x{\cdot}c^{-1}$ shows the position of the aerodynamic center of the airfoil.

Measurements were performed at zero incidence angle and at $Re \approx 2.6\times10^5$. The Reynolds number was based on the length of the chord. The airfoil model setup in the test section and the coordinate system are shown in Fig. 1 b). The aerodynamic center of the airfoil was located at $\Delta z\cdot c^{-1}\approx1.9$ and $\Delta x\cdot c^{-1}\approx2.8$. Where Δz and Δx indicate the distance between the wall and the entrance to the test section. Measurements were performed in three cross-sections from $z\cdot c^{-1}\approx$ -1.5 to 1.5 at $x\cdot c^{-1}\approx0.2$, 0.4, 1.0, and at the height $y\cdot c^{-1}\approx1.8$. Thus the measuring field corresponded to the center of the airfoil span. The distance k_z between the measuring points was different depending on the measurement location. Thus, according to methodology [6], in the wake region $k_z\approx$ 3 mm, and outside $k_z\approx$ 5 mm.

Experimental studies were performed using Hot-wire anemometry and two types of probes. Namely, it was a split-fiber film probe 55R55 and a miniature X-wire probe 55P64. It should note that they allow investigation of the two-dimensional characteristics of the turbulent flow (u and υ components). The split-fiber probe was used for wake topology investigation, while the wire probe for spectral and scaling analysis of vortices. In both cases, the sampling frequency and measurement time at one point were 70 kHz and 10 s, respectively. The filtering of the received signal was also applied. Namely, the High Pass filter was 10 Hz, while the Low Pass filter was 30 kHz.

3 Results and discussion

First of all, with the help of split-fiber probe 55R55, we evaluated the patterns of velocity distribution behind different types of profiles and the position of the measuring cross-section $x\cdot c^{-1}$ (Fig. 2). For convenience, the resulting curves are located relative to the position of the trailing edge at zero angle of attack and normalized by the average flow velocity U. For data analysis, we selected the range from $z\cdot c^{-1}\approx$-0.5 to 05. The obtained results allowed us to evaluate the wake characteristics in streamwise and spanwise directions. In general, we can observe classical dependencies, where with increasing distance from the trailing edge, the wake depth decreases, and its width increases. According to the data, the highest and lowest velocity deficits correspond to the S817 and S803 airfoils, respectively. Especially this pattern is detected close to the trailing edge at $x\cdot c^{-1}\approx0.2$ in a streamwise direction. Namely, under these conditions, the minimum value of $u\ U^{-1}$ for S817 is 13% lower than for S803. It should also note some features of the velocity distribution for S807 and S803 in the spanwise direction. The geometry of these airfoils has a more curved trailing edge than for others, which leads to some inflection of their velocity distribution at $z\cdot c^{-1}\approx0.2$.

We also found distributions of normalized standard deviation in different measuring cross-sections behind the various types of airfoils (see Fig. 3). As we can see, the largest values of $\sigma_u\cdot U^{-1}$ and $\sigma_\upsilon\cdot U^{-1}$ are observed for the S817 profile. Moreover, the obtained graphs clearly reflect the influence of the airfoil shape on the overall distribution of the standard deviation. Namely, in the streamwise direction, the difference in the formation of double peaks of the data curve is clearly visible. That phenomenon occurs due to the vortices interactions from pressure and suction sides of the airfoil [7]. Thus, it is logical that this effect is better observed for a S813 profile with a more symmetrical profile shape. While in other cases, the flow contains mainly a single dominant structure.

The Reynolds shear stress distribution was applied to estimate of the turbulent flow anisotropy (see Fig. 4). Obtained data clearly show the range where the turbulence structure is anisotropic. Usually, it is displayed as a graphical distribution asymmetry with a combination of positive and negative numbers through the wake centerline. Thus greatest width of this range, regardless of $x\cdot c^{-1}$, is observed for S817, while the smallest one for S803 airfoil.

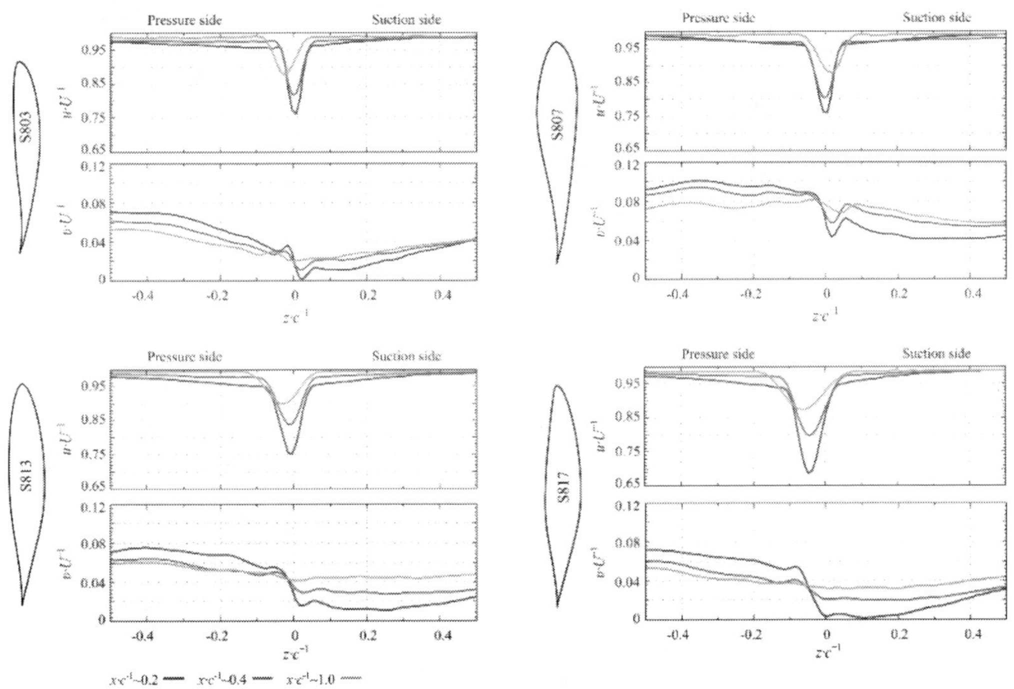

Fig. 2. The normalized profiles of streamwise $u \cdot U^{-1}$ and spanwise $v \cdot U^{-1}$ velocity components depending on the airfoil types and the distance of the measuring cross-sections $x \cdot c^{-1}$. Measurements were performed at zero incident angle and at $Re \approx 2.6 \times 10^5$. Violet, green and yellow curves characterize the position of the measuring cross-section behind the trailing edge at $x \cdot c^{-1} \approx 0.2$ and 0.4 and 1, respectively.

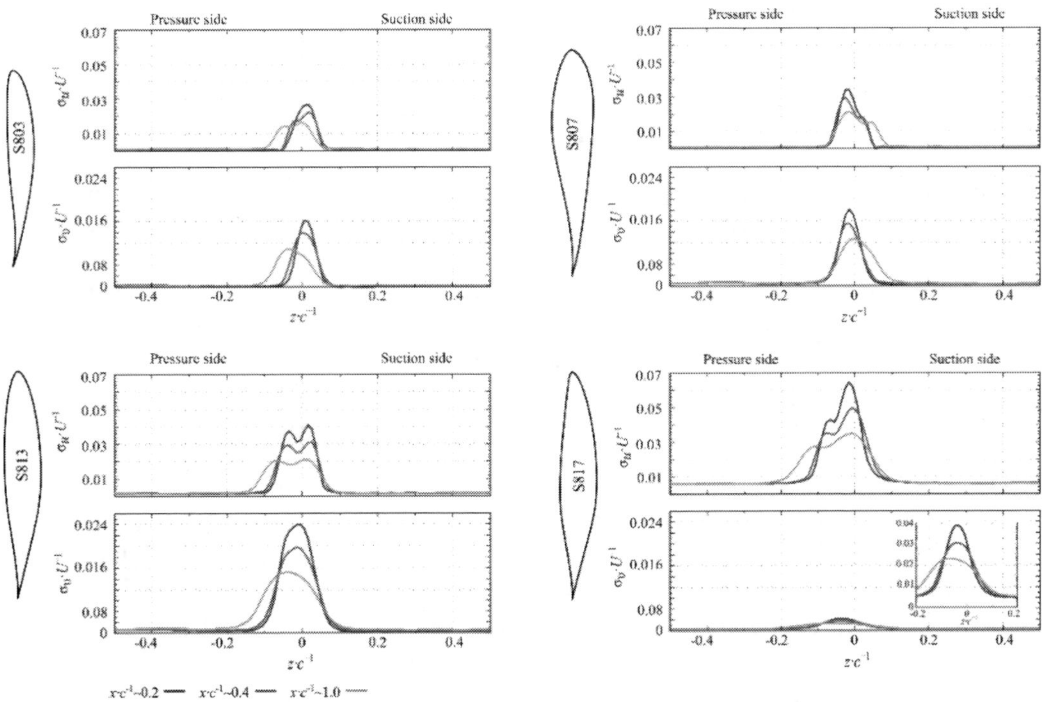

Fig. 3. Profiles of the normalized standard deviation of the streamwise $\sigma_u \cdot U^{-1}$ and spanwise $\sigma_v \cdot U^{-1}$ components depending on the airfoil types and the distance of the measuring cross-sections $x \cdot c^{-1}$. Measurements were performed at zero incident angle and at $Re \approx 2.6 \times 10^5$. Violet, green and yellow curves characterize the position of the measuring cross-section behind the trailing edge at $x \cdot c^{-1} \approx 0.2$, 0.4 and 1, respectively.

In addition, among the entire composition of the blade, exactly these two profiles have the largest and smallest value of the Reynolds shear stress component. Accordingly to results, the magnitude of $\overline{u'v'} \cdot U^{-2}$ for S817 at $x{\cdot}c^{-1}{\approx}0.2$ is 15 times larger than for S803.

Moreover, Fig.4 shows that the perturbation region of shear stress with a negative value is located directly on the suction side. This pattern is due to the intensity of turbulence, which increases the frictional resistance in the middle flow. As a result, turbulence receives energy, while the average flow loses. Usually, this trend is especially noticeable in the stream-wise direction. Accordingly, to Fig. 2 and Fig. 3, the highest magnitude of the velocity deficit and standard deviation was for the S817, while S803 has the opposite trend.

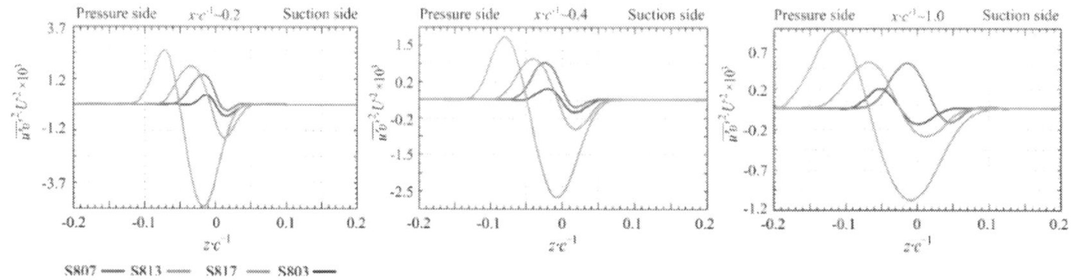

Fig. 4. Normalized Reynolds shear stress distribution at $Re \approx 1.7{\times}10^5$ depending on airfoil types and various positions of the measuring cross-sections $x{\cdot}c^{-1}$. Green, light blue, yellow and violet curves characterize the distributions for S807, S813, S817 and S803 airfoils respectively.

As the distance from the trailing edge increases up to $x{\cdot}c^{-1}{\approx}1$, the shear stress decreases almost fourfold and the wake with increases almost twice. Thus, we can see some relationship between the amplitude of the Reynolds shear stress component and the energy characteristic of the vortices. According to Kolmogorov's theory [8-11], the greatest energy is contained in the largest vortices, the size of which can be described by an integral length scale L_E. To determine it, we applied the methodology well described in [12, 13]. Where the authors used correlation time series analysis to estimate the spatial scales of turbulent flows. Accordingly, the autocorrelation function for stream-wise component R_{uu} can be interpreted as:

$$R_{uu}(t) = \frac{\langle u(t+\Delta t)\cdot u(t)\rangle}{\sigma_u^2} \tag{1}$$

where the angular brackets indicate the average value u and Δt is the shift time of the signal, while $\sigma_u = \sqrt{N^{-1}\sum_{i=0}^{N-1}(u_i - \underline{u})^2}$ is standard deviation. Accordingly, N and u_i are the number of the data set, and g is the instantaneous velocity at the measurement set.

Then, using the cumulative integral of the autocorrelation function, it is easy to calculate the time scale T. That from a physical point of view interprets the average lifetime of vortices.

$$T = \int_0^\infty R_{uu}(t)dt \tag{2}$$

It is clear that the integral length L and time scales T are interrelated. Thus integral length can be evaluated by the equation:

$$L = T/\sigma_u \tag{3}$$

For scaling analysis, we performed measured by 55R55 probe at cross-section with coordinates $x{\cdot}c^{-1}\approx 1$, $y{\cdot}c^{-1}\approx\approx 1.8$ and $z{\cdot}c^{-1}$ from -0.5 to 0.5. The obtained distributions of the integral length are shown in Fig. 5 a). As we can see, the values of the integral length scale for different types of airfoils in the spanwise direction are almost the same, where in the wake region the $L_v \approx$ 1.1-1.3 mm. While in the stream-wise direction, as for the distribution of shear stresses (see Fig. 4), this parameter varies greatly. Namely, the smallest eddies are observed at S803 airfoil $L_u \approx$ 1.3 mm, while the biggest at S817 are $L_u \approx$ 2.6 mm. It should note that the configuration of the airfoil geometry significantly changes the evolution of the vortices. That is especially noticeable from the suction side, where the behavior of obtained distributions, in contrast to the pressure side, does not have a common trend.

After that, we analyzed the effect of the geometry of each profile on the Taylor micro-scale characteristics. From a physical point of view, that micro-scale shows the size of the vortices at which energy dissipation begins. To calculate the Taylor micro-scale, we used the classical equation:

$$\lambda = \sigma_u\sqrt{15 \cdot \nu/\epsilon} \tag{4}$$

were ν is kinematic viscosity velocity of air.

Graphical interpretation of the distribution λ depending on the type of profile shown in Fig. 5. b). Accordingly, the highest value of λ observe for S817 - $\lambda \approx 0.6$ mm, whereas the smallest one is for S803 - $\lambda \approx 0.2$ mm. The relation of these values is approximately the same in streamwise and spanwise directions.

Another important parameter of turbulent flow is the Kolmogorov microscale η. What characterizes the smallest size of vortices which cause the dissipation of kinetic energy of the flow into heat. According to paper [14], the value of η can be found from the equation:

$$\eta = (\nu^3/\epsilon)^{1/4} \tag{5}$$

where ϵ is the dissipation rate determined by integration of the second-order structure-function and ν is kinematic viscosity of air.

Graphical interpretation of Kolmogorov microscale depending on the airfoil types is shown in Fig. 5 c).

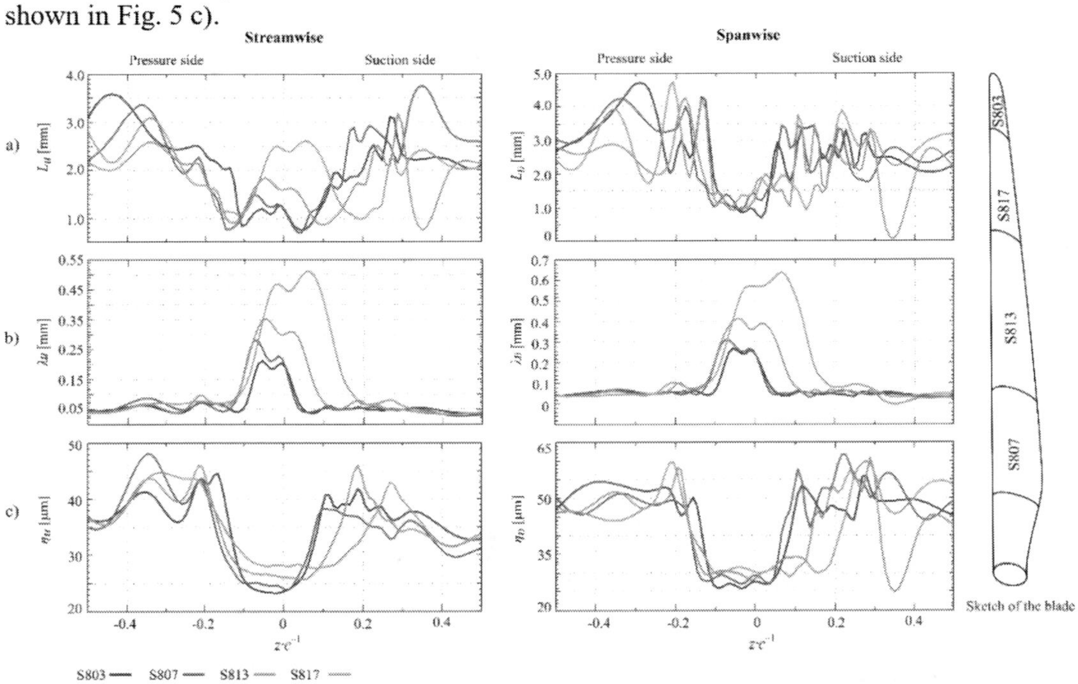

Fig. 5. Distributions of the integral length scale L, Taylor microscale λ and Kolmogorov microscale η in streamwise and spanwise directions depending on airfoils type. Measurements were carried out at $Re \approx 2.6\times10^5$, zero incident angle and $x{\cdot}c^{-1} \approx 1$. Green, light blue, yellow and violet curves characterize the distributions for S807, S813, S817 and S803 airfoils respectively.

The analysis of the obtained data revealed that regardless of the measurement direction and profile type, the value of the Kolmogorov micro-scale in the wake region does not change significantly. Thus, the maximum and minimum dimensions of the smallest vortices observe for S817 and S807. Accordingly, in both measuring directions, their values correspond to $\eta \approx 30$ μm and $\eta \approx 25$ μm.

Power spectral density is used to estimate the distribution of flow energy between different scales of turbulence. To perform this analysis, we chose the positions of measurement points at $x{\cdot}c^{-1} \approx 1$, where the value of standard deviation in the streamwise direction is highest (see Fig. 3). Spectral interpretation of velocity fluctuations depending on the airfoil types is shown in Fig. 6. Usually, according to Kolmogorov's theory [12, 13], isotropic turbulent flow is characterized by a spectral distribution with a slope of the curve in the inertial range $f = k^{-5/3}$ (see Fig. 6 a)). To better identify this range, the obtained spectrum was normalized $E(k)\cdot k^{5/3}$ and the wave number $k\cdot\eta$ (Fig. 6 b)). The obtained data show that depending on the type of profile, the width of the inertial range varies. Accordingly, we can say that the characteristics of the energy cascade, where the kinetic energy of turbulence is transmitted without loss from large scales to smaller, for different parts of the wind turbine blades change. Thus largest and smallest width of the inertial subrange is observed for S817 and S803 profiles, respectively. This pattern is manifested equally in the streamwise and spanwise directions.

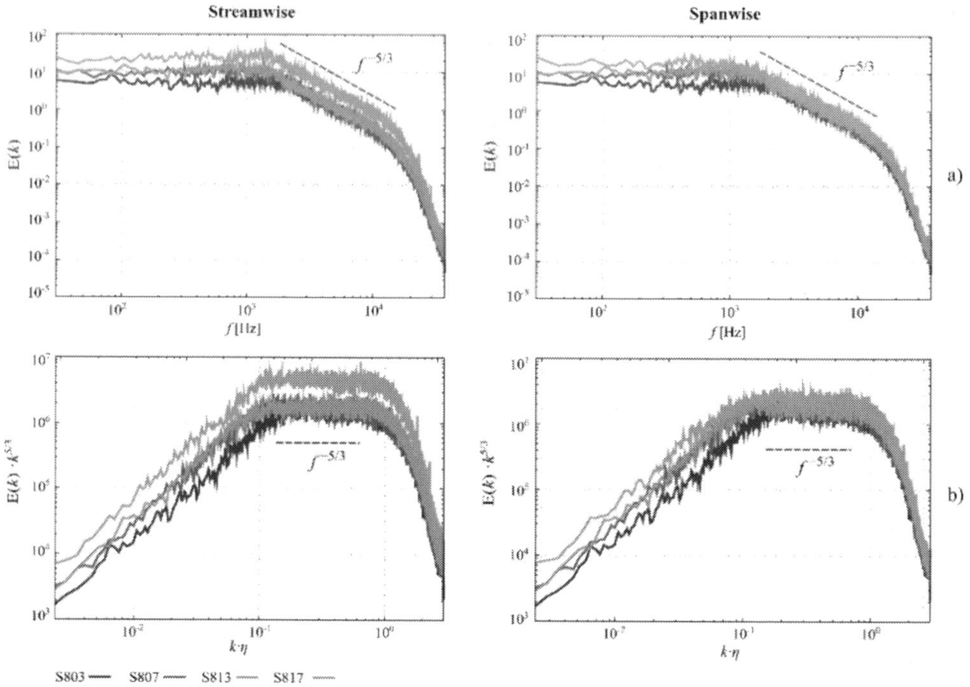

Fig. 6. Power spectral densities of stream-wise and span-wise velocity fluctuations at $Re \approx 2.6\times10^5$ and $x{\cdot}c^{-1} \approx 1$ depending on airfoils type. The upper graphs show the spectrum in the classical form, while the lower two are normalized by $k^{-5/3}$. Green, light blue, yellow and violet curves characterize the distributions for S807, S813, S817 and S803 airfoils respectively

4 Conclusion

We used Hot-wire anemometry to investigate the turbulence topology behind different parts of the wind turbine blade. Airfoil profiles S807, S813, S817, and S803, which correspond to selected parts of the blade, were selected for experimental studies. We investigated the dependence of wake parameters and turbulent flow characteristics at zero angles of attack and at the chord-based Reynolds number 2.6×10^5. While the measuring cross-sections were placed behind the airfoil at a distance of $x{\cdot}c^{-1}\approx0.2$, 0.4 and 1.0.

Initially, using the split-fiber probe 55R55, we obtained patterns of velocity distribution and standard deviation depending on the airfoil type in the stream-wise and span-wise directions. The data show that the highest and lowest velocity deficits correspond to airfoils S817 and S803, respectively. Moreover, a similar trend persists for the standard deviation distributions.

After that, using miniature X-wire probe 55P63, we performed scaling analyses of the turbulence structure $x \cdot c^{-1} \approx 1.0$. Namely, we determined the value of the Integral length scale, the Taylor and Kolmogorov microscale in the points that correspond to the maximum values of the standard deviation. The maximum values of these parameters, regardless of the measurement direction, are observed behind S817, while the minimum is behind S803. This trend is particularly noticeable in a streamwise direction. It should be noted that there is some relationship between the behavior of the Reynolds shear stress component and the energy characteristic of vortices. Because the values of shear stress and Integral length scale vary equivalently depending on the type of profile.

Finally, we obtained the spectral power density of the flow energy distribution between different scales of turbulence for each of the profiles. The obtained graphs show that the inertial interval is the widest and narrowest for S817 and S803, respectively.

Acknowledgments

This work was financial supported by "IDEG Programme", as part of the project Improving the Quality of Internal Grant Schemes at the University of West Bohemia in Pilsen, project reg. No. CZ.02.2.69/0.0/0.0/19_073/0016931.

References

1. N. Tenguria, N. D. Mittal and S. Ahmed. Investigation of blade performance of horizontal axis wind turbine based on blade element momentum theory (BEMT) using naca airfoils. *International journal of engineering, science and technology* **2(12)**, 25–35 (2010).
2. J. Tangler, B. Smith, D. Jager and T. Olsen. *Atmospheric Performance of the Special-Purpose SERI Thin-Airfoil Family: Final Results*, in Proceedings of the European Wind Energy Conference, 10-14 September, Madrid, Spain (1990).
3. J. Tangler and D. Somers, (1995). *NREL airfoil families for HAWT's*, report of the National Renewable Energy Laboratory (January 1995, Cole Boulevard Golden, Colorado, USA).
4. M. S. Alsoufi, A. E. Elsayed. Surface roughness quality and dimensional accuracy - a comprehensive analysis of 100% infill printed parts fabricated by a personal/desktop cost-effective FDM 3D printer. *Materials Sciences and Applications*. **09**, 11–40 (2018).
5. V. Yanovych, D. Duda, V. Horáček and V. Uruba. *Research of a wind tunnel parameters by means of cross-section analysis of air flow profiles*, in Proceedings of the Power System Engineering, AIP conference 2189, Pilsen, Czech Republic (2019).
6. V. Yanovych, D. Duda, V. Horáček and V. Uruba. *Creation of recombination corrective algorithm for research of a wind tunnel parameters* in Proceedings of the Power System Engineering, AIP conference 2118, Pilsen, Czech Republic (2019).
7. A. Zobeiri, P. Ausoni, F. Avellan, M. Farhat. How oblique trailing edge of a hydrofoil reduces the vortex-induced vibration. *Journal of Fluids and Structures* **32**, 78–89 (2012).
8. K. Choi, J. L Lumley. The return to isotropy of homogeneous turbulence. *J. Fluid Meach.* **436**, 59–84 (2001).
9. A. N. Kolmogorov, V. Levin, J. C. R. Hunt, O.M. Phillips, and D. Williams. Dissipation of energy in the locally isotropic turbulence. *Proceedings of the Royal Society of London. Series A: Mathematical and Physical Sciences* **434**, 15–17 (1991)

10. G. K. Batchelor. The Theory of Homogeneous Turbulence (Cambridge University Press, New York, US, 1982).
11. J. L. Lumley, G. Newman. The return to isotropy of homogeneous turbulence. *J. Fluid Meach.* **82**, 161–178 (1977).
12. M. Bourgoin et al. Investigation of the small-scale statistics of turbulence in the modane S1MA wind tunnel. *CEAS Aeronautical Journal* **9**, 269–281 (2018).
13. V. Yanovych, D. Duda, V. Uruba. Structure turbulent flow behind a square cylinder with an angle of incidence, *European Journal of Mechanics, B/Fluids* **85**, 110–123 (2021).
14. V. Yanovych, D. Duda, V. Uruba and T. Tomášková. Hot-Wire Investigation of Turbulence Topology behind Blades at Different Shape Qualities, *Processes* **10(3)**, 522 (2022).
15. S. B. Pope. Turbulent Flows, 1st ed. (Cambridge University Press, New York, US, 2000).
16. P. Kundu, I. Cohen and D. Dowling. Fluid Mechanics. 6th ed. (Academic Press, London, UK, 2016).

The Alternative concept - external steam-gas heat exchanger (DaGaVo)

Tomas Zizka[1]

[1]Technical University of Liberec, Faculty of Mechanical Engineering, Studenstká 1402/2, 461 17 Liberec, CZ

Abstract. In Waste to Energy plants ((WtE)) is treatment of municipal solid waste a complex process that places strictly demands on design and operation of technology. As part of the general renovation of the WtE plant in Prague Malesice was installed an alternative concept for flue gas heating in front of Selective Catalytic Reduction System (SCR) DeNOx. This concept replaced the original flue gas heating with natural gas burners with steam heating using an external steam-gas heat exchanger (DaGAVo). This alternative concept aims to reduce natural gas consumption, reduce secondary emissions and increase the energy efficiency of the installation.

1 Introduction – Waste to Energy Prague – Malesice

The Company Pražské služby a.s. according to a 10-year contract with the City of Prague shall to thermally treated of approx. 260 000 t/a of residual waste. The current approved capacity of the plant is 330 000 t/a. T*he plant is located on west industrial part of city Prague as you can see in Figure 1 and is operated by Pražské služby a.s.. The plant is equipped with four combustion lines.

Fig. 1. The WtE plant in Prague – Malesice.

*Corresponding author: tomas.zizka1@tul.cz

1.1 Project description

WtE Prague – Malešice - GOLEM

Pražské služby a.s. signed on 9 November 2017 the contract for work designated "GENERAL RENEWAL AND ECOLOGIZATION MALEŠICE - THERMIC TREATMENT OF WASTE" called "Project GOLEM". The primary purpose of the project was to renovate the technology and related equipment after 20 years in operation. The execution of renovation follows the basic objective of ensuring the safe and reliable operation of the WTE Prague at least until 2045, while achieving higher annual utilization. [1]

Alternative concept

As a part of general renovation was an Alternative concept that shall improve the ecologization and make the energy management of the WTE Prague more efficient to an even greater extent than it was originally agreed in the Basic concept and achieve a significant decrease in operating costs. The Alternative concept distinguishes itself from the boiler concept by virtue of additional super heater surfaces in the fourth boiler path and of an external steam-gas heat exchanger (DaGAVo) positioned upstream of the SCR system to reheat the flue gases as you can see in Figure 2.

Due to the additional heating surfaces in the boiler, the steam temperature increases to 320° C, 15 bar(a). This steam is taken from the boiler into an external steam-gas heat exchanger positioned upstream of the SCR in order to increase the operating temperature of the flue gases in the ceramic catalysts (SCR DeDiox/DeNOx) from 235°C to 275-280°C. The steam parameters suited for supplying steam to the existing turbine are achieved by means of spray cooling downstream of the heat exchanger.

Position of external steam-gas heat exchanger in the flue gas path you can see in Figure 3.

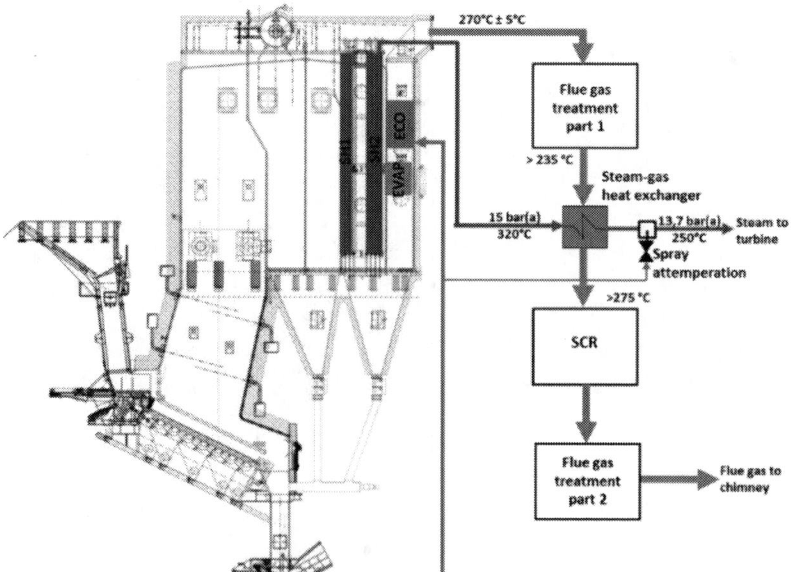

Fig. 2. Steam boiler with heat displacement system (MARTIN).

Fig. 3. Position of external steam-gas heat exchanger in FGT system.

Own natural gas consumption

WtE plant use the natural gas for the burners in 4 boilers and also in flue gas cleaning system for increasing flue gas temperature. In the boilers are burners used to put in the operation after shut down or for stabilization of the burning process (wet waste etc.) [4]. With installation the honeycomb catalyst in 2007 for reducing PCDD/PCDF emissions also was installed in flue gas cleaning part of the technology additional natural gas burners for regularly monthly 8 hours' regeneration process with the temperature 290°C. Ceramic catalysts were installed in the purification of raw flue gases immediately after dedusting. From an energy point of view, this location is more suitable, which is in line with the information in reference [3].

In 2010 we added the ammonia water dosing and started to use it as a Selective Catalytic Reduction System (SCR) with reducing NOx emissions from the flue gases. For ammonia water dosing we need permanent operation of the burners due to the flue gas temperature. It had to negative influence to operation costs. For this reason, we gradually optimized the SCR System to keep balance between costs for the natural gas and the NOx emission. Currently we operate the plant with average 110-120 mg.Nm-3 NOx emission, the limit is 200 mg.Nm-3. Annual consumption natural gas from 2009 to 2021 you can see on the Figure 4 bellow. You can see the reduction in consumption since 2019 due to the gradual replacement of flue gas heaters with flue gas heater (DaGaVo) on Figure 4.

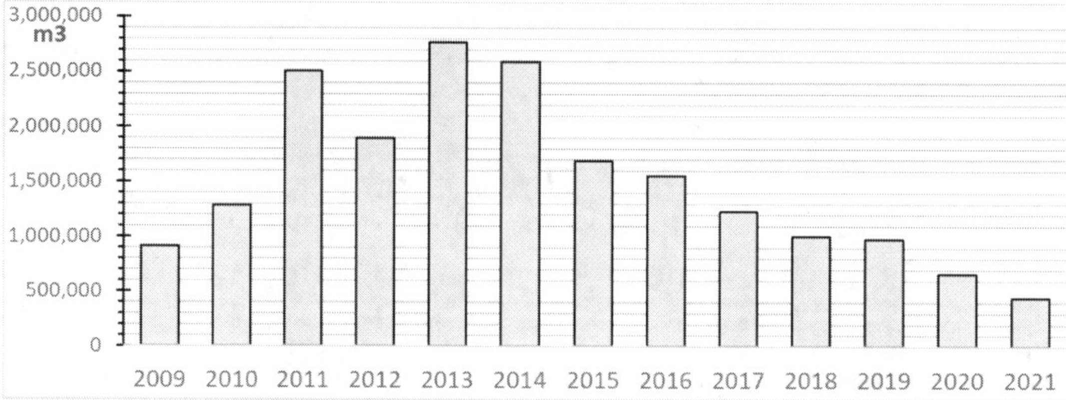

Fig. 4. Annually natural gas consumption.

2 Conclusion

The goal of the measures facilitate was considerably reducing the operating hours of the burner used to heat up the flue gas upstream of the SCR system and significantly lowering the associated natural gas consumption. After 3 years in operation with the Alternative concept we could decide to remove and store the natural gas burners for SCR System. The considerably reducing is done with combination of the high work availability. December 2021 was our historically first month, when we did not have to start a natural gas burner for a whole month.

The results of replacing natural gas burners with a steam exchanger clearly show that natural gas consumption are significantly reduced in 2021 to less than 30% of long-term average annual consumption. The Alternative concept proved to be operationally reliable and economical after 3 years in operation.

Acknowledgement:

The work has been financially supported by the project "Hybrid materials for hierarchical structures", research goal: Composite materials and structures, research program: Materials and structures on the metal basis, reg. no. CZ.02.1./0.0/0.0/16_019/0000843 provided by the European Union and the Czech government and by the Student Grant Scheme at the Technical University of Liberec through project no. SGS-2022-5040.

References

1. Blaha A., Eickhoff N.: Project GOLEM: New Construction and Modernization of an Existing Plant in Prague/Czech Republic, Waste Management Volume 8 (2018), p. 203-208,
2. Keunecke, M, Mengede, F.-J., Energie aus Abfall – Steigerung der Energieeffizienz als Treiber für Neu und Umbau von Energie-aus-Abfall-Anlagen, in Strategie, Planung, Umweltrecht, K.J. Thome-Kozmiensky und M. Beckmann, S. 29-39, Neuruppin: TK Verlag, 2016
3. Lemann, M.F., Verfahrenstechnik und Umwelt / Martin F.Lemann, Bern; Berlin; Bruxelles; Franfurt am Main; New York; Oxford; Wien: Lang, ISBN 3-906764-66-4 (2000)
4. Operational Manual Waste to Energy plant Malešice, Pražské služby a.s., (2019)

AUTHOR INDEX

EDP Sciences – Web of Conferences
17, Avenue du Hoggar
Parc d'Activité de Courtabouf
BP 112
F-91944 Les Ulis Cedex A
France

ISBN 978-1-7138-6275-8

CURRAN ASSOCIATES INC. proceedings.com

Eastern Snow Conference 2018

College Park, Maryland, USA
5 - 8 June 2018

Eastern Snow
Conference